Catalysis

Volume 14

A Specialist Periodical Report

Catalysis

Volume 14

A Review of Recent Literature

Senior Reporter
James J. Spivey, *Research Triangle Institute,*
Research Triangle Park, North Carolina, USA

Reporters
Pehr H. Björnbom, *Royal Institute of Technology (KTH), Stockholm, Sweden*
Michel Boudart, *Stanford University, Stanford, CA, USA*
Linda J. Broadbelt, *Northwestern University, Evanston, IL, USA*
Ya-Huei Chin, *University of Oklahoma, Norman, OK, USA*
Anthony G. Dixon, *Worcester Polytechnic Institute, Worcester, MA, USA*
Anders G. Ersson, *Royal Institute of Technology (KTH), Stockholm, Sweden*
Rolf Gabrielsson, *Royal Institute of Technology (KTH), Stockholm, Sweden*
D. Heinz, *Rheinisch-Westfalische Technische Hochschule Aachen, Aachen, Germany*
Wolfgang F. Hölderich, *Rheinisch-Westfalische Technische Hochschule Aachen, Aachen, Germany*
Christina Hörnell, *Royal Institute of Technology (KTH), Stockholm, Sweden*
Sven G. Järås, *Royal Institute of Technology (KTH), Stockholm, Sweden*
E. Magnus Johansson, *Royal Institute of Technology (KTH), Stockholm, Sweden*
P. Govind Menon, *Royal Institute of Technology (KTH), Stockholm, Sweden*
Dennis Papadias, *Royal Institute of Technology (KTH), Stockholm, Sweden*
Daniel E. Resasco, *University of Oklahoma, Norman, OK, USA*
Mehri Sanati, *Royal Institute of Technology (KTH), Stockholm, Sweden*
Philippe O. Thevenin, *Royal Institute of Technology (KTH), Stockholm, Sweden*

RS•C
ROYAL SOCIETY OF CHEMISTRY

ISSN 0140-0568
ISBN 0-85404-214-8

Published by The Royal Society of Chemistry,
Thomas Graham House, Science Park, Milton Road, Cambridge CB4 0WF, UK

For further information see our web site at www.rsc.org

Typeset by Computape (Pickering) Ltd, Pickering, North Yorkshire, UK
Printed and bound by Athenaeum Press Ltd, Gateshead, Tyne & Wear

Preface

Applying the principles of catalysis to new challenges is one of the most rewarding parts of our profession. As I look at the changes in content of the Specialist Periodical Reports over the last ten years, it is clear that although the problems that must be solved change, many of the principles used to solve those problems do not. It is my hope to bring to you in this series some of both – the principles, and the problems to which they are applied.

I am especially pleased with the response of the authors that have been invited to contribute to this volume. Their attention to schedules, responsiveness to reviewers' comments, and enthusiasm has been very encouraging. I have had the opportunity to meet each of the authors over the last few years, and that has been one of the most rewarding elements of my job as Senior Reporter.

Michel Boudart (Stanford University) illustrates a principle – how kinetic coupling in and between catalytic cycles can help in the discovery process. He provides examples from both heterogeneous and homogeneous catalysis to demonstrate this principle.

Wolfgang Hölderich and D. Heinz (Technische Hochschule Aachen) illustrate a problem – the application of heterogeneous catalysts to the environmentally benign synthesis of fine chemicals, such as fragrances. This is one example of how catalysts can be developed to synthesize products efficiently and cleanly. This type of 'green catalysis' is one of the challenges we face, especially with large volume commodity chemicals.

Antony Dixon (Worcester Polytechnic Institute) describes one of the most innovative areas of research in catalysis – catalytic membranes. His review focuses on the emerging area of inorganic membranes which extends the use of these materials to the higher tempertatures needed for many synthesis processes.

Linda Broadbelt (Northwestern University) addresses a very specific and important problem – the use of catalysts to recover polymeric wastes into value-added products. This review concentrates on several high-volume polymers (such as polyethylene), and also discusses the use of catalysts in 'coprocessing' strategies in which wastes are processed with conventional feedstocks such as coal or petroleum.

Magnus Johansson, Dennis Papadias, Philippe Thevenin, Anders Ersson, Rolf Gabrielsson, Govind Menon, Pehr Björnbom, and Sven Järås (Kungl Tekniska Hogskolan) present a summary of the use of catalysts for combustion in gas turbines. The use of catalysts has tremendous promise in lowering NO_x emissions from this increasingly used power source. However, there are difficult challenges in developing thermally stable catalysts that can withstand the demanding conditions of a modern gas turbine.

Their colleagues at KTH, Mehri Sanati, Christina Hörnell, together with Sven Järås, also present a review of alkene oligomerization using heterogeneous catalysts. This is an important reaction in the production of gasoline and diesel blending stocks, and other widely used commodity chemicals such as detergents.

Finally, Ya-Huei Chin and Daniel Resasco (University of Oklahoma) review the catalytic oxidation of methane under lean-burn conditions. They focus on palladium-based catalysts, which are the most active for methane oxidation. They examine both the low temperature region (<800 °C), which is most relevant to exhaust control, and the high temperature region (>800 °C), which is applicable to gas turbines.

Volume 15 is underway and I look forward to bringing it to you. As always, comments are welcome.

James J. Spivey
Research Triangle Institute
Research Triangle Park, NC
USA
Email 'jjs@rti.org'

Contents

Catalysis, Volume 14

1

Catalytic Oxidation of Methane on Supported Palladium Under Lean Conditions: Kinetics, Structure and Properties

BY YA-HUEI CHIN AND DANIEL E. RESASCO

1 Introduction

The catalytic total oxidation of hydrocarbons is generally considered as an effective method to generate power and reduce emissions. In recent years, the interest towards catalytic combustion of methane has increased considerably. Methane has several advantages as an energy source. It has a high H/C ratio, and therefore the heat of combustion per mole of 'greenhouse' CO_2 generated is significantly higher for methane than that for other fuels. For example, while the combustion of methane generates 890 kJ mol^{-1} of CO_2 produced, the corresponding values for *n*-decane and coal are 680 and 390 kJ, respectively. At the same time, the level of sulfur and nitrogen impurities in natural gas is much lower than in other fuel sources. The two main applications of the catalytic combustion of methane are:

(1) catalytic combustion as an alternative to conventional thermal combustion in gas turbine combustors used for power generation;
(2) abatement of methane emissions from compressed natural-gas vehicles (NGVs).

In the first application, the use of a catalyst results in minimization of NO_x emissions. Owing to the presence of the catalyst, the combustor can operate at air/fuel ratios higher than those of a flammable mixture. The role of the catalyst is to initiate the reaction at the relatively low temperatures typical of the inlet of the combustor. As the exothermic combustion accelerates, the temperature along the combustor rises until the mass transfer limitation conditions are reached. At about this point, homogeneous gas phase reaction occurs, completing the combustion process.

The maximum temperature attainable in the combustor can be controlled by varying the air/fuel ratio.[1] This is a unique feature of the catalytic combustor, since without a catalyst a flame can only be sustained in a narrow air/fuel ratio range. By using an appropriate catalyst, instead of operating at the typical temperature of conventional flame combustion (*i.e.* 1500 °C), the combustor can operate under flameless conditions below 1300 °C. It is well known that the

Catalysis, Volume 14

emissions of nitrogen oxides can be minimized by reduction of the average temperature at which the combustion takes place[2–4] and by elimination of hot spots (1700–1800 °C) that are sites of rapid NO_x production. In a typical combustion process, 95% of NO_x is generated *via* the Zeldovich radical chain mechanism,[5] which is kinetically limited below 1500 °C. Above 1500 °C, the thermal NO_x production is doubled every time the temperature increases by 40 °C.[6] The use of a catalytic combustor not only increases the fuel efficiency, but also eliminates the possibility of local hot spots. In this way, most of the 'thermal' NO_x[3] is eliminated, resulting in a dramatic decrease in the NO_x emissions.[7] While a standard diffusion flame combustion turbine produces exhausts with NO_x concentrations higher than 150 ppm, a flameless catalytic combustor achieves concentrations of the order of 3 ppm.[8] The search for efficient catalysts, active over a wide temperature range, and able to withstand the severe conditions under which combustion takes place has generated a large body of information on catalytic materials and the mechanisms by which they operate.

In the second type of application, the intrinsic knock resistance of natural gas as a motor vehicle fuel makes it very attractive, particularly because this property is maintained over a wide range of air/fuel ratios.[9] Therefore, natural-gas engines can operate under lean conditions, thus increasing their fuel efficiency and minimizing the typical products of incomplete combustion, such as soot, CO, and volatile organic compounds (VOCs). In addition to this advantage, the lower cost and lower emissions associated with natural gas are important benefits that have greatly increased the potential of natural-gas vehicles. However, one of the concerns about the use of methane as a fuel is that it is a greenhouse gas with a global warming potential much higher than that of CO_2.[10,11] Thus, large emissions of unburned methane would become an environmental problem. Therefore, there has been considerable interest in the study of catalytic materials for methane combustion under typical exhaust conditions, with low CH_4 concentrations and in the presence of varying concentrations of O_2, H_2O, CO_2, SO_2, and NO_x. In this case, the desired characteristic is high activity in the presence of these components and at temperatures typical of exhausts (*i.e.* < 500 °C).

Among a large number of formulations investigated, the superiority of Pd-based catalysts has been widely recognized.[12,13] In this contribution, we will review the current ideas about the nature of the active sites and the effects of particle size, state of Pd, and metal–support interactions on the catalytic properties. We will analyse these effects for two different regimes. The low-temperature region (below 800 °C) is most relevant for combustion catalysts employed in catalytic converters for the abatement of unburned methane in exhausts.[14] Pd-based catalysts can also be potentially useful for the simultaneous elimination of unburned CH_4 and NO_x.[15] At the same time, low temperature combustion studies are also important to light-off the gas feed in catalytic combustors. The high temperature region usually refers to reactions carried out at or above the temperature of PdO decomposition in air (*i.e.* 800 °C). This region is particularly important for applications in gas turbine combustors. There are significant differences between the two temperature regimes. Therefore, we will analyse them separately.

2 Catalytic Combustion of Methane at Low Temperatures (below 800 °C)

2.1 Effects of Particle Size – The question of whether the methane oxidation is structure sensitive or structure insensitive has generated some controversy in the scientific literature. Some studies have demonstrated a strong dependence of activity on particle size.[16,17] For example, Figure 1 summarizes literature data of the variation of turnover frequency (TOF) with Pd dispersion for catalysts preconditioned in He or H_2. Despite some scattering in the data, a clear correlation is observed, indicating that the specific activity of Pd does in fact increase with particle size.[18]

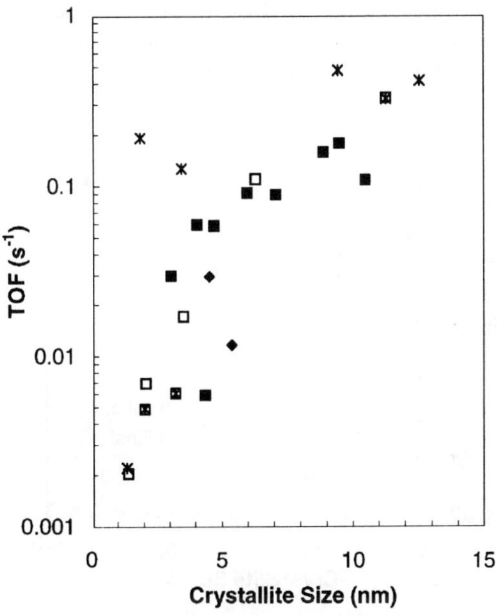

Figure 1 *Turnover frequency (TOF) at* 280 °C *as a function of Pd crystallite size on various supports. Data from the literature as follows: solid squares* (ref. 18); *open squares* (ref. 23); *diamonds* (ref. 29); *crosses* (ref. 16). *When the temperature was different from* 280 °C, *the TOFs were extrapolated using the reported activation energy*

The correlation is not so clear when the comparison is made with catalysts that have been pre-oxidized before the reaction. However, most studies seem to agree that the TOF increases with particle size. For example, Hicks *et al.*,[16] working at 335 °C on pre-reduced Pd/Al_2O_3 and $Pd/Y–ZrO_2$ samples, observed a steady state TOF of about 1.3 s^{-1} for large particles and 0.02 s^{-1} for small particles. Similarly, Muller *et al.*[19] examined the effect of crystallite size on Pd/ZrO_2 catalysts. The crystallite size, as measured by XRD, was varied by using different reduction temperatures. This study showed that the intrinsic activity increased with Pd crystallite size. At 330 °C, the TOF went from 0.035 to 0.17 s^{-1} as the crystallite size was varied from 6 to 12 nm.

By contrast, some investigations have found no clear correlation between particle size and TOF. For example, experiments conducted by Baldwin and Burch[20] on Pd/Al_2O_3 catalysts calcined in air at different temperatures and time periods showed a wide variation in activity without any clear relationship between the TOF and particle size. Similar data with little correlation with particle size was reported by Muto *et al.*[21] Ribeiro *et al.*[22] have found relatively modest increases in TOF as the particle size was varied. They observed that when the particle size increased from 2 to 110 nm, the TOF increased from 2×10^{-2} s^{-1} to 8×10^{-2} s^{-1} (see Figure 2). Despite this increase, the authors have suggested that methane oxidation is structure insensitive and the observed changes in TOF should not be ascribed to particle size effects.

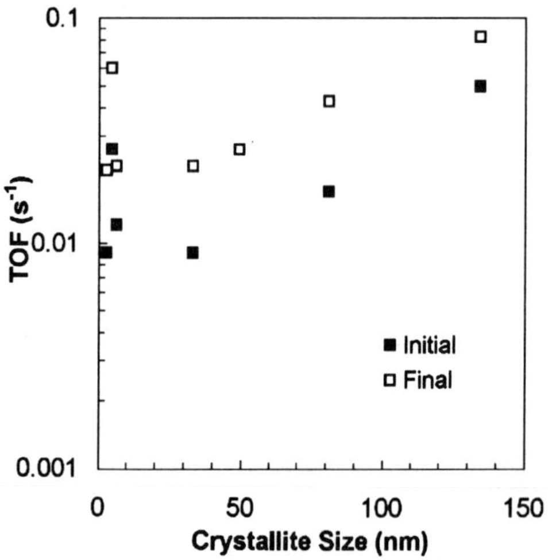

Figure 2 *Turnover frequency (TOF) as a function of Pd crystallite size at* 277 °C. *Initial: 5 min on stream; final: 24 h on stream*
(After ref. 22)

Fujimoto *et al.*[18] have identified several possible causes for the wide range of TOF values and, in some cases, the lack of correlation found in the literature. In the first place, they have pointed out that, in most studies, the TOF values are based on H_2 or CO chemisorption measurements conducted on reduced catalysts. However, these initial dispersion measurements may not represent the actual dispersion present under reaction conditions. Both oxidation state and morphology of the Pd particles may be altered in the presence of the reaction mixture. The second important consideration necessary to analyse the wide variation of TOF values is the role of product inhibition,[22] which, obviously, would greatly vary as the conversion varies. To make a proper comparison one needs to operate at similar conversions, and this aspect has not been considered in many of the studies. A third complication in comparing TOF values is the presence of long induction and activation periods exhibited by some of these catalysts. In most cases, the initial activity is very different from that at steady state and, as described in the next section, in some cases, steady state conversions are only achieved after several hours. Therefore, in making a comparison of activity data, one must be sure that steady state values are used. When all of these conditions were carefully taken into consideration,[18] a clear correlation between TOF and particle size was obtained, showing that the TOF indeed increases with particle size, at least for particles smaller than about 15 nm.

2.2 Time-dependent Catalytic Activity – The strong variation of activity as a function of time on stream is a typical feature of the methane combustion reaction on most Pd catalysts. Several different transient phenomena have been reported. In some cases, the activity is initially low, or even zero, but then it increases with time on stream. In other cases, the activity starts high but then it drops to a lower steady state value. The approach to the steady state has also been found to vary greatly. In some cases, it reaches a relatively constant value in a few minutes. In other cases, the activity still changes after several hours on stream.

It has been proposed[23] that the increases in activity as a function of time on stream may be due to the decomposition of inactive Pd oxy-chloride species, which attenuate the oxygen adsorption capacity. In line with this concept, previous investigations[24] have shown that the presence of halogenated hydrocarbons and organosiloxanes greatly inhibited the methane oxidation reaction. It was found that, above 350 °C, the presence of chlorine- or bromine-containing compounds has a strong deactivating effect, and thus the deactivation could be ascribed to the adsorption of halogens, which block sites needed for oxygen activation. Similarly, Simone *et al.*[25] observed higher activity on Pd catalysts prepared from nitrate than on those prepared from chlorides.

However, the presence of halogenated compounds may not be the only cause for the low initial activity of Pd catalysts, since halogen-free catalysts also exhibit time-dependent activity increases. Peri and Lund[26] have made a clear distinction between induction periods (*i.e.* time during which no activity is observed after exposing the catalyst to reaction conditions) and activation periods (*i.e.* the activity steadily increases with time on stream). Figure 3 illustrates these two types of transient behavior. According to Peri and Lund, the induction periods are related to the presence of residual chlorine, but the activation periods should not necessarily be linked to chlorine. They arrived at this conclusion from their observation that induction periods were observed only on samples containing chlorine. Those samples, from which chlorine had been eliminated by either washing the sample or pre-treating it under reaction conditions, did not exhibit induction periods, but they did exhibit activation periods. Similarly, chlorine-free samples prepared from nitrates and employed in other studies did not exhibit an induction period, while they still presented activation periods.[20,30]

If the presence of impurities is not the cause of the observed activation periods,

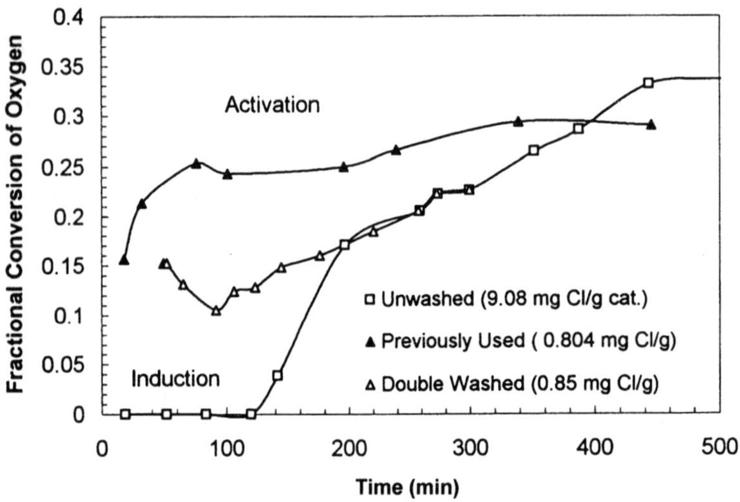

Figure 3 *Time-dependent activity of previously used, unwashed, and double washed Pd/SiO$_2$ catalysts with indicated amounts of residual chlorine* (After ref. 26)

a reconstruction or morphological change of the Pd particles must be responsible for the observed transients. In a combined characterization study, using FTIR, TEM, and nano-diffraction techniques, Garbowski *et al.*[27,28] observed clear reconstruction of the Pd particles supported on Al_2O_3. Before the methane combustion period, the reduced catalyst had metal particles in epitaxy with the support and exhibited mainly Pd (111) planes. After exposure to the reaction mixture, the (111) planes were exchanged by (100) and (110) planes. These authors speculated that these more open structures, generated during the reaction, are able to accommodate oxygen and form the PdO needed for the reaction, lowering the activation energy for the whole process.

The activity transients are also affected by the type of support used. As shown in Figure 4, the activation periods were very different for a series of 0.5 wt% Pd catalysts on various supports.[21] It was found that the activity of the alumina-supported catalyst first increased, reached a maximum, and then decreased with time on stream. By contrast, the silica- and silica–alumina-supported catalysts did not have an activation period and the conversion decreased with time from the start. Finally, with the mordenite support, the activity slowly increased and did not reach a maximum during the period of the study, about 7 h.

2.3 Effects of Thermal Pretreatments – The transient and steady state activities of Pd catalysts are highly sensitive to the pretreatment procedures. For example, the activities of Pd/Al_2O_3 samples were compared after different pretreatments with dry or wet air, H_2, N_2, or methane.[29] Table 1 shows the dispersion, the initial and steady state conversions on a 4% Pd/Al_2O_3 catalyst, preconditioned

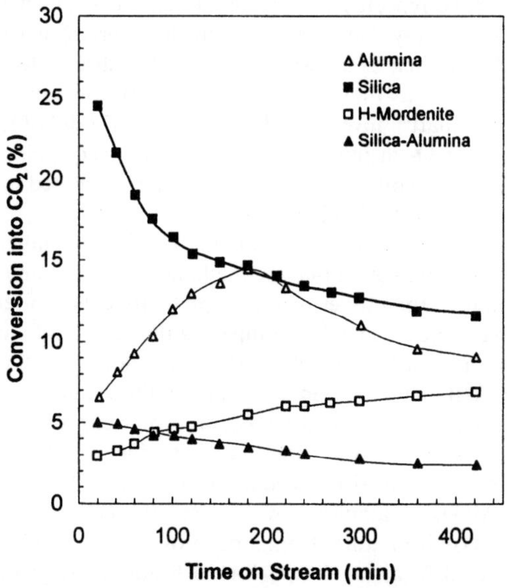

Figure 4 *Time-dependent activity curves for several supported Pd catalysts* (After ref. 21)

under various environments during either 1 h or 16 h. Although the dispersion was relatively constant for all samples, ranging from 18 to 25%, a very broad range of time-dependent conversions was obtained for the pretreatments conducted under different gases or for different periods of time. For example, the catalyst pretreated in H_2 for 1 h exhibited a low initial activity, but it increased by a factor of about 4.5 before reaching the steady state activity. On the other hand, the samples pretreated in dry air or N_2 only exhibited modest changes.

Table 1 *Effects of thermal pretreatment at* 500 °C *on catalytic activity of* 4% *Pd/Al$_2$O$_3$* (After ref. 29)

Pretreatment	*Initial dispersion (%)*	*Methane conversion(%)*	
		Initial	*Final*
1h/H_2	25	7.5	37.5
16 h/H_2	21	2.5	12.0
1 h/H_2	20	30.0	27.0
16 h/H_2	22	32.0	25.5
1 h/dry air	18	31.5	25.0
16 h/dry air	19	27.0	23.0
1 h/1% CH_4/air	21	17.5	19.5
16 h/1% CH_4/air	10	7.0	12.5

Reaction conditions: 100 $cm^3\ min^{-1}$ 1% CH_4/air, 0.10 g catalyst, at 280 °C.

Several studies have reported that catalysts pretreated under reducing conditions have a higher steady state activity than those pretreated under oxidizing atmospheres.[19,20,30] In fact, as shown in Table 1, the steady state activity of the catalyst reduced in H_2 was the highest among the various samples. In a similar study, Cullis and Willatt[30] investigated the activity of Pd/Al_2O_3 after pretreatments H_2+He, He, or O_2 in pulse measurements. In every pulse (except for the first one), the catalyst conditioned under H_2+He exhibited higher transient activities than those pretreated in He or O_2.

Although in Table 1, the samples activated in 1% CH_4/air show a relatively low activity, most studies have found that the methane combustion mixture is a better activating environment than other oxidizing pretreatments. For example, important differences have been found[17] in comparing the activity of a catalyst after 14 h at 600 °C on stream in a 1% CH_4/4% O_2/N_2 environment, with that of the same catalyst pretreated at the same temperature for the same period of time, but without CH_4. The catalyst aged under reaction conditions was significantly more active. Similarly, to study the variation in activity due to different pretreatments while keeping the crystallite size constant, Baldwin and Burch[20] compared the activities of two Pd/Al_2O_3 samples with the same particle size, conditioned either in air or in a 1%CH_4/air mixture. Comparison of activity at 350 °C showed that the sample pretreated in air at 500 °C for 16 h and subsequently treated in a reaction mixture at 405 °C for 60 h was two orders of magnitude more active than another sample only exposed to air at 600 °C for 16 h.

From all of the results reviewed in this section, it is clear that particle size may have an important effect on activity, but measurement of particle size alone cannot be used to predict the combustion activity. It is apparent that the structure and composition of the surface plays a decisive role in activity, and it greatly depends on the thermal history of the sample. As a result, catalysts with similar particle size may exhibit very different activities. Therefore, plots like Figures 1 and 2 are only valid when the samples have been pretreated under similar conditions.

3 Structure of Pd Species and its Effects on Catalytic Activity

3.1 Sintering and Redispersion of Pd Species under Different Environments – To understand the great differences in activity observed after different pretreatments, one needs to analyse the dramatic effects that the nature of the gaseous environment has on the morphology of the supported Pd particles. Ruckenstein and Chen[31,32] have investigated the morphological changes of Pd crystallites supported on thin alumina films during heating in oxygen or hydrogen. Their transmission electron microscope studies showed that, below 870 °C, oxidation of Pd crystallites with sizes between 3 and 35 nm under an oxygen treatment led to the migration and spreading of the Pd species. This phenomenon was explained in terms of interfacial forces. It was proposed that the oxidation started on specific sites, generating a porous structure that resulted in the appearance of pits and cavities. In a subsequent step, the oxidized Pd would have a tendency to migrate. Spreading of PdO was proposed to occur through an Ostwald ripening mechanism,[33] where PdO was emitted from large crystallites to the substrate. Extensive spreading, accompanied by rupture and fragmentation was observed at 500 °C. The authors proposed that the extent of spreading was determined by the equilibrium wetting angle (θ), obtained from interfacial free energies of gas–substrate (σ_{gs}), crystallite–gas (σ_{cg}), and crystallite–substrate (σ_{cs}) through the Young's equation.

$$\sigma_{gs} = \sigma_{cs} + \sigma_{cg} \cos \theta \qquad (1)$$

The spreading occurs because both, σ_{cs} and σ_{cg} decrease after the oxidation of the crystallite while σ_{gs} does not change significantly. This lowers the wetting angle (θ) and thus favors the dispersion of the crystallite onto the substrate. This wetting phenomenon results in the formation of a two-dimensional surface complex, which has also been proposed by Lieske and Volter[34] after their temperature program reduction (TPR) studies on Pd/Al_2O_3 treated in oxygen between 800 and 900 °C.

The redispersion phenomenon is not favorable with metallic Pd because it has a much larger wetting angle than the oxide. Therefore, contrary to the redispersion phenomenon observed under oxidation conditions, sintering of Pd particles has been generally observed upon reduction in H_2 above 400 °C.[31–36] In the sintering process, small metal crystallites migrate and merge with larger crystallites. Agglomeration is associated with the atomic mobility of Pd metals, which can be expressed in terms of the Tammann temperature, *i.e.* half of the melting point (in degrees K) of the bulk metal. In the case of Pd, the Tammann temperature is 641 °C. When the reduction temperature approaches this value, a rapid increase in surface mobility and, consequently, in particle size may be expected.[37] This trend is experimentally verified for the case of Pd/ZrO_2 catalysts, as illustrated in Figure 5.

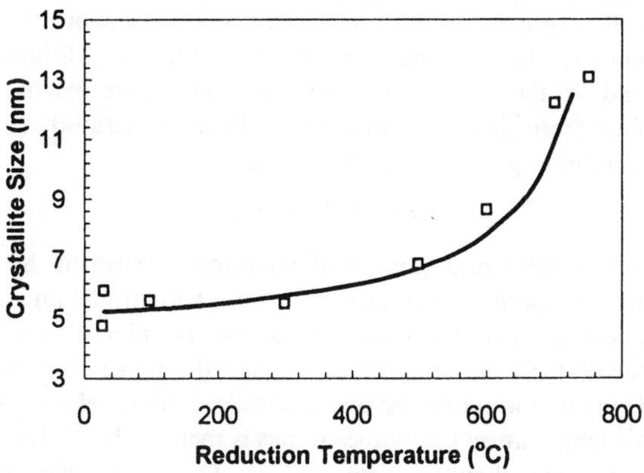

Figure 5 *Pd crystallite size on Pd/ZrO_2 samples after various 1 h reduction treatments in H_2 at varying temperatures*
(After ref. 19)

Owing to its relatively low Tammann temperature, Pd is much more sensitive to sintering than other noble metals, such as Rh, Pt, or Ir. The strength of bonding for these metals, as well as their Tammann temperatures follows the order Ir > Rh > Pt > Pd, which is the same order of stability to sintering observed experimentally.[38,39] Accordingly, even in the absence of any effects of the support, one may expect a higher propensity to sintering in Pd metal particles than in PdO particles, simply due to the higher volatility of the metal.[40]

Based on these trends, one may explain the higher activity generally observed with pre-reduced catalyst in terms of the particle size. It could be expected that during a pretreatment under hydrogen, coalescence of metal crystallites may occur. By contrast, under an oxidizing pretreatment and below the PdO decomposition temperature, Pd would redisperse. As a result, the pre-reduced catalyst should have larger particles and a lesser extent of interaction with the support than the pre-oxidized catalyst. As described above, the former would lead to high activity and the latter to low activity. However, these proposed trends have some limitations. For example, prolonged reduction treatments in H_2 have been found to suppress the overall activity greatly while the metal dispersion only shows a modest decay. In contrast to the high steady state activities of samples treated for 1 h in H_2, those treated for 16 h in H_2 exhibited a very low activity (see Table 1). It is possible that prolonged hydrogen treatments result in the generation of closed planes on the surface, such as Pd (111). As proposed by Garbowski *et al.*,[27,28] the presence of these closed structures would retard the oxygen uptake and increase the activation energy for the oxidation reaction. Interestingly, the length of pretreatment in other gases had no effect on the steady state activity.[29]

3.2 Interaction of Pd with Oxygen and its Effects on Activity – The high affinity of Pd to oxygen can be linked to its high combustion activity, superior to that of other noble metals such as Pt. For example, the oxygen uptake of a Pd catalyst at 400 °C is typically more than 100 times higher than a similar Pt catalyst,[30] in good correlation with their relative activities under lean conditions.[21] The rate and extent of Pd oxidation depends upon the temperature, partial pressure of oxygen, crystallite grain size, and heating rate. From a thermodynamic point of view, the dissociation of palladium oxide, expressed as,

$$PdO(s) \rightleftharpoons Pd(s) + \tfrac{1}{2}O_2 \qquad (2)$$

can be described in terms of a van't Hoff relationship between the dissociation temperature and the equilibrium oxygen pressure. Theoretical[41] and experimental data[42,43] have been used to construct the equilibrium plot shown in Figure 6, which shows the dissociation line that determines the range of stability of PdO. Therefore, PdO in air at atmospheric pressure will be stable below around 800 °C, while above this temperature the stable species is metallic Pd.[44] The shaded area indicated in the figure shows a typical 2–4% O_2 in the temperature range 500–600 °C. Therefore, under these conditions, it is clear that PdO is the thermodynamically stable phase. However, as described below the kinetics of the oxidation process may limit the extent of oxidation.

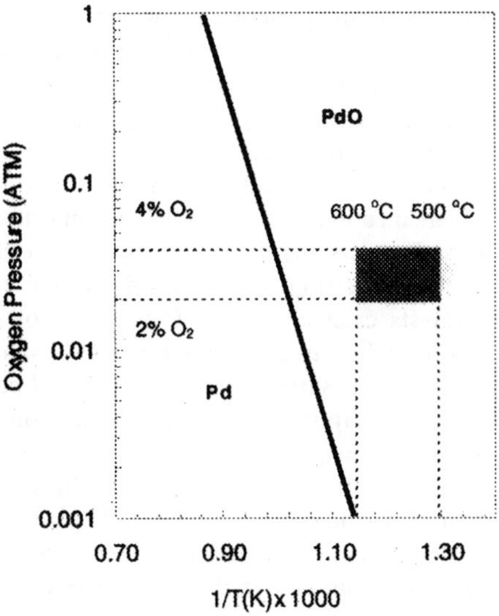

Figure 6 *Van't Hoff plot for equilibrium PdO decomposition. The shaded area indicates conditions between* 2 and 4% O_2 *and* 500 and 600 °C

Studies of oxygen uptake on polycrystalline Pd by White *et al.*[45] indicated that, below 740 °C and within the range 0.002–0.38 torr, the oxygen pressure has no effect on the rate of uptake. These authors concluded that the oxidation of Pd started from an oxygen-saturated surface, and the rate-limiting step was the transport of chemisorbed oxygen from the surface to the bulk. Under these conditions, the rate of oxygen adsorption is fast and the surface is rapidly covered. The ability of oxygen to penetrate into the bulk of Pd was previously proposed by Ertl and Koch.[46] They showed that, when an oxygen-covered Pd(111) surface was heated, changes in the LEED pattern and the work function indicated that oxygen had disappeared from the surface. However, no oxygen desorption was detected, denoting that it had penetrated into the bulk. The activation energy values measured in both studies for the oxygen penetration process were very similar, at 70–80 kJ mol^{-1}.[45,46]

Even though it is generally accepted that Pd oxide is formed under oxygen-rich reaction conditions,[16,17,23] depending on the reaction temperature, pretreatment history, air/fuel ratio, and metal dispersion, both PdO and Pd species may be present on the catalysts. For example, it has been found[42] that oxidation of Pd powder upon 8 C min^{-1} heating in air starts at 200 °C and it is only completed at about 700 °C. Similarly, Damiani *et al.*[47] have found that the extent of Pd oxidation increases with increasing temperature in the range 200–500 °C. They also showed that incomplete oxidation of Pd occurs under methane combustion conditions. Their XRD investigation of samples used in a 1% CH_4/2% O_2 mixture for 5 h at 350–500 °C revealed the presence of both PdO and Pd phases. From the thermodynamic and kinetics arguments discussed above, it may be expected that, when these two phases are present at low temperatures, they may not coexist on the surface, but they would rather form a complex structure in which an oxygen-rich layer covers a metallic Pd core. The oxidation of this core may be inhibited by the slow rate of oxygen transport through the solid.[45]

Using volumetric adsorption techniques, Hicks *et al.*[23] examined the equilibrium extent of oxidation of Pd/Al_2O_3 catalysts at 300 °C and 110 torr of oxygen after the catalysts were calcined in air at 700 or 900 °C, and subsequently reduced in H_2 at 300 °C. Figure 7 shows the extent of oxidation (bulk fraction) as a function of the initial Pd dispersion. In the Pd dispersion range 3–55%, the oxygen uptake for catalysts calcined at 900 °C was half of that for catalysts calcined at 700 °C. This difference in oxygen uptake at the same Pd particle size was ascribed to differences in the surface structure of the Pd particles. As shown by Ruckenstein *et al.*,[32] Pd oxidation starts on specific sites on the metal surface. These sites could be imperfections, such as kinks, steps, edges, or low-coordination sites. The sample calcined at 700 °C is expected to have a higher density of such imperfections than that calcined at 900 °C. The former may have experienced the cracking and pit formation process described in Section 3.1. The latter has been exposed to a temperature above the PdO dissociation and, as a result, the Pd metal may have agglomerated and annealed into highly ordered structures. Therefore, for the same particle size, the sample exhibiting a more irregular surface will result with a higher extent of oxidation.

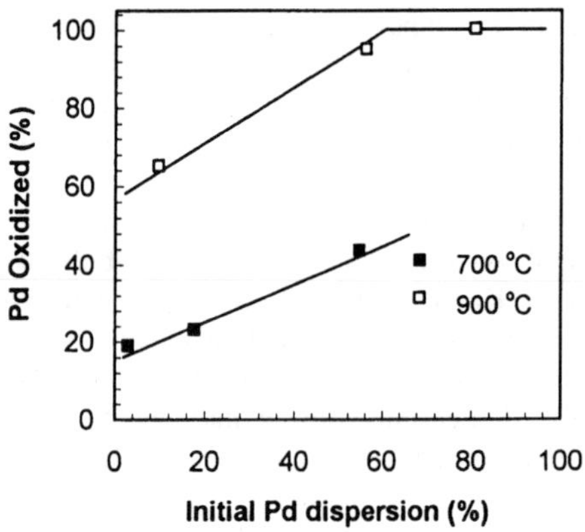

Figure 7 *Percent of bulk Pd oxidized at* 300 °C *as a function of the initial Pd dispersion after two different calcination temperatures,* 700 *and* 900 °C (After ref. 23)

Burch and Urbano[29] have attempted to correlate the combustion activity in pulse experiments with the rate and extent of oxygen uptake at 300 °C for 4% Pd/Al_2O_3 catalysts exposed to various pretreatments. The shape of the oxygen uptake curves as a function of time was dependent on the type of pretreatment performed on the catalysts. However, a common characteristic in all cases was that the initial oxygen uptake was much faster than after a monolayer had been completed. As a result, clear breaks were observed on all the uptake curves. In a separate set of experiments, the samples were exposed to pulses of $CH_4/O_2/N_2$ mixtures and the conversion was followed as a function of time after the pulse injection.

A quantitative comparison between these two sets of experiments cannot be made since they were performed under very different conditions, one under static 10 torr O_2, the other after the injection of a pulse containing 1% CH_4 and 10% O_2. However, a qualitative comparison of the shapes of the two curves implies important differences. The most important one is that while the oxygen uptake curve presented a clear break when a monolayer was completed, no break was present in the conversion curve. From this behavior the authors have concluded that chemisorbed oxygen was not the active species. They proposed that oxidized Pd is much more active and that the activity continuously increases with the extent of oxidation until a plateau is reached after formation of several oxygen monolayers.

In a more recent study, Carstens *et al.*[48] have examined the relationship between the methane conversion and the fraction of oxidized Pd by simultaneously measuring the extent of oxidation and the activity of a pre-reduced Pd/ZrO_2 catalyst. The extent of Pd oxidation was determined by TPR (by CH_4) at various points during the activation period after rapidly quenching from the reaction temperature of 500 °C. The instantaneous methane conversion was zero until the extent of Pd oxidation was 0.05, then it increased almost linearly with the fraction of oxidized Pd until a saturation point was reached at an oxidation extent of 0.35, which corresponds to about 6–7 ML. After this point, the extent of Pd oxidation continued to increase while, in agreement with Burch and Urbano, the catalytic activity remained unchanged.

The activity of pre-dosed oxygen towards methane oxidation was investigated in a temperature-programmed reaction experiment.[23] Oxygen was pre-dosed at 21 °C or at 300 °C and at an oxygen pressure of 110 torr, followed by exposure to methane, as the temperature was linearly increased with time. It was found that over a range of bulk oxidation fraction of 18–25%, methane reacted stoichiometrically with all the oxygen dosed at either temperature in a CH_4/O(ad) ratio of 0.4. According to the following expression, the expected stoichiometric ratio for CH_4 reacting with pre-dosed oxygen atoms would be 0.25:

$$CH_4 + 4\,Pd_sO \rightarrow 4\,Pd_s + CO_2 + 2\,H_2O \quad (3)$$

The authors have proposed that the observed ratio of 0.4 can be explained if deposition of carbon on the clean Pd surface occurs after removal of the adsorbed oxygen in a ratio $CH_4/Pd_s = 0.15$. To support this explanation, in a separate experiment it was confirmed that this number is in agreement with the methane uptake by reduced Pd. It is important to note that all of the pre-dosed oxygen reacted with methane, keeping the same CH_4/O ratio, even for samples with different degrees of surface and bulk oxygen contents. The authors explain this fact by proposing that during the oxidation process the Pd crystallites are broken apart, exposing a large fraction of inner Pd atoms for further oxidation. However, the possibility of bulk oxygen migrating to the surface during the reaction with methane cannot be ruled out.

A few surface science studies[49–51] have investigated the structure of adsorbed oxygen on Pd at temperatures relevant to the methane oxidation reaction. For example, it has been found that oxygen dissociatively adsorbs on Pd (100) and occupies the four-fold hollow sites below 0.25 monolayers, forming an ordered p (2X2) phase below 330 °C. At higher coverages[52] a mixture of a dense $(\sqrt{5} \times \sqrt{5})R27^\circ$ phase and a less dense '(2X2)' phase are formed at a temperature near the oxygen desorption temperature, 280 °C, regardless of the original phase present. This high-density surface structure has been linked to the formation of PdO overlayers oriented with the Pd (100) substrate, where the $(\sqrt{5} \times \sqrt{5})R27^\circ$ and '(2X2)' phases correspond to PdO(110)//Pd(100) and PdO(001)//Pd(100), respectively.[53]

Klier *et al.*[52] have a different interpretation for the phase transformation observed at high coverages. By modelling the experimental LEED and HREELS spectra with kinematic theory and simulated vibration frequencies, they showed that, above half monolayer, oxygen occupies some of the nearest-neighbor four-fold sites. The repulsive interaction between oxygens would induce a lateral shift of the surface Pd atoms, and would give rise to distorted four-fold sites, which would be stabilized by an increase in the Pd–O bond strength. From this analysis it was speculated that highly active oxidation catalysts could be achieved by maintaining a low surface oxygen density, and avoiding the phase transformation that would result in strong Pd–O bonds.

Although the explanations of the role of oxygen in the methane combustion reaction are diverse, a common concept that emerges in most studies is the realization that different forms of oxygen can be present in the catalyst and that they have different reactivities. The following sections analyse the methodology that has been used to characterize the various Pd–O species and their role in activity.

3.3 Characterization of the Pd–O Bond Strength – It is generally agreed that at least two types of PdO species may exist under methane oxidation conditions:

(1) PdO interacting with the PdO bulk oxide or metallic Pd crystallites; and
(2) stable PdO interacting with the support,[23,54,55] perhaps forming a two-dimensional surface complex, *e.g.* isolated Pd–aluminate over alumina support.[55]

The former species is typical of samples with low Pd dispersion where the majority of the PdO interacts with other Pd species. By contrast, the latter would be present on well dispersed samples, where the metal–support interaction is dominant. The relative amount of these Pd species is related to the crystallite size and the types of support used. These Pd species have been identified by their distinct oxygen adsorption–desorption characteristics. For example, microcalorimetric studies revealed the presence of strong Pd–O bonds in small Pd metal clusters.[56,57] The integral isothermal heat of adsorption (Q_{ad}) of oxygen on Pd powder, and Pd supported on SiO_2, Al_2O_3 or SiO_2–Al_2O_3, was about 210 kJ mol^{-1} and independent of supports and crystallite size above 3 nm.[57] However, below 3 nm, the strength of Pd–O bond was strongly dependent on the crystallite size, with a sharp increase in Q_{ad} up to 290–335 kJ mol^{-1} as the crystallite size decreased. The authors attributed this sudden change to the deviation of electronic properties from bulk Pd. This deviation was explained in terms of an increase in the d-orbital energy gap due to the lower average coordination number in small Pd crystallites. Such an increase in energy gap has been experimentally observed by X-ray absorption spectroscopy.[58] Of course, electronic and structural effects are always inter-related and it is generally difficult to distinguish which one is more relevant. Small metal clusters (below 3–5 nm) have a large amount of defective sites with low-coordination features such as kinks, steps, and edges. They may present adsorption sites structurally different than those typically found on the low-index planes that abound on the larger particles.[59] Whatever the exact reason behind the observed changes in bond strength, the important point for this discussion is that the Pd–O bond strength significantly increases when the particles become very small. Also, as the extent of Pd oxidation increases, one may expect a decrease in the Pd–O bond strength.[60] If oxygen removal is part of the reaction mechanism, a lower TOF can be expected for smaller particles. This concept will be further discussed in Section 4.2.

The state of Pd in methane oxidation catalysts has been characterized by a wide variety of techniques. For example, in agreement with the calorimetric results mentioned above, temperature program desorption (TPD) of oxygen adsorbed on Pd/Al_2O_3,[61] as well as TPR studies on Pd/ZrO_2,[18] have shown modifications of the Pd–O bond strength at high dispersions. The temperatures corresponding to the maximum desorption rate in TPD and maximum reduction rate in TPR were found to decrease significantly with increasing crystallite size. Likewise, XPS experiments[54] on oxidized Pd/Al_2O_3 catalysts have shown that the low-loading catalysts had Pd $3d_{5/2}$ binding energies 1.6 eV higher than those for high-loading samples and the bulk PdO reference, at 336.7 eV. The higher binding energy values are consistent with an increase in the Pd–O bond strength as the Pd loading decreases, which is normally accompanied with higher dispersion and smaller particles. Similar XPS results were obtained by Widjaja *et al.*[62] on a series of Pd/Al_2O_3 catalysts with crystallite size ranging from 10 to 80 nm. A gradual decrease in the Pd $3d_{5/2}$ core level binding energy was found with increasing crystallite size.

Temperature program oxidation (TPO) studies on highly dispersed Pd/Al_2O_3 catalyst[55] have shown a single oxidation peak with uptake of excess oxygen at an O/Pd ratio greater than two. By contrast, on catalysts with high Pd loading and low dispersion, the oxidation peak splits into two peaks with nearly complete oxidation (O/Pd *ca.* 0.91). The appearance of the double peaks has been explained as follows. The low temperature oxidation peak would correspond to the oxidation of bulk Pd particles, while that at higher temperatures was assigned to Pd species modified by the support. The important point, however, is that these support-modified species have a different oxidation behavior in low and high Pd loading catalysts. On the low loading, highly dispersed catalysts, the oxygen uptake per Pd atom is much higher than on the low dispersion catalyst. This difference could be due to a different environment of the Pd atoms on the highly dispersed catalysts. EXAFS data on Pd K edge,[63] support the idea that, on highly dispersed catalysts, oxygen-rich Pd occupies the octahedral aluminum-vacant sites, with an average of 5–6 neighboring oxygen atoms. The extra lattice oxygen would be responsible for the excess oxygen uptake during TPO of low-loading Pd/Al_2O_3 catalysts.[55]

The presence of two distinct states of Pd has also been demonstrated by TPR studies[21,34] which usually show two or more H_2 consumption peaks. The consumption peak observed below room temperature has been ascribed to the reduction of bulk PdO clusters, while consumptions at higher temperatures have generally been related to metal–support interactions. The negative peak typically observed near 60 °C is due to H_2 release from the decomposition of β Pd–hydride.[34,64,65] TPR has been used to correlate the amount of H_2 consumption with methane oxidation activity.[21] It has been observed that the fresh catalyst that exhibited low activity had a H_2/Pd ratio of 0.3. As the catalyst activated and reached its maximum rate the H_2/Pd ratio increased to 1.6, and finally after 7 h on stream the partially deactivated catalyst exhibited a H_2/Pd ratio of 1.25.

The role of the different types of PdO species in methane oxidation has been investigated by several authors. Ribeiro *et al.*[22] compared two 8.5% Pd/Al_2O_3 catalysts, calcined in air below and above the dissociation temperature of PdO (*i.e.* 700 and 900 °C, respectively). H_2–O_2 titration of the catalyst after calcination at 700 °C (sample A) gave a Pd surface site density almost three times higher than that of the one calcined at 900 °C and immediately quenched to liquid nitrogen temperature (sample B). Despite a much larger density of Pd sites, sample A showed the same rate per gram as sample B, indicating that a large fraction of Pd sites on sample A had very low or no activity. From the discussion in the previous sections, it can be expected that on sample A, a large fraction of Pd species is in the form of highly dispersed PdO (inactive) species, interacting with the Al_2O_3 support. By contrast, in sample B the most common species would be PdO clusters (active), or as proposed by some authors a PdO layer spread on a metallic Pd core.

Similar conclusions can be obtained from FTIR studies[16] of adsorbed CO on Pd/Al_2O_3 catalysts. The linear/bridge intensity ratio was used as a semi-quantitative way to determine the relative amounts of Pd on low-coordination sites and on crystal planes. The linear/bridge ratio would be high for small particles and low for large crystallites. In agreement with the concepts discussed above, it was found that the catalysts that exhibited a low linear/bridge ratio after the reaction period had the high catalytic activity. The samples that exhibited the highest TOF were those with either a low initial dispersion, or those with initially moderately high dispersion, but the linear/bridge ratio decreased significantly during reaction.

3.4 Stability of Pd Species on Different Supports – Most studies on Pd–support interactions have focused on alumina-supported catalysts. The situation may be somewhat different on other supports. However, it is important to note that when comparing catalysts with different supports, the effects of particle size and metal–support interactions are inter-related and sometimes cannot be distinguished. Keeping this in mind, a simple criterion to determine the ability of a support for preventing metal sintering is to use the interfacial free energies (σ) involved in the Young's equation mentioned in Section 3.1. The degree of metal spreading onto the support is favored as the substrate–gas interfacial free energy (σ_{sg}) increases. Accordingly, the ability of a support to inhibit sintering based on the σ_{sg} values should be,

$$ZrO_2\,(1130) > Al_2O_3\,(925) > SiO_2\,(605)$$

where the experimental σ_{sg} values, expressed in ergs cm^{-2} are indicated for each support.[66] This trend is supported by experimental data. For example, it is generally observed that metal particles supported on silica sinter faster than on alumina.[67] At the same time, Pd/SiO_2 catalysts exhibit higher activity[21] and shorter activation periods[20] than Pd/Al_2O_3 catalysts. These differences can be explained in terms of a greater extent of sintering on the silica-supported catalysts, which leads to larger and more active Pd particles.

Zirconia and alumina have high σ_{sg} values, inhibiting the mobility of Pd species. However, when they are modified by the addition of promoters such as La or Y, their stabilization of Pd species may be greatly altered. For example, the use of Y-stabilized zirconia as a support for Pd resulted in a catalyst that started with a high density of low-coordination sites, as indicated by a high linear/bridge adsorbed CO ratio, but drastically dropped during the reaction.[16] This change was accompanied by a large increase in activity, which could be related to the agglomeration of Pd species. Similarly, TPO studies[55] have shown that the addition of La to alumina results in an inhibition in the metal–support interactions between Pd and alumina and the elimination of the highly dispersed, oxygen-rich Pd.

When some zeolites are used as supports, a very high and stable dispersion of Pd can be achieved.[68] In particular, when the zeolite contains protons, they become effective anchoring sites for the stabilization of positively charged Pd ions. From the conclusions presented in the previous section, it is reasonable to expect that Pd supported on zeolites should exhibit very low methane combustion activity. This expectation is confirmed by the trend shown in Figure 8, which presents a comparison of the steady state conversions as a function of temperature on two zeolite-supported catalysts and a silica-supported one.[69] The light-off temperature noticeably increased in the sequence 0.3% Pd/SiO_2 < 0.3% Pd/H-Mordenite < 0.1% Pd/H-Mordenite. This trend, which was also observed for a similar Pd/H-ZSM-5 series, is in agreement with the concepts proposed above. That is, the combustion activity was markedly reduced on the zeolite-supported, low-loading catalysts, where Pd was highly dispersed and in close interaction with the support, most probably in the form of isolated ions, stabilized inside the zeolite. This situation is certainly most favorable in the 0.1% Pd/H-Mor, less favorable in the 0.3% Pd/H-Mor, and not possible on the silica-supported catalyst. In fact, EXAFS characterization of these catalysts showed that, following the CH_4 combustion reaction, Pd was in the form of PdO on the silica-supported catalyst, but as highly dispersed Pd ions in the case of the low-loading catalysts supported on H-Mordenite. At higher Pd loading, the amount of PdO clusters increases, correlating with the increase in activity.

In apparent contrast with these results, Li and Armor[70] have observed high methane combustion activity on Pd catalysts supported on ZSM-5. However, the non-acidic zeolites employed in that study contained a relatively high Pd loading (*i.e.* about 3.4 wt%), exceeding the capacity of the zeolite to stabilize isolated Pd^{n+} ions.[71] More recent studies have shown that when the Pd loading is greater than about 0.3 wt%, palladium oxide clusters begin to form and the combustion activity greatly increases.[15,71]

A series of Pd catalysts supported on mordenite and ZSM-5 zeolites has exhibited a clear relationship between the activity and the $Al_2O_3/(Al_2O_3 + SiO_2)$ ratio.[72] When this ratio and, consequently, the anchoring capacity of the zeolite increases, the activity of the catalyst decreases, *i.e.*, it appears as a general trend that as the interaction with the support increases, Pd becomes less active for combustion.

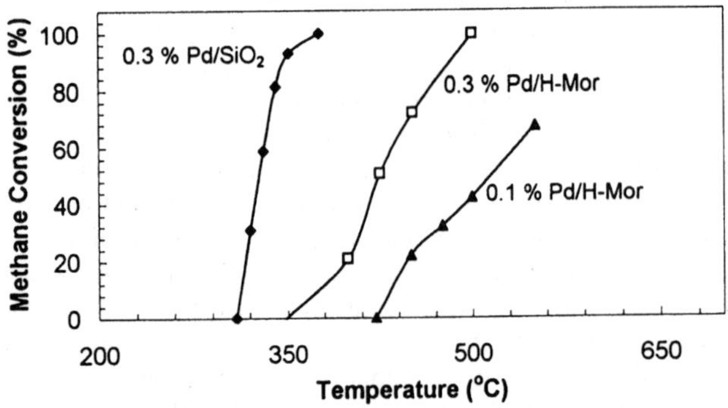

Figure 8 *Methane combustion as a function of temperature over Pd catalysts. Reaction conditions: GHSV* 10 000 h^{-1}, 1% CH_4, 2.5% O_2 (After ref. 69)

4 Reaction Kinetics and Possible Mechanisms

4.1 Mass and Heat Transfer Limitations – The methane oxidation reaction is very exothermic and relatively fast. Therefore, it presents the possibility of heat and mass transfer limitations during the measurement of reaction rates. One of the methods that can be used to check for heat and mass transfer limitations is the Koros–Nowak test.[73] Ribeiro *et al.*[22] have employed this test to demonstrate that their data were not affected by heat or mass transfer limitations. In the Koros–Nowak test, rate measurements are conducted on catalysts with similar dispersions but different metal loadings. The comparison should be done at the same conversion, since, as mentioned above, the combustion products inhibit the reaction. If the observed TOFs are the same, it can be concluded that, under the tested conditions, those samples are not subject to heat or mass transfer limitations. This has been the case for the samples tested by Ribeiro *et al.* As shown in Table 2, the same TOF was obtained on two samples with the same Pd dispersions, but with loadings varying by an order of magnitude.

Table 2 *Test for heat and mass transfer limitations* (After ref. 22)

Catalyst	*Rate* (10^{-3} mol s^{-1} g^{-1})	*TOF* (10^{-2} s^{-1})	*Methane conversion* (%)	*Percentaage of metal exposed* (%)	*Crystallite size* (nm)
7.7% Pd/Si–Al_2O_3	220	1.7	1.6	18.1	6.2
0.77% Pd/Si–Al_2O_3	20	1.7	1.52	16.1	7

The apparent activation energies (E_{app}) may also be a good indicator for heat and mass transfer limitations. For example, most studies have found a constant activation energy of between 75 and 180 kJ mol^{-1} throughout the temperature range of 250–650 °C. However, earlier studies from Firth[74] and Cullis and Willatt[30] have independently reported a drastic drop in E_{app} above *ca.* 470 °C, while Firth and Holland[75] have shown a break in the Arrhenius plot at about 290 °C. In these contributions, the authors attributed the transitions to a change from kinetic control at low temperature to the mass transfer control at high temperature. The mass transfer control could be due to intraparticle pore diffusion limitation, as well as due to the interparticle transfer of methane to the catalyst surface. Cullis and Willatt[30] have provided another interpretation. They suggested that, at low temperature, the E_{app} included the sum of the activation energy for methane combustion and the activation energy for Pd oxidation. On the other hand, oxidation of Pd is no longer kinetically controlled at high temperatures, so the E_{app} would be similar to that for methane combustion without the Pd oxidation. A similar explanation was given by Firth and Holland[75] to the changes in E_{app} at 290 °C in their microcalorimetric studies of pre-reduced Pd over alumina beads, where the drop in E_{app} (87 kJ mol^{-1}) was found to be comparable to the heat of formation of PdO (85 kJ mol^{-1}). This also implied that the rate-determining step for low temperature combustion includes the oxidation of Pd to PdO, while at temperatures high enough so that Pd oxidation is rapid, the reaction is not controlled by the formation of PdO species. However, other studies either did not observe a drastic drop in E_{app} across the transitional temperature[20,21,76] or found a E_{app} at high temperature which falls within the range of that found at low temperatures.[72]

4.2 Kinetics and Mechanisms – In most kinetic studies, the methane oxidation rate exhibited a first order to half order in methane[18,21,22,30,77] but a very weak dependence on oxygen concentration.[18,76,78] Firth[74] first discovered the constancy of rate with oxygen concentration, over a 15-fold increase in oxygen pressure. He proposed that methane does not compete with the strongly chemisorbed oxygen on a Pd surface. Although it is accepted that the oxidation reaction involves surface oxygen species, either chemisorbed or as lattice oxide, the interaction of methane with Pd under a CH_4/O_2 mixture is not well established. Earlier studies invoked an Eley–Rideal model to explain the methane oxidation reaction.[79] In this model, gaseous methane would react with adsorbed atomic oxygen.

More recently, a redox (Mars and van Krevelen) mechanism has been favored by several authors. Participation of lattice oxygen was confirmed by isotopic labelling experiments working with $Pd^{18}O/ZrO_2$.[80] It was found that at 300 °C, at least 20% of the CO_2 produced from pulses of $CH_4/^{16}O_2$ contained oxygen from the lattice. At higher temperatures, the exchange of oxygen from CO_2 and the lattice is so rapid that it was not possible to determine the fraction of CO_2 produced by the redox mechanism.

In the Mars and van Krevelen mechanism, gas phase methane interacts with lattice oxygen and reduces the Pd. The Pd metal is subsequently oxidized by chemisorbed oxygen, completing the redox cycle. Fujimoto *et al.*[18] have also proposed a redox mechanism in which methane first adsorbs on a coordinatively unsaturated Pd site on the surface of a PdO crystallite (see Figure 9). Subsequently, a H atom is abstracted by interaction with an adjacent Pd–O species to form a Pd–OH species. This hydrogen abstraction would be the rate limiting step. Combinative desorption of two OH species recovers the catalytic sites to yield an unsaturated Pd and a Pd–O species. In this mechanism, the ability of a catalyst to create oxygen vacancies is essential. Consequently, the strength of the Pd–O bond has a direct effect on the activity, a conclusion that is in direct agreement with the concepts on Pd–O interactions discussed in previous sections. Based on this concept, it is then reasonable to expect that a reduced catalyst has initially a low activity, due the high strength of the initial Pd–O bonds. Although metallic Pd can activate methane,[81–83] if the Pd–O bonds are too strong, the rate would still be low. As the reaction progresses and the oxygen content increases, the Pd–O bonds become weaker and the activity increases. A similar idea invoking the role of surface oxygen vacancies as active sites for methane activation has also been proposed for the methane oxidation reaction on a Cu/Al_2O_3 catalyst.[84]

This mechanism is also consistent with the higher reactivity of Pd compared to other noble metal catalysts. For example, Pd is more easily reduced than Pt,[85] with a reduction temperature 120–200 °C lower, which correlates with its higher methane combustion activity under lean conditions.[86,87] For example, at 500 °C and at the same metal loading, the rate of methane combustion on a 0.5 wt% Pd/Al_2O_3 is almost 50 times higher than that on Pt/Al_2O_3.[21] However, the opposite is true under fuel-rich conditions.[87] The relative activity could be linked to the heat of oxygen adsorption on Pd and Pt in their oxide or metallic forms. Higher activities are obtained on surfaces with lower heat of adsorption. In fact, quantum-chemical calculations reveal that under an oxidizing atmosphere, Pd, most probably in the form of PdO, has a lower heat of oxygen adsorption than PtO.[88] However, metallic Pd, which may be present under fuel-rich conditions, has a higher heat of oxygen adsorption than metallic Pt. In agreement with these trends, Machida *et al.*[89] have also found that metallic Pd, generated by high-energy ion beam implantation, is less active than PdO for low temperature combustion. As opposed to Pd, comparable Pt and Rh catalysts, prepared in a similar fashion, showed an increase in activity after the high-energy implantation.

The lower activity exhibited by the small Pd particles can also be explained in

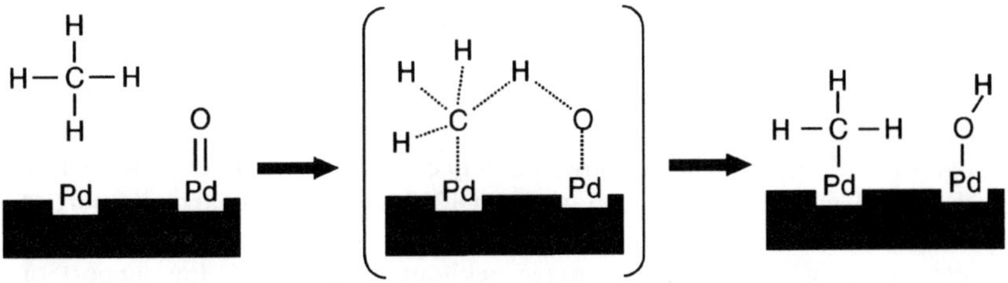

Figure 9 *Reaction scheme for the activation of methane on a surface Pd–PdO pair. The methane interaction with the surface starts at the coordinatively unsaturated Pd site. The activation occurs by H abstraction on the neighboring Pd–O site*
(After ref. 18)

terms of the proposed role of the oxygen vacancies. Spectroscopic evidence obtained by Weber,[90] shows that the band gap energies increase as the size of the Pd clusters decreases. This increase in energy, which can also be inferred from the XPS data,[54] would make the generation of oxygen vacancies more difficult, thus inhibiting the redox process.

Fujimoto *et al.*[18] oppose the idea of the coexistence of metallic and oxidized Pd on the surface of a crystallite in the low temperature region. They argue that the coexistence of these unstable structures would cause oscillations, which are not detected under lean conditions and below the PdO dissociation temperature. Oscillations in the reaction rate, coupled with oscillations in the temperature of the catalyst bed, have been observed under rich conditions on Pd/TiO_2[91] as well as Pd and Pt supported on Al_2O_3.[92] The appearance of oscillations between 400 and 600 °C was attributed to periodic phase changes on the Pd surface. It was proposed that metallic and oxidic sites were present on the surface; the latter would be active for methane conversion to CH_x species, while the former would be responsible for direct oxidation.[91] However, these oscillations are only observed at CH_4/O_2 ratios higher than the stoichiometric ratio. Under these reducing conditions, the PdO phase may not be stable, and metallic Pd may be present on the surface. It is interesting to note that, even under fuel-rich conditions, PdO phases are still present and are completely eliminated only when oxygen has been totally consumed.[93] It has been observed that the addition of small pulses of H_2 during reaction causes a temporary increase in the methane combustion rate.[48] This increase was ascribed to the ability of metallic Pd to activate CH_4. However, it was stated that the presence of metallic Pd could not be sustained under oxygen-rich, low-temperature, steady state combustion. Under these conditions the most probable structures are PdO species interacting with the support, PdO clusters, and Pd particles covered by a layer of PdO. Oxygen vacancies on this layer are certainly possible and may be directly linked to the catalyst activity. As discussed in the following section, this situation may change at high temperatures.

5 Catalytic Combustion of Methane at High Temperatures (above 800 °C)

Studies in methane combustion at elevated temperatures have focused on stabilizing PdO against decomposition and severe sintering.[41] At temperatures above 1000 °C, Pd has relatively low volatility compared with other noble metals, such as Pt, Ru, Os, or Ir.[94] For example, at 1230 °C, the metal vapor pressure of Pd is about 10 times lower than the corresponding value for Pt.[95] Also, Pd is stable over a wide temperature range, without reacting with the support or forming inactive ionic solid solutions, such as the aluminates that form in the case of Rh.[96] At temperatures above 1000 °C, even inert materials such as quartz are active for hydrocarbon combustion.[97] Therefore, in that temperature range, high activity is not so critically important. However, long-term operability and the ability of maintaining a stable/gradual change in conversion with a temperature ramp are of great importance. In that sense, Pd-based catalysts are also the material of choice for high temperature operations.

5.1 Hysteresis in Activity and Oxygen Content – Upon heating and cooling under air or reaction conditions within a temperature range of 500–1000 °C, a hysteresis between the decomposition and re-formation of PdO has often been observed.[98,99,103] Using gravimetric analysis (TGA), Farrauto *et al.*[98] have shown that PdO supported on alumina rapidly loses oxygen at about 800 °C upon heating in air at 20 °C min^{-1}. By contrast, during cooling, reoxidation does not start until 650 °C (see Figure 10). In addition to this hysteresis, the Pd/Al_2O_3 exhibited a peculiar lack of reversibility upon reoxidation. The TGA profile showed that the final weight after cooling was greater than the initial weight. When similar temperature cycles were performed in CH_4/air mixtures, while measuring the catalytic activity for methane oxidation on Pd catalysts, hysteresis in reaction rates were also observed.[99] The appearance of hysteresis in both activity and oxygen content in the catalyst is a strong indication that they are directly related. As illustrated in Figure 11, when the temperature is increased above the decomposition temperature and then cooled to a point at which Pd is still metallic (*e.g.* 700 °C), the activity of a Pd/Al_2O_3 catalyst is the same as that of the bare alumina support. This result would suggest that, at this point, metallic Pd is inactive. When the temperature is further cooled below the point at which oxygen starts to be taken up by Pd (see Figure 10), the activity increases and approaches that of the original Pd/Al_2O_3 catalyst.

The above experiments were conducted at conversions close to 100%. Further information about the relationship between PdO decomposition and activity has been recently obtained by Lyubovsky and Pfefferle[100] in an experiment conducted at lower conversions, with the temperature first increasing and then decreasing. As the temperature was increased from 800 to 870 °C, which may have caused the decomposition of PdO, the activity rapidly increased. At that point, the temperature was reduced, which caused a rapid decrease in activity. As the temperature kept decreasing, the activity increased again with a negative apparent activation energy similar to that described by Farrauto.[98] The main difference with Farrauto's work is the sudden increase in activity observed during the PdO

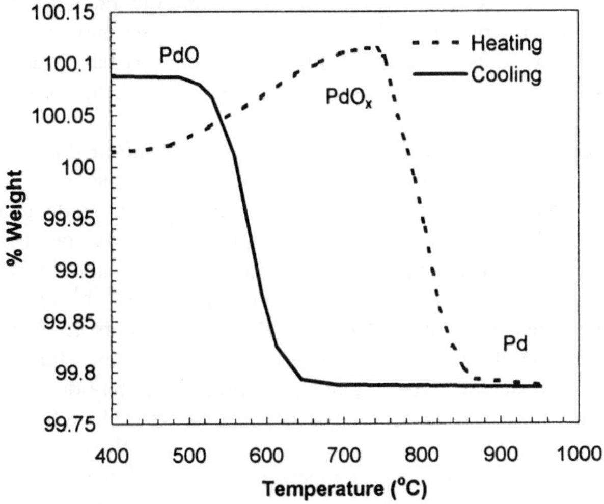

Figure 10 *TGA profile for thermal decomposition of Pd/Al_2O_3 in air*
(After ref. 98)

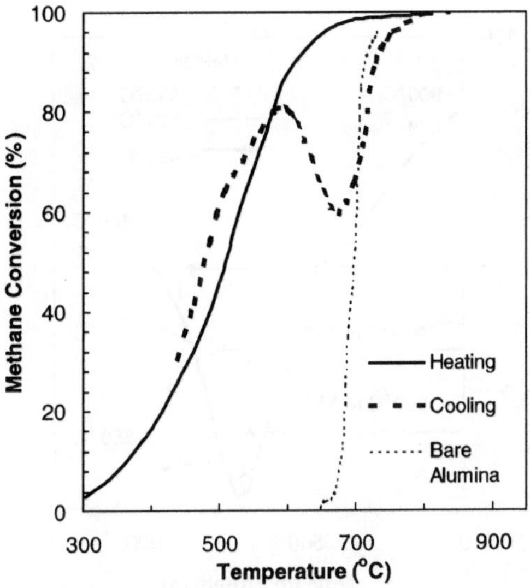

Figure 11 *Hysteresis behavior observed on methane combustion activity on Pd/Al_2O_3 during a heating/cooling cycle*
(After ref. 98)

decomposition. This sudden increase observed upon the PdO decomposition would indicate that the presence of both PdO and Pd phases may result in a high combustion activity. In agreement with this idea, when a partial reduction of the PdO was effected at 800 °C by decreasing the O_2 in the feed, the combustion activity was found to increase.[101]

In situ XRD studies during the temperature cycles confirm that the activity hysteresis is closely related to the dissociation of PdO to metallic Pd and the subsequent regeneration of PdO. When the intensity of the diffraction peak corresponding to Pd (111) was followed as a function of temperature in the heating and cooling ramps, a sudden increase was observed at about 900 °C in the heating ramp, and a slow decrease in the cooling ramp.[41,102] This behavior indicates that the decomposition of PdO to Pd readily occurs at about 900 °C. However, the regeneration of the PdO phase at a lower temperature is a much slower process. The activity recovery in the cooling ramp may depend on both temperature and time, but in most studies these two variables have changed simultaneously. To discriminate between the role of these two variables McCarty[103] has conducted time- and temperature-transient kinetics. Although, as shown in Figure 12, the activity recovery is both a function of time and temperature, the dependence on temperature is the more important. The author has ascribed this strong dependence on temperature to an inhibition by chemisorbed oxygen to the nucleation and growth of the active PdO phase. As the temperature decreases, nucleation and growth of active PdO will increase.

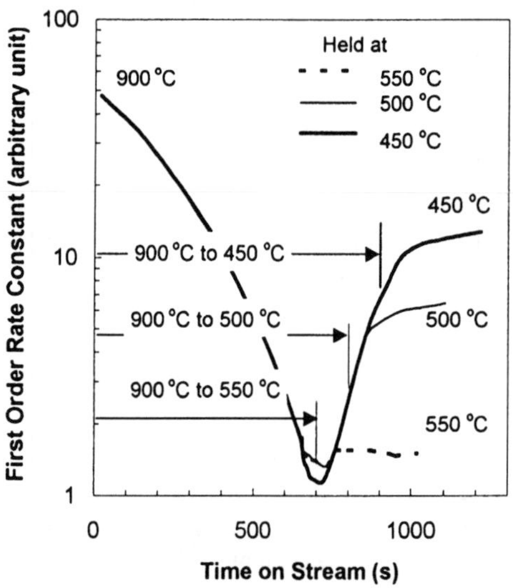

Figure 12 *Temperature–time dependence of the rate constant upon cooling from* 900 °C *to* 450, 500, *or* 550 °C *at* 0.5 °C s^{-1}, *then holding at the indicated temperatures*
(After ref. 103)

Lyubovsky and Pfefferle[100] have noted that the negative activation energy region is only observed on cooling after the catalyst has been heated above the PdO decomposition temperature. They propose that the high temperature treatment results in the formation of large Pd particles exhibiting low-index planes, which upon cooling in the reaction mixture begin to be covered by small patches of PdO. This structure has been proved by high resolution TEM analysis.[104] The presence of discrete PdO patches on the metallic Pd surface would generate a very active catalyst in which both species participate in the reaction. The metallic Pd would be very efficient in activating CH_4, while the PdO species would be responsible for the subsequent oxidation.

We could ask why, at low temperatures, the most probable structure under the reaction conditions is a core of Pd covered by a layer of PdO while, when cooling from high temperatures, small patches of PdO can be present over the metallic Pd surface? One can explain this difference in terms of the overall transformation rate of Pd to the thermodynamically more stable PdO. Depending on temperature, this transformation can be limited by either the nucleation rate or the growth (diffusional) rate, which have opposing temperature dependences. At low temperatures (*e.g.* 300–550 °C), the nucleation rate is fast and the diffusion is slow. Consequently, the oxidation results in a thin oxide shell that inhibits further oxidation of the metallic Pd core. On the other hand, when the particles are cooled from a temperature at which the metallic Pd is the thermodynamically stable phase (>800 °C) to one slightly below (*e.g.* 700–800 °C), the nucleation rate is slow while diffusion is fast. Under these conditions, nucleation can only start on surface defects or impurities. As a result, nucleation would be more difficult on particles that have been annealed at high temperature and exhibit smooth low-index planes. The resulting structure in this case will be Pd metal partially covered by islands of PdO. Of course, when the temperature is lowered further, nucleation becomes more favorable and the whole particle is again covered by a PdO layer.

In conclusion, although the most active catalyst would be one that contains both PdO and Pd sites, this situation can only be achieved close to the PdO decomposition temperature, in a metastable condition, which is limited by kinetics (nucleation rate). At both extremes the combustion activity is low. Without PdO species present, metallic Pd is not active. Similarly, completely oxidized PdO would not be active unless it contained oxygen vacancies.

5.2 Catalyst Improvement for High Temperature Operation – Increasing the PdO dissociation temperature by strengthening the Pd–O bond through metal–support interactions has been used to prevent loss of activity at high temperatures.[105] Interestingly, this strategy would result in lower combustion activities in the low-temperature region. Enhanced metal–support interactions have been achieved by incorporating additives, such as Mn, Ni or rare earths, onto the supports. For example, although the activity of Pd over Mn-promoted hexa-aluminate supports was lower than the unpromoted catalyst at low temperature, the activity drop normally seen at about 800 °C was not observed.[102] On the other hand, a Ni-promoted-Al_2O_3 support was found to exhibit an overall increase in activity as well as to eliminate the drastic drop at high temperature.[102] Addition of rare earth oxides has also been found to be useful for the stabilization of PdO against decomposition. For example, in recent patents, Farrauto *et al.*[106,107] have presented examples in which the addition of La, Ce, or Pr to PdO/Al_2O_3 resulted in increases in the temperature of decomposition. Also, the use of other supports has a strong effect on the PdO decomposition/reoxidation process. For example, PdO decomposition on ZrO_2 occurs at much lower temperatures than on Al_2O_3, while reoxidation is also lowered in a similar way.[105] Consequently, the extent of hysteresis is about the same on ZrO_2 as it is on Al_2O_3, but shifted to lower temperatures by about 200 °C. On CeO_2, the situation is more interesting since the PdO decomposition temperature decreases while the reoxidation increases, resulting in a relatively narrow hysteresis, *i.e.* about 45 °C.

Preparation of Pd/Al_2O_3 catalysts by special methods, such as hot isostatic pressing (HIP) under an oxygen-containing environment, may also reduce the loss of activity at high temperatures.[108] The improvement exhibited by the HIPed catalysts has been ascribed to the enhanced crystallinity of the resulting PdO structure. These conclusions were supported by a clear sharpening of the PdO (101) XRD peak, observed for the HIPed catalyst, in comparison to that for the conventional catalyst. This enhanced diffraction contrasted with the smaller particle size determined by CO chemisorption, and suggested that the particles on these novel catalysts were smaller but more ordered.

It should be kept in mind that industrial gas turbines operate at a much higher overall pressure than that of most reaction studies. The oxygen pressure may have a very important effect on the PdO decomposition as well as on the activity hysteresis that may be relevant under industrial combustion applications.

6 Rate Inhibition by Impurities and Products

6.1 Inhibition by H_2S and SO_2 – Natural gas typically has a low sulfur content, of less than a few ppm. However, the presence of even trace amounts of sulfur, derived not only from the gas but also from lubricating oils,[109] can severely poison a methane combustion catalyst. This poisoning is particularly critical for natural-gas vehicles (NGVs) because sulfur affects the methane combustion much more drastically than combustion of other hydrocarbons.[110] As a result, NGVs equipped with Pd catalysts can meet the standards for non-methane hydrocarbon emissions, but fail the total hydrocarbon emission tests, which require methane elimination at lower temperatures.

Depending on temperature, and whether the environment is oxidizing or reducing, sulfur species may exist in the form of sulfur dioxide, sulfate, or sulfide. Khairulin *et al.*[111] have shown that the presence of 100 ppm H_2S raised the light-off curve for Pd/Al_2O_3 by 150 °C. When compared at the same temperature, the loss of activity in the poisoned catalyst was more than 90%. This activity loss was attributed to the irreversible adsorption of sulfate species on the metal. Adsorption of sulfate species on Pd surface has been demonstrated by XPS studies, which showed the presence of a S_{2p} peak after the Pd/Al_2O_3 catalyst had been poisoned by 200 ppm H_2S in a 1% CH_4/air mixture.[112] The binding energy for S_{2p} observed in the range 169.0–169.4 eV suggests the formation of sulfate species rather than sulfides on the Pd surface.[113] At the same time, the adsorption of sulfate resulted in a binding energy increase for Pd $3d_{5/2}$ of about 0.4 eV.[112]

In most cases, sulfur poisoning is at least partially reversible. For example, by treating a poisoned Pd/SiO_2 catalyst in N_2 from 350 to 600 °C, a gradual recovery of the activity was observed, reaching a final recovery of more than 90%.[114] FTIR characterization of the poisoned catalyst during the heating process, in N_2 or in vacuum, showed that the intensity of the band ascribed to the Pd-sulfated species subsided as the temperature reached 400 °C and was completely eliminated at 600 °C. The latter temperature corresponds to the decomposition of $PdSO_4$. Removal of sulfate from the Pd surface was verified by the appearance of characteristic Pd–CO bands upon a mild reduction treatment followed by CO adsorption. By contrast, the attempt of regenerating the poisoned catalyst in H_2 was not successful. Although the intensity of the band corresponding to Pd–sulfate species disappeared at 350 °C under heating in H_2, the activity was not recovered. The authors related the irreversibility in activity by the H_2 treatment to the reduction, as opposed to decomposition, of the sulfate groups on the Pd surface, forming PdS, according to the reaction;

$$PdSO_4 + 4\,H_2 \rightarrow PdS + 4\,H_2O \tag{4}$$

The PdS species formed after the H_2 treatment can be converted back to $PdSO_4$ upon exposure to the CH_4/O_2 mixture, preventing the regeneration.

The support has also been found to play an important role in the sulfur poisoning process. Alumina-supported catalysts deactivate slowly in comparison with silica-supported catalysts. In the presence of 100 ppm of H_2S, both catalysts exhibited an initial methane conversion of about 80% and a final conversion of about 6%. However, on the Pd/SiO_2 catalyst, the activity rapidly dropped to 6% within 50 min on stream, as opposed to alumina, which had a much longer deactivation period of about 200 min. The slow poisoning rate on Pd/Al_2O_3 is due to the sulfur storage capacity of alumina. IR studies have shown that, under oxidizing conditions, alumina is able to bind sulfate species strongly.[115–117] The higher sulfur storage in alumina was further confirmed by chemical analysis, which showed that the poisoned alumina contained almost 10 times more sulfur than the silica support.[114]

In a recent study, Lampert *et al.*[110] have reported that although deactivation on alumina-supported catalysts is slower than on other supports, the activity is not totally recovered upon removal of S from the feed. This is due to spillover of S-species from the support onto the PdO surface. They compared the reversibility of several Pd catalysts over non-sulfating (SiO_2 and ZrO_2–SiO_2), and sulfating (alumina) supports over a temperature range between 150 and 650 °C. Upon removal of SO_2, a higher activity was obtained for the Pd/Al_2O_3 catalyst due to the spillover of sulfate from Pd to the alumina, which results in the cleaning of the PdO surface. However, the activity was higher for the non-sulfating support at temperatures above 600 °C, the decomposition temperature of Pd–sulfate. Subsequent tests showed a nearly complete restoration of activity on poisoned Pd over the non-sulfating supports while on the sulfating support, S-species continuously spilled over the PdO surface, inhibiting the recovery of activity. Kinetics studies on Pd/Al_2O_3 and Pd/ZrO_2–SiO_2 catalysts, in clean conditions and after ageing in SO_2, showed that the activation energies for both catalysts increased from 75 and 82 kJ mol^{-1}, respectively, to a same value of 105 kJ mol^{-1}.[110] This increase was ascribed to the presence of identical catalytic sites left after ageing in SO_2. In a similar study[114] conducted in the presence of sulfur at a much higher concentration (200 ppm H_2S instead of 1 ppm SO_2), the activation energy under sulfur contamination was lower than under clean conditions. In this case, pore diffusion resistance, generated from deposition of sulfur onto alumina, was responsible for the observed changes in activation energy. It is interesting to note that on the non-sulfating support, no modification in activation energy was observed.

6.2 Inhibition by Water and CO_2 – The products of the methane oxidation reaction, water and carbon dioxide, may also act as rate inhibitors. Water is formed not only by the methane oxidation reaction, but it is already present at high concentrations (up to 15%) in typical exhausts. It is generally agreed that the inhibiting effect of water is more severe than that of CO_2. The reaction order with respect to H_2O has been found to be -1, while that for CO_2 depends upon concentration.[22] Below 0.5 vol%, CO_2 exhibited no effect, but at higher concentrations it strongly suppressed the rate with an order dependency of -2. However, other studies have found a weak correlation or no inhibition above a CO_2 concentration of 10%.[118]

Cullis *et al.*[120] were the first to report that, in the presence of 1.3% water, the rate suppression was independent of the added water. It was proposed that water interacts with the PdO surface, forming thermodynamically favored Pd–hydroxide species according to the reaction,

$$PdO(s) + H_2O(g) \rightarrow Pd(OH)_2(s) \tag{5}$$

although TGA experiments under heating in dry nitrogen showed that bulk $Pd(OH)_2$ decomposes at 250 °C.[119] It may be that in the presence of excess water vapor the decomposition may be shifted to higher temperatures.[121] The way that the inactive $Pd(OH)_2$ inhibits the rate can be by blocking sites for methane dissociation on the PdO surface.[120] Recent work[18] has refined the idea of active site blocking, by invoking a quasi-equilibrium adsorption–desorption step of water, as shown below:

$$2\,OH^* \rightleftharpoons H_2O\,(g) + O^* + * \tag{6}$$

An increase in water concentration in the gas phase would shift the equilibrium towards higher OH concentration on the surface, titrating the unsaturated Pd and PdO sites required for the combustion reaction. A similar argument was used in this work to explain the inhibition by CO_2. From a detailed kinetics analysis, a complex rate expression was obtained. This expression contains both water and CO_2 concentrations in the denominator. Under low CO_2 concentrations, the most abundant species is OH, so the rate appears to be of -1 in water and independent of CO_2. At higher CO_2 concentrations, CO_2 or carbonate surface species become dominant and the rate is greatly affected by the CO_2 concentration.

Burch *et al.*[121] studied the inhibition and reversibility of simultaneous poisoning by water and CO_2. Both water (2.7%) and CO_2 (20%) poison the catalyst, but the poisoning by water alone is more severe than that of CO_2 alone. In the presence of both water and CO_2, the degree of inhibition is the same as that of water alone. This suggests that water and CO_2 poison the same type of sites, and water displaces CO_2 from the poisoned sites. The magnitude of inhibition was found to vary with reaction temperature. The drops in conversion due to CO_2 or H_2O were found to decrease with increasing temperature (of course, accompanied by an increase in methane conversion). The authors related this decrease in inhibition to the instability of the inactive $Pd(OH)_2$ at high temperatures. Also, since the conversion varies as the temperature is increased, at higher conversions, the amount of water formed from the combustion reaction increased and so the effects of added water vapor were less pronounced.

7 Methane Combustion in the Presence of NO on Low-loading Pd/Zeolite Catalysts

In the introduction we mentioned that the two applications for the catalytic methane combustion were in power generation and in the abatement of methane emissions in engine exhausts. In the first application, the amount of NO produced is in fact very small because the presence of a catalyst reduces the operating temperature. However, in the second type of application, the amounts of NO present can be substantial (*e.g.* several thousand ppm). In addition, the so-called selective catalytic reduction of NO can be effected by using methane as a reductant. In this case, the methane combustion takes place simultaneously with the NO reduction. Therefore, it is important to understand how the presence of NO affects the methane combustion under typical exhaust conditions. In the presence of NO, methane can undergo two different reactions, the reduction of NO and the total oxidation:

$$CH_4 + 2\,NO + O_2 \rightarrow CO_2 + 2\,H_2O + N_2 \tag{7}$$

$$CH_4 + 2\,O_2 \rightarrow CO_2 + 2\,H_2O \tag{8}$$

7.1 Suppression of Combustion Activity – A remarkable alteration of the typically high combustion activity of Pd occurs in the presence of low concentrations of NO, *e.g.* a few thousand ppm, typical of exhausts. In a recent study,[69] the methane combustion activity with and without NO in the feed has been compared on a series of Pd catalysts on various supports and with different Pd loadings. It was observed that supports containing Brønsted acidity exhibited a high selectivity towards the selective reduction of NO by methane, and that this selectivity enhancement was accompanied by a drastic decrease of the methane combustion rate.

A comparison of the methane conversions obtained in the presence and absence of NO in the feed over Pd/H-Mor catalysts with 0.1 wt% and 0.3 wt% Pd, respectively, is made in Figure 13. The data for CH_4 combustion, in the absence of NO, are also included. In this case, the 0.3% Pd sample shows a higher combustion activity than the 0.1% Pd sample. A remarkable contrast is observed in the presence of NO. While on the 0.1% Pd, a much higher overall CH_4 conversion was obtained with than without NO, on the 0.3% Pd, the increase was only apparent at lower temperatures. In fact, at high temperatures, the overall CH_4 conversion was lower in the presence of NO than without NO. As a result, the computed combustion of CH_4 in the presence of NO (after subtracting the fraction of methane used in the NO reduction), greatly decreases on the 0.3% Pd

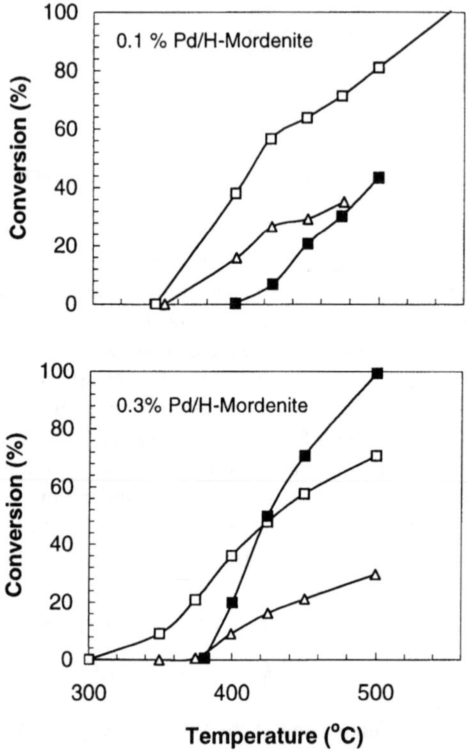

Figure 13 *Methane conversion as a function of reaction temperature in the presence and absence of NO over Pd/H-Mordenite catalysts. Note that in the presence of NO, overall conversion of methane includes combustion and NO reduction. Open squares: overall methane conversion in the presence of NO (see reactions 7 and 8); solid squares: combustion in the absence of NO; open triangles: combustion in the presence of NO* (After ref. 69)

catalyst as compared to the $CH_4 + O_2$ alone, while it does not decrease on the 0.1% Pd catalyst. These changes, occurring in the presence of NO, make the combustion in the presence of NO about the same for both catalysts, even though in the absence of NO their combustion activities were very different.

The change in combustion activity, from the high rate in the absence of NO, to the low rate in the presence of NO, was monitored in several consecutive cycles. The methane conversion was measured as a function of time over a 0.3 wt% Pd/H-ZSM-5 sample while the NO flow was switched on and off. The CH_4 and O_2 concentrations were kept constant at 7440 ppm and 1.87%, respectively, while the NO concentration was switched from 3560 ppm to zero in several cycles. As shown in Figure 14, at 6900 GHSV, the catalyst had a high combustion activity in the absence of NO, but when the NO flow was turned on, the combustion activity drastically dropped to about 20%, as the NO conversion increased to 60% in 20 min. The catalyst regained its combustion activity as the NO flow was interrupted again, and it dropped once more when NO was reintroduced. This cyclic behavior indicates that the transformation associated with the presence of NO is reversible.

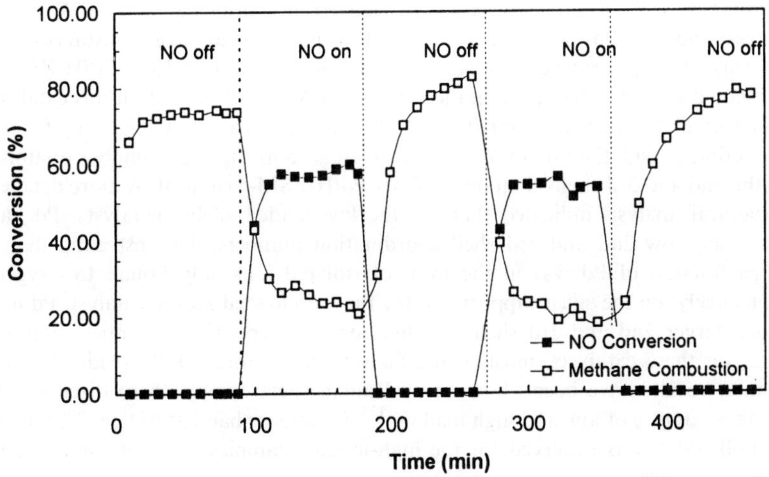

Figure 14 *Methane combustion and NO conversion as a function of time on stream over 0.3% Pd/H-ZSM-5 while turning NO on and off. GHSV 6900 h^{-1}, 0.74% CH_4, 1.87% O_2, with or without [NO] = 3560 ppm, balance He* (After ref. 69)

7.2 Characterization of Pd Species Stabilized by the Support in the Presence of NO – In order to explain the drastic effects of the presence of NO in combustion activity, all of the catalysts were characterized by X-ray absorption after reduction in H_2 at 300 °C or after exposure to a mixture of $CH_4/NO/O_2$ at 500 °C, and cooled down in the same mixture.[122] An XANES analysis of the Pd K edge for the reduced 0.3 wt% Pd/H-ZSM-5 and 0.3 wt% Pd/SiO_2 catalysts indicates that, before reaction, both catalysts contained Pd in the metallic state. By contrast, the catalysts that were exposed to the CH_4/NO/O_2 mixture presented XANES spectra clearly different from that of metallic Pd, but very similar to that of a PdO powder used as a reference. These results indicate that, during the first few minutes in contact with the CH_4/NO/O_2 mixture, Pd is oxidized regardless of the support used. Therefore, XANES indicates that a strong alteration in the state of Pd occurs under reaction conditions, but did not show differences that justify the dramatically different catalytic behavior observed on the low-loading catalysts supported on materials containing Brønsted sites from that of the Pd/SiO_2. On the other hand, EXAFS showed important differences in the state of Pd on the different catalysts. In the first place, in agreement with the XANES results, the EXAFS data for all reduced catalysts showed that, before the reaction, Pd was in the form of small metallic clusters.

On the other hand, the EXAFS data for the various used catalysts presented important differences depending on the Pd loading and the type of support. The Fourier transforms of the $k^1\chi$ data for 0.3% Pd and 1.0% Pd on H-ZSM-5 are shown in Figures 15 and 16, respectively. The r-space data for the 0.3% Pd catalyst showed a single peak at an uncorrected distance of 1.53 Å which corresponds to Pd–O bond, and very low intensity at larger distances. By contrast, the spectrum corresponding to the higher loading 1.0% Pd/H-ZSM-5 presented a much larger peak in the region 2–4 Å. It is clear that on this catalyst, which exhibited very high activity for CH_4 combustion and low selectivity for the reduction of NO, the Pd atoms have, on average, a much larger number of atoms in the 2nd and 3rd shells than on the 0.3% Pd/H-ZSM-5 catalyst. A more detailed structural analysis indicated that, on the low loaded zeolite catalysts, Pd had extremely low 2nd and 3rd shell coordination numbers, demonstrating that a large fraction of Pd was in the form of isolated ions only bound to oxygen. Conversely, on the silica support and for the high loaded zeolite catalyst, Pd had much larger 2nd and 3rd shell coordination numbers. These numbers indicate that, on these catalysts, most of the Pd was in the form of Pd oxide clusters. Similar results have been observed by Raman spectroscopy on oxidized Pd/H-ZSM-5 samples of low and high loading.[123] An intense band at 651 cm^{-1}, typical of bulk PdO, was observed for the high-loading samples, but not for the low-loading sample.

These results show that, under the $CH_4 + NO + O_2$ reaction, Pd is oxidized to a different morphological state depending on the support and its loading. On the low-loading H-ZSM-5 and H-Mor catalysts, a large fraction of Pd becomes isolated Pd^{n+} ions, while on SiO_2 or at higher Pd loadings it forms clusters of PdO. The transformation of Pd particles into isolated ions stabilized by a zeolite is a rather general phenomenon, which has been previously observed on several systems. Sachtler and Zhang[68] have reviewed the phenomenon and proposed that the oxidative redispersion occurs in two steps. In the first step, the metal is oxidized to oxide particles, and in the second step, the oxide reacts with the protons of the zeolite forming water. This transformation has been observed when Pd or Cu particles supported on NaHY zeolites were oxidized at tempera-

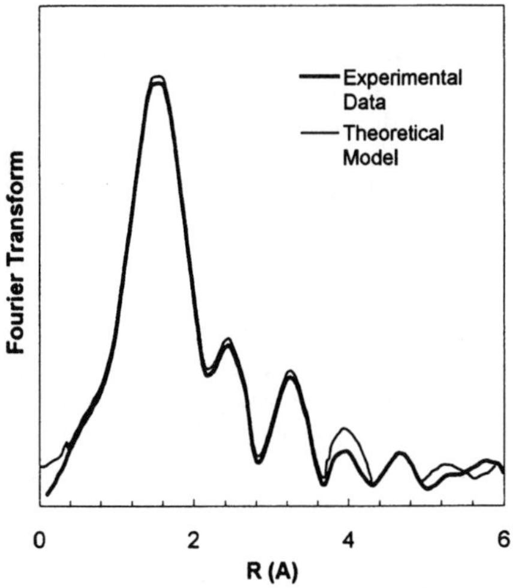

Figure 15 *Experimental EXAFS data obtained for the* 0.3% *Pd/H-ZSM-5 catalyst exposed to the SCR reaction at* 500 °C *for* 15 *min, compared with the fitted theoretical model*
(After ref. 122)

Figure 16 *Experimental EXAFS data obtained for the* 1.0% *Pd/H-ZSM-5 catalyst exposed to the SCR reaction at* 500 °C *for* 15 min, *compared with the fitted theoretical model*
(After ref. 122)

tures higher than 400 °C.[124,125] In a combined EPR, IR, and XRD study conducted by Che *et al.*[126] several years ago, it was found that when metallic Pd particles located inside the zeolite Y were exposed to NO, the XRD signal corresponding to metallic Pd disappeared while isolated Pd^{2+} ions were spectroscopically detected. In agreement with these reports, Garbowski *et al.*[127] found a similar oxidative dissolution of Ni particles in the presence of NO at 200 °C over a mordenite support.

In excellent agreement with these results, Adelman and Sachtler[128] conducted TPR studies that demonstrated that, after exposure to $NO_2 + O_2$ in a low-loading Pd/H-ZSM-5 sample, the majority of the Pd species are in the form of Pd^{n+} ions, stabilized by the zeolite. TPR is a good technique to quantify the amounts of PdO clusters, which reduce at temperatures below that of room temperature, and Pd^{2+} ions, which require much higher temperatures for reduction, usually appearing in the TPR as a broad hydrogen consumption in the range 100–250 °C.[129] When the sample was not exposed to NO_2, but only reduced and then oxidized, it did not show any Pd^{2+} ions, but only PdO clusters. Finally, when Pd was supported on Na-ZSM-5 instead of H-ZSM-5, the amount of Pd^{2+} ions was much lower, and the amount of PdO clusters much higher. Although in the absence of NO, Pd supported on Na-ZSM-5 or H-ZSM-5 exhibits similar methane combustion rates, in the presence of NO, only the one supported on H-ZSM-5 loses its combustion activity and becomes selective for NO reduction.[15,130,131]

Also, in good agreement with these results, Aylor *et al.*[71] have described the redispersion and stabilization process from their IR data. It was found that following an initial oxidation in O_2, a low-loading Pd/H-ZSM-5 catalyst had most of the Pd species as isolated Pd^{n+} ions. Following a reduction process in CO, the subsequent treatment in NO resulted again in the formation of Pd^{n+} ions, together with a decrease in the density of Brønsted acid sites in an amount that corresponds to 2 H^+/Pd. This correspondence indicates that the Brønsted sites participate in the stabilization of the Pd ions by formation of species of the type $Z^-H^+(PdO)H^+Z^-$. When a higher-loading sample was used, the initial treatment in O_2 resulted in the formation of both Pd^{n+} ions and PdO clusters. When the treatment in NO after reduction was performed again on this sample, all of the Pd was redispersed in the form of Pd^{n+} ions. However, only a fraction of them remained as such after removal of NO from the gas phase. The rest of the Pd reverted to PdO clusters.

8 Concluding Remarks

Palladium-based catalysts are the most active materials for the total oxidation of methane in the presence of excess oxygen. The structure and chemical identity of the species present on the surface of these materials have a decisive influence on their specific activity and time-dependent behavior. Since the structure and composition of these species strongly depend on preparation, pretreatment, thermal history, and reaction conditions, a very broad range of specific activities can be obtained. The several orders of magnitude differences observed in the literature for relatively similar materials illustrate this concept. With all other variables constant, the TOF generally increases with particle size. This particle size effect may be an indirect one, since smaller particles have stronger interactions with the support. Highly dispersed PdO species, interacting with the support, have much lower methane combustion activity than PdO clusters. The reason behind the low activity of the highly dispersed PdO species can be linked to the Pd–O bond strength, which in turn increases with decreasing the size of the PdO cluster.

The Pd–O bond also varies with the extent of oxidation of Pd. During the methane combustion reaction, the catalyst surface is a non-equilibrium, kinetically controlled structure. The oxygen concentration profile in the particle results from a combination of particle reconstruction, oxygen adsorption, bulk diffusion, and oxygen removal. This concentration profile varies as a function of time, and as the oxygen content increases, the Pd–O bond strength decreases. This increase is accompanied by an increase in the specific activity. The most widely accepted reaction pathway is the Mars and van Krevelen redox mechanism, which involves lattice oxygen and uncoordinated Pd centers as active species. Inhibition by products (H_2O and CO_2) and impurities (SO_2) is a major drawback for low temperature combustion. The effect of sulfur is particularly important for catalytic converters for NGV applications because it drastically reduces the methane combustion activity.

In the high temperature regime, the problem of poisoning or site blockage is not as severe for Pd catalysts as that of PdO decomposition and metallic sintering. When PdO decomposes by exceeding the decomposition temperature, the catalytic activity suffers a drastic decay, which is not regained in cooling until a much lower temperature is reached. This hysteresis is paralleled by the oxygen uptake, which indicates that metallic Pd is much less active than PdO. However, when both phases are simultaneously present on the surface the catalyst becomes very active. This is a metastable kinetically controlled structure, which can only occur close to the PdO decomposition temperature.

The presence of NO in the gas phase may interfere with the methane combustion activity. Low-loading Pd zeolites may stabilize Pd^{n+} species, inactive for methane combustion, when NO is present. These catalysts exhibit very low combustion activity, but are effective for the selective reduction of NO.

Acknowledgement

This work has been supported by the National Science Foundation by grant number CTS-97266465.

References

1 R.M. Heck and R.J. Farrauto, in *Catalytic Air Pollution Control: Commercial Technology*, Van Nostrand Reinhold, New York, 1995, p. 197.

2 J.A. Barnard and J.N. Bradley, in *Flame and Combustion*, Chapman and Hall, London and New York, 1985, p. 270.
3 Z.R. Ismagilov and M.A. Kerzhentsev, *Catal. Rev.-Sci. Eng.*, 1990, **32**, 51.
4 R. Prasad, L.A. Kennedy and E. Ruckenstein, *Catal. Rev.-Sci. Eng.*, 1984, **26**, 1.
5 K. Wark and C.F. Warner, *Air Pollution: Its Origin and Control*, IEP-Dun-Donnely, Harper and Row, New York, 1976.
6 A.F. Sarofim and R.C. Flagan, 'NO_x Control for Stationary Combustion Sources', *Prog. Energy Combust. Sci.*, 1976, **2**, 1.
7 L.D. Pfefferle and W.C. Pfefferle, *Catal. Rev.-Sci. Eng.* 1987, **29**, 219.
8 J.C. Schlatter, R.A. Dalla Betta, S.G. Nickolas, M.B. Cutrone, K.W. Beebe and T. Tsuchiya, *Proc. Intern. Gas Turbine and Aeroengine Congr.*, Orlando, FL (1997) paper 97-GT-57.
9 C.S. Weaver, SAE Paper No. 892133.
10 B. Hillemann, *Chem. Eng. News*, 1989, **67**, 25.
11 D. Albritton, R. Derwent, I. Isaksen, M. Lal and D. Wuebbles, in *Climate change 1995*, eds. J.T. Houghton, L.G. Meira Filho, B.A. Callander, N. Harris, A. Kattenberg and K. Maskell, Cambridge University Press, Cambridge, UK, 1996.
12 R.B. Anderson, K.C. Stein, J.J. Feenan and L.J.E. Hofer, *Ind. Eng. Chem.* 1961, **53**, 809.
13 S.H. Oh, P.J. Mitchell and R.M. Siewert, *J. Catal.*, 1991, **132**, 287.
14 K. Eguchi, *Catalysis*, Specialist Periodical Reports, vol. 12, The Royal Society of Chemistry, Cambridge, UK, 1996, p. 132.
15 C.J. Loughran and D.E. Resasco, *Appl. Catal., B: Environ.*, 1995, **7**, 113.
16 R.F. Hicks, H.Qi, M.L. Young and R.G. Lee, *J. Catal.*, 1990, **122**, 280.
17 P. Briot and M. Primet, *Appl. Catal.*, 1991, **68**, 301.
18 K. Fujimoto, F.H. Ribeiro, M. Avalos-Borja and E. Iglesia, *J. Catal.*, 1998, **179**, 431.
19 C.A. Muller, M. Maciejewski, R.A. Koeppel and A. Baiker, *J. Catal.*, 1997, **166**, 36.
20 T.R. Baldwin and R. Burch, *Appl. Catal.*, 1990, **66**, 337; *ibid.* 1990, **66**, 359.
21 K. Muto, N. Katada and M. Niwa, *Appl. Catal., A: General*, 1996, **134**, 203.
22 F.H. Ribeiro, M. Chow and R.A. Dalla Betta, *J. Catal.*, 1994, **146**, 537.
23 R.F. Hicks, H. Qi, M.L. Young and R.G. Lee, *J. Catal.*, 1990, **122**, 295.
24 C.F. Cullis and B.M. Willatt, *J. Catal.*, 1984, **86**, 187.
25 D.O. Simone, T. Kennelly, N.L. Brungard and R.J. Farrauto, *Appl. Catal.*, 1991, **70**, 87.
26 S.S. Peri and C.R.F. Lund, *J. Catal.*, 1995, **152**, 410.
27 E. Garbowski, C. Feumi-Jantou, N. Mouaddib and M. Primet, *Appl. Catal.*, 1994, **109**, 277.
28 E. Garbowski and M. Primet, *Appl. Catal.*, 1995, **125**, 185.
29 R. Burch and F.J. Urbano, *Appl. Catal.*, 1995, **124**, 121.
30 C.F. Cullis and B.M. Willatt, *J. Catal.*, 1983, **83**, 267.
31 J.J. Chen and E. Ruckenstein, *J. Phys. Chem.*, 1981, **85**, 1606.
32 E. Ruckenstein and J.J. Chen, *J. Colloid Interface Sci.*, 1982, **86**, 1.
33 B.K. Chakraverty, *J. Phys. Chem. Solids*, 1967, **28**, 2401.
34 H. Lieske and J. Volter, *J. Phys. Chem.*, 1985, **89**, 1841.
35 R.J. Farrauto, J.K. Lampert, M.C. Hobson and E.M. Waterman, *Appl. Catal. B*, 1995, **6**, 263.
36 E. Ruckenstein and I. Sushumna, *Chem. Ind.*, 1988, **31**, 259.
37 E. Ruckenstein, in *Metal–Support Interactions in Catalysis, Sintering and Redisper-*

sion, eds. S. Stevenson, J.A. Dumesic, R.T.K. Baker and E. Ruckenstein, Van Nostrand Reinhold Catalysis Series, New York, 1987, p. 153.
38 T. Wang and L.D. Schmidt, *J. Catal.*, 1980, **66**, 301.
39 A. Graham and S.E. Wanke, *J. Catal.*, 1981, **68**, 1.
40 H. Jehn *J. Less-Common Metals*, 1984, **100**, 321.
41 K. Sekizawa, M. Machida, K. Eguchi and H. Arai, *J. Catal.*, 1993, **142**, 655.
42 G. Bayer and H.G. Wiedemann, *Thermo. Acta*, 1975, **11**, 79.
43 W.E. Bell, R.E. Inyard and M. Tagami, *J. Phys. Chem.*, 1966, **70**, 3735.
44 C. Mallika, O.M. Sreedharan and J.B. Gnanamoorthy, *J. Less-Common Metals*, 1983, **95**, 213.
45 C.T. Campbell, D.C. Foyt and J.M. White *J. Phys. Chem.*, 1977, **81**, 491.
46 G. Ertl and J. Koch, in *Adsorption–Desorption Phenomena*, ed. F. Ricca, Academic Press, New York, 1972, p. 345.
47 L.W. Konopny, A. Juan and D.E. Damiani, *Appl. Catal., B: Environ.*, 1998, **15**, 115.
48 J.N. Carstens, S.C. Su and A.T. Bell, *J. Catal.*, 1998, **176**, 136.
49 C. Nyberg and C.G. Tengstal, *Solid State Commun.*, 1982, **44**, 251.
50 K.H. Reider and W. Stocker, *Surf. Sci.*, 1985, **150**, L66.
51 E.M. Stuve, R.J. Madix and C.R. Brundle, *Surf. Sci.*, 1984, **146**, 144.
52 G.W. Simmons, Y. Wang, J. Marcos and K. Klier, *J. Phys Chem.*, 1991, **95**, 4522.
53 T.W. Orent and S.D. Bader, *Surf. Sci.*, 1982, **115**, 323.
54 K. Otto, L.P. Haack and J.E. deVries, *Appl. Catal., B*, 1992, **1**, 1.
55 T.E. Hoost and K. Otto, *Appl. Catal., A*, 1992, **92**, 39.
56 M.A. Vannice and P. Chou, in *Strong Metal–Support Interactions*, ed. R.T.K. Baker, S.J. Tauster and J.A. Dumesic, American Chemical Society, Washington, 1986, ch. 8, pp. 77–88.
57 P. Chou and M.A. Vannice, *J. Catal.*, 1987, **105**, 342.
58 M.G. Mason, *Phys. Rev. B*, 1983, **27**, 748.
59 V.I. Bukhtiyarov, in *Interfacial Science*, ed. M.W. Roberts, Int. Union of Pure and Appl. Chemistry, UK, 1997, ch. 5, pp. 109–127.
60 G.D. Zakumbaeva and S.V. Artamonov, *React. Kinet. Catal. Lett.*, 1979, **10**, 183.
61 T.Y. Chou, C.P. Hwang and C.T. Yeh, *J. Therm. Anal.*, 1996, **46**, 305.
62 H. Widjaja, K. Sekizawa, K. Eguchi and H. Arai, *Catal. Today*, 1997, **35**, 197.
63 E. Lesage-Rosenberg, G. Vlaic, H. Dexpert, P. Lagarde and E. Freund, *Appl. Catal.*, 1986, **22**, 211.
64 G. Chen, W.T. Chon and C. Yeh, *Appl. Catal.*, 1983, **8**, 389.
65 V. Ragaini, R. Giannantonio, P. Magni, L. Lucarelli and G. Leofanti, *J. Catal.*, 1994, **146**, 116.
66 S.H. Overbury, P.A. Bertrand and G.A. Somorjai, *Chem. Rev.* 1975, **75**, 547.
67 M. Chen and L.D. Schmidt, *J. Catal* 1978, **55**, 348.
68 W.M.H. Sachtler and Z. Zhang, in *Advances in Catalysis*, eds. D.D. Eley, H. Pines and P.B. Weisz, Academic Press Inc., San Diego, 1993, vol. 39, p. 129.
69 A. Ali, Y.-H. Chin and D.E. Resasco, *Catal. Lett.*, 1998, **56**, 111.
70 Y. Li and J.N. Armor, *Appl. Catal., B: Environ.*, 1994, **3**, 275.
71 A.W. Aylor, L.J. Lobree, J.A. Reimer and A.T. Bell, *J. Catal.*, 1997, **172**, 453.
72 H. Maeda, Y. Kinoshita, K.R. Reddy, K. Muto, S. Komai, N. Katada and M. Niwa, *Appl. Catal., A*, 1997, **163**, 59.
73 R.M. Koros and E.J. Nowak, *Chem. Eng. Sci.*, 1967, **22**, 470.
74 J.G. Firth, *Trans. Faraday Soc.*, 1966, **62**, 2566.
75 J.G. Firth and H.B. Holland, *Trans. Faraday Soc.*, 1969, **65**, 1121.
76 J.G. Firth and H.B. Holland, *Trans. Faraday Soc.*, 1969, **65**, 1891.

77 Y.F.Y. Yao, *Ind. Eng. Chem. Prod. Res. Dev.*, 1980, **19**, 293.
78 A.K. Neyestanaki, N. Kumar and L.-E. Lindfors, *Appl. Catal., B*, 1995, **7**, 95.
79 S. Seimanieds and M. Stoukides, *J. Catal.*, 1986, **98**, 540.
80 C.A. Muller, M. Maciejewski, R.A. Koeppel, R. Tschan and A. Baiker, *J. Phys. Chem.*, 1996, **100**, 20006.
81 H. Dropsch and M. Baerns, *Appl. Catal.*, 1997, **165**, 159.
82 E. Broclawik, R. Yamauchi, A. Endou, M. Kubo and A. Miyamoto, *Int. J. Quantum. Chem.*, 1997, **61**, 673.
83 M.R.A. Blomberg, P.E.M. Siegbahn and M. Svensson, *J. Phys. Chem.*, 1992, **96**, 5783.
84 A.Q.M. Boon, H.M. Huisman and J.W. Geus, *J. Mol. Catal.*, 1992, **75**, 293.
85 M. Niwa, K. Awano and Y. Murakami, *Appl. Catal.*, 1983, **7**, 317.
86 Y. Mizushima and M. Hori, *Appl. Catal., A*, 1992, **88**, 137.
87 R. Burch and P.K. Loader, *Appl. Catal., B: Environ.*, 1994, **5**, 149.
88 V.A. Drozdov, P.G. Tsyrulnikov, V.V. Popovskii, N.N. Bulgakov, E.M. Moroz and T.G. Gäleev, *React. Kinet. Catal. Lett.*, 1985, **27**, 425.
89 M. Machida, H. Taniguchi, T. Kihima and J. Nakatani, *J. Mater. Chem.*, 1998, **8**, 781.
90 R.S. Weber, *J. Catal.*, 1995, **151**, 470.
91 U.S. Ozkan, M.W. Kumthekar and G. Karakas, *J. Catal.*, 1997, **171**, 67.
92 Y. Deng and T.G. Nevell, *Faraday Discuss.*, 1996, **105**, 33.
93 T. Maillet, C. Solleau, H. Barbier, Jr. and D. Duprez, *Appl. Catal., B: Environ.*, 1997, **14**, 85.
94 F.A. Cotton and G. Wilkinson, in *Advanced Inorganic Chemistry*, Interscience, New York, 5th edn., 1988, p. 880.
95 J.G. McCarty, D.L. Hildenbrand, K.H. Lau, B.J. Wood, J.A. Kezerle and D.A. Scarpiello, in *Moderation of Surface Temperatures in Adiabatic Combustion Catalysts*, Proc. Int. Gas Research Conf., Cannes, France, Nov. 6–9, 1995, p. 607.
96 C. Wong and R.W. McCabe, *J. Catal.*, 1989, **87**, 117.
97 R.W. Schefer and F.A. Roben, Proc., in *Fourth Workshop on Catalytic Combustion*, 1980, EPA Report 600/9-80-035, p. 266.
98 R.J. Farrauto, M.C. Hobson, T. Kennelly and E.M. Waterman, *Appl. Catal.*, 1992, **81**, 227.
99 P. Salomonsson, S. Johansson and B. Kasemo, *Catal. Lett.*, 1995, **33**, 1.
100 M. Lyubovsky and L. Pfefferle, *Appl. Catal., A*, 1999, **173**, 107.
101 M. Lyubovsky, R. Weber and L. Pfefferle, *Proc. 26th Int. Symp. On Combustion*, 1996, p. 1779.
102 K. Sekizawa, K. Eguchi, H. Widjaja, M. Machida and H. Arai, *Catal. Today*, 1996, **28**, 245.
103 J. McCarty, *Catal. Today*, 1995, **26**, 283.
104 M. Lyubovsky, L. Pfefferle, A. Datye and J. Bravo, unpublished data.
105 R.J. Farrauto, J.K. Lampert, M.C. Hobson and E.M. Waterman, *Appl. Catal., B: Environ.*, 1995, **6**, 263.
106 R. Farrauto, T. Kennelly, M. Hobson and E. Waterman, US Pat. 5,214,912.
107 T. Kennelly and R. Farrauto, US Pat. 5,216,875.
108 K. Furuta, A. Nishida, H. Rata, K. Narui and Y. Kohtoku, *Appl. Catal., A*, 1997, **164**, 149.
109 J.K. Lampert, M.S. Kazi and R.J. Farrauto, *SAE* 961971, 1996.
110 J.K. Lampert, M.S. Kazi and R.J. Farrauto, *Appl. Catal., B*, 1997, **14**, 211.

111 S. Khairulin, B. Beguin, E. Garbowski and M. Primet, *J. Chem. Soc., Faraday Trans.*, 1997, **93**, 2217.
112 Y. Deng, T.G. Nevell, R.J. Ewen, C.L. Honeybourne and M.G. Jones, *Appl. Catal., A*, 1993, **101**, 51.
113 C.D. Wagner, in *Practical Surface Analysis*, ed. D. Briggs and M.P. Seah, Wiley, New York, 1985, p. 477.
114 L.J. Hoyos, H. Praliaud and M. Primet, *Appl. Catal., A*, 1993, **98**, 125.
115 C.C. Chang, *J. Catal.*, 1978, **53**, 374.
116 M. Waqif, J. Bachelier, O. Saur and J.C. Lavalley, *J. Mol. Catal.*, 1992, **72**, 127.
117 M. Waqif, O. Saur, J.C. Lavalley, S. Perathouer and G. Centi, *J. Phys. Chem.*, 1991, **95**, 4051.
118 A.F. Ahlstrom-Silversand and C.U.I. Odenbrand, *Appl. Catal., A*, 1997, **153**, 157.
119 R.J. Card, J.L. Schmitt and J.M. Simpson, *J. Catal.*, 1983, **79**, 13.
120 C.F. Cullis, T.G. Nevell and D.L. Trimm, *J. Chem. Soc., Faraday Trans. 1*, 1972, **68**, 1406.
121 R. Burch, F.J. Urbano and P.K. Loader, *Appl. Catal., A*, 1995, 123, 173.
122 A. Ali, W. Alvarez, C.J. Loughran and D.E. Resasco, *Appl. Catal., B: Environ.*, 1997, **14**, 13.
123 W.H. Weber, R.J. Baird and G.W. Graham, *Raman Spectrosc.*, 1988, **19**, 239.
124 G. Bergeret, P. Gallezot and B. Imelik, *J. Phys. Chem.*, 1981, **85**, 411.
125 S.T. Homeyer and W.M.H. Sachtler, *Appl. Catal.*, 1989, **54**, 189.
126 M. Che, J.F. Dutel, P. Gallezot and M. Primet, *J. Phys. Chem.*, 1976, **80**, 2371.
127 E.D. Garbowski, C. Mirodatos, M. Primet and M.V. Mathieu, *J. Phys. Chem.*, 1983, **87**, 303.
128 B.J. Adelman and W.M.H. Sachtler, *Appl. Catal., B: Environ.*, 1997, **14**, 1.
129 B.T. Carvill, B.A. Lerner, Z.C. Zhang and W.M.H. Sachtler, *J. Catal.*, 1993, **143**, 314.
130 M. Misono, Y. Nishizaka, M. Kawamoto and H. Kato, *Stud. Surf. Sci. Catal.*, 1997, **105**, 1501.
131 Y. Nishizaka and M. Misono, *Chem. Lett.*, 1994, 2237.

2
Innovations in Catalytic Inorganic Membrane Reactors

BY ANTHONY G. DIXON

1 Introduction

Chemical reaction and the separation of the resulting products are the two most important, and often the two most costly, unit operation steps in a chemical process. The combination of these two operations into a single processing step is a desirable goal to improve process economics and to affect waste minimization favorably. Various types of multifunctional reactors have been suggested,[1] such as catalytic distillation and the chromatographic reactor. The efficiency of biological systems, however, suggests the use of membranes integrated into the reactor.[2] Until recently, the use of membranes was restricted to low-temperature processes with mild chemical environments which could be tolerated by polymeric materials. One historical exception was the separation of hydrogen by palladium membranes. Recent advances in materials, to high temperatures and chemically harsh environments have broadened the scope of membrane use. This has allowed the consideration of membranes for integration into reactors for catalytic reactions.

Section 2 of this review gives a brief survey of well-established uses of membrane reactors. The original idea was to use a membrane to remove a product from a reaction limited by chemical equilibrium, thus allowing higher conversions than would be possible in a conventional fixed bed catalytic reactor. The integration of catalytic reactions and membranes has developed beyond the original goals to include the introduction of reactor feed species throughout the reactor volume, and the control of how the reactants contact. Both of these allow greater control of the species' concentrations in contact with the catalytic surface.

Section 3 presents a look at the current state of membrane reactor evaluation. Despite the advances of recent years, many challenges and difficulties still face catalytic membrane reactors. Some are technical: the materials available, although much improved, still suffer from problems of stability. The difficulty of obtaining high permeation rate and permselectivity in the same material still confronts us. Some problems are commercial or competitive in nature. For example, membrane reactors may be too expensive. In certain applications, reactors that are more conventional can outperform them. An active area of research answers the question of how to make such comparisons. Simple measures, such as conversion or yield, do not always give the whole story.

Catalysis, Volume 14

The heart of the review is in Section 4. Two main considerations drive innovations in the membrane reactor field. One is the need to solve the technological challenges that prevent the incorporation of membrane reactor technology into the plant. The second is the need to find new 'niche' applications where there really is an economic, environmental or safety advantage in using a membrane reactor. The resulting innovations can be separated into those that will originate from advances in materials science, and those that will originate from advances in catalysis and reaction engineering. The purpose of this contribution is to provide an *interpretative review* of the recent advances in the latter category, to place them in context, and to identify trends for the future.

The scope of the review is limited to inorganic membrane reactors under chemically harsh environments at high temperatures (> 100 °C). We will not discuss bioreactors, biological membranes, and enzyme catalysis or organic/polymeric membranes. We refer readers with interests in these areas to other reviews.[3,4] We will discuss only configurations involving catalysts. The membrane itself may be catalytically active. It may contain a catalyst, or be in proximity to a catalyst. This last statement means that the membrane and the catalytic surface *directly* influence each other. That is, the reactor and membrane are integrated into one unit. This excludes staged reactor/membrane configurations where these are two separate operations. An additional limitation of this review is that it will focus on innovations in reactor type or configuration. A comprehensive review of new developments in the field of membrane materials and their properties is not possible within the page limits of the present contribution. Membrane reactor materials will be mentioned only as necessary for the understanding of new catalytic reactor types, or as the reasons for new developments.

A comprehensive review of this ever-widening field is today hardly possible within a single journal article, even with the above restrictions. In recent years, a series of excellent reviews has appeared at regular intervals[5-16] that has comprehensively documented and interpreted the work of the previous 20 years. In the present article, therefore, the focus is on those innovative uses of catalytic membrane reactors that either are very new or have not previously been covered. The theme of the review is that a *diversification of ends*, and an *integration of means*, has taken place. Researchers are investigating membrane reactors for a greater variety of purposes than was apparent from earlier work. Many investigators have studied the integration of the separation and reaction tasks in a membrane reactor. They are now studying these ideas together with control of other phenomena in order to exploit the flexibility of the membrane reactor concept more fully.

2 Background on Catalytic Inorganic Membrane Reactors

To put recent developments in catalytic membrane reactors into perspective, providing some background for the reader who is not an expert in this area is necessary. Those for whom this is superfluous may wish to skip to Section 3. The present section provides a brief overview of well-established developments in catalytic inorganic membrane reactors. For the sake of brevity, many original papers will not be cited and the reader is referred to the reviews mentioned above. In particular, the review of Zaman and Chakma[11] that lists 221 references, and that of Saracco and Specchia[12] with approximately 400 references, provide comprehensive coverage of the literature up to late 1993. More recent developments are covered by Sanchez and Tsotsis,[15] whose article is closest in spirit to the present work, and Coronas and Santamaría,[16] who present a clear discussion of porous membranes, especially microporous. For a thorough discussion of mathematical modelling principles for catalytic membrane reactors, the reader may consult Tsotsis *et al.*[10] for work before 1993, and Sanchez and Tsotsis[15] for an update to 1995.

2.1 Types of Membrane and Reactor Configurations – Membrane reactors may be classified by the geometry and materials of the membrane, and by the configuration of the reactor.

The two most common membrane geometries are the flat plate and the tube. The flat plate has been used frequently in laboratory scale investigations due to its ease of fabrication and incorporation into a diffusion-cell type of apparatus. Commercial-scale operation will require much higher ratios of membrane surface area to equipment volume. The most likely configuration is presently thought to be multitubular, and for that reason a great deal of research work has focused on the single tube-and-shell configuration.

The materials of membrane construction can be classified as either dense or porous. Dense metal materials include palladium membranes that are semipermeable to hydrogen,[8] and silver membranes that are semipermeable to oxygen. The low permeation rates for silver membranes have led to the more recent use of solid oxide electrolyte dense membranes such as modified zirconias and perovskites, which have higher O_2 permeation rates at high temperatures.[17,18]

Porous inorganic membranes can be divided into macroporous ($d_p > 50$ nm), mesoporous ($50 > d_p > 2$ nm) and microporous ($d_p < 2$ nm). Macroporous materials, such as α-alumina membranes, provide no separative function, but may be used to support layers of smaller pore size to form composite membranes, or in applications where a well-controlled reactive interface is required. Mesoporous materials for membranes have generally had pore sizes in the 4–5 nm range, so that permeation is governed by Knudsen diffusion. The ideal separation factor α_{ij} of two species is then proportional to the square root of the inverse of the ratio of their molecular weights. For most systems studied, this implies a separation factor in the range 1.5 to 10, which is low. Typical materials are Vycor glass, and composite membranes of γ-alumina supported on successively larger-pore layers of α-alumina support. Microporous membranes offer the potential for molecular sieving effects, with very high separation factors. They are currently the subject of much research, and various materials such as carbon molecular sieves,[19] microporous glasses[20] and zeolites[21] have been studied. Of these, the zeolites appear to have the most promise, due to their higher stability under processing conditions.

Some materials have been investigated to obtain fundamental information regarding transport mechanisms. Anodic aluminas were studied by Tsotsis and co-workers[22] for the diffusion of asphaltenes. These materials can be synthesized with straight, parallel pores, thus easing the interpretation of experimental data. Microporous glass membranes were investigated by Shelekhin *et al.*[23] to obtain understanding of the roles of the adsorbed phase and the gas phase in small pore transport.

The most active areas of development for membrane materials are currently synthesis of supported thin films, and pore modification. The complete selectivity to one species provided by dense membranes is very attractive, but is accompanied by low permeation rates if the membrane is composed entirely of the dense material. To maintain structural stability, thinner dense films must be supported by materials that are strong but that offer no additional resistance to permeation. Similar principles apply to the use of microporous materials with high permselectivities or molecular sieving effects. Some examples of these developments are supported Pd films on porous aluminas,[24] or on porous stainless steel,[25] and supported zeolite films.[26–30] Pore modification has been used to deposit materials inside mesoporous materials, an example being the deposition of SiO_2 films in porous glass.[31]

Classification of membrane reactors with catalysts is based on the location of the catalyst. The notation introduced by Tsotsis *et al.*[10] has been widely adopted, and will be used here. Table 1 gives a list of the acronyms used in this paper.

Table 1 *Explanation of acronyms used throughout the review. Alternative acronyms used by some other authors are given in parentheses*

Acronym	*Explanation*
CMR	Catalytic Membrane Reactor
PBMR (IMR, IMRCF)	Packed Bed Membrane Reactor (Inert Membrane Reactor, Inert Membrane Reactor with Catalyst on Feed side)
PBCMR	Packed Bed Catalytic Membrane Reactor
FBMR	Fluidized Bed Membrane Reactor
FBCMR	Fluidized Bed Catalytic Membrane Reactor
CNMR	Catalytic Non-permselective Membrane Reactor
SLPCMRS	Supported Liquid-phase Catalytic Membrane Reactor–Separator
PFR (PBR, FBR)	Plug Flow Reactor (Packed Bed Reactor, Fixed Bed Reactor)

The catalyst may be deposited within the membrane, including the case where the membrane itself is intrinsically catalytically active (CMR), catalyst particles may be packed inside or outside a membrane tube (PBMR) or both (PBCMR). These configurations are illustrated in Figure 1.

Similar terminology is used if the membrane is a flat plate, and by extension, if the membrane is used with a fluidized bed then FBMR and FBCMR are used for the two possible configurations.

The PBMR acts to change the fluid phase composition in contact with the catalyst particles, either by removing intermediate or product species, or by gradually supplying reactant species. The CMR provides an alternative way to

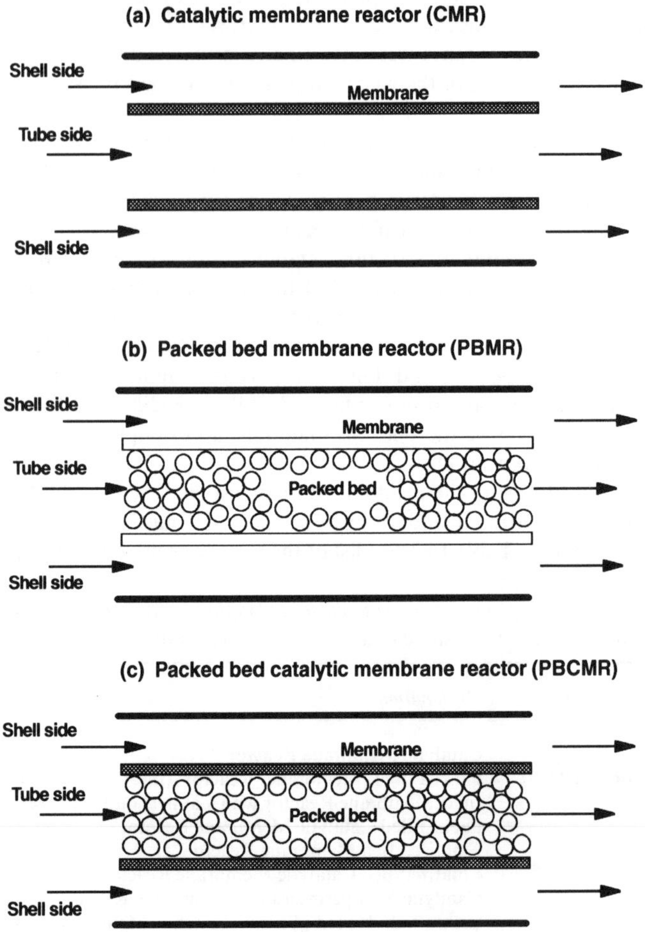

Figure 1 *Schematic of three membrane reactor configurations*

contact the reactants and the catalyst. Microporous membranes especially have a high surface area per gram, and if a catalyst can be well-dispersed within the pores they can be highly effective. The membrane geometry allows products to leave the catalytic site without having to diffuse against reactants, and possibly react further. In addition, the membrane provides effective contact since all molecules can be forced to pass through the membrane, as opposed to less efficient contacting in a packed bed. Drawbacks include operating in a diffusion-controlled regime, and difficulties in obtaining sufficient catalyst loading in the membrane.

The following three sections present further details on three well-established areas of application of catalytic membrane reactors: product removal, reactant feed and control of reactant contact in a non-permselective membrane. An excellent discussion can also be found in Harold *et al.*,[14] who present a review of the principles behind some successful applications of catalytic inorganic membranes.

2.2 Membrane Reactors for Preferential Removal of a Species – The most studied type of membrane reactor is illustrated schematically in Figure 2.

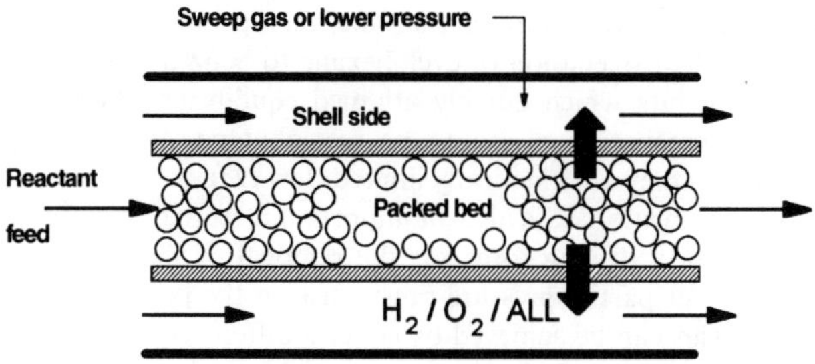

Figure 2 *Schematic of a membrane reactor for product removal*

The membrane may be porous or dense, and is most frequently in a tube-and-shell configuration. All three types, PBMR, CMR and PBCMR have been used. The reactant gases may be fed either to the tube or to the shell side (the reaction side). The other side (the permeate side) is either at a lower pressure or has an inert, sweep gas flowing through.

2.2.1 Principles – The function of the membrane is preferentially to remove a product from the reaction mixture. If the reaction is equilibrium limited, the decreased activity of the species being removed will permit further conversion to occur, beyond that which would be possible if no species were removed. This phenomenon has been called 'equilibrium shifting' or sometimes 'exceeding equilibrium'. It must be noted that the equilibrium being referenced in these expressions is the hypothetical equilibrium of the closed system formed by a fixed mass of the reaction side feed only. It is not the equilibrium of the actual membrane reactor system, which would include both the feeds to the reaction and permeate sides. This point is discussed further in Section 3.2 below.

In a conventional isothermal plug-flow reactor (PFR), two important rates govern its performance – the rate of reaction and the rate of reactant feed per catalyst volume to the reactor. The ratio of these gives the Damköhler number, *Da* = (reactor volume)(maximum reaction rate per volume)/(inlet flow rate), which also involves reactor tube dimensions. The membrane reactor brings in at least one additional rate, the permeation rate of the fastest gas. The ratio of these has been labelled differently by various authors; we will follow Bernstein and Lund[32] and term it the Damköhler–Peclet product, *DaPe* = (maximum reaction rate per volume)/(maximum permeation rate per volume). For proper performance of a membrane reactor, it is important that these three rates are correctly balanced. This has been clearly explained by Harold *et al.*[14] If the permeation rate is too low, the membrane has little effect and the reactor behaves like a PFR with the reactant gas feed. If the permeation rate is too high, the shell and tube sides will equilibrate rapidly. Bernstein and Lund[32] recommend $0.1 < DaPe < 10$ as covering the optimal range.

The activity of the catalyst is also important, as reflected in the value of *Da*. If the reaction rate is slow, equilibrium will not be approached and the removal of a product by a membrane will not affect the ultimate yield. This was demonstrated by Raich and Foley,[33] who showed that the very promising results of early studies using the dehydrogenation of cyclohexane to benzene were achieved due to the fast reaction rate which quickly attained equilibrium. Slower reactions, such as the dehydrogenation of isobutane to isobutene, are not helped by a membrane reactor as they are kinetically limited (low *Da*).

The role of the permeate side is also an important feature of membrane reactor operation. To obtain a driving force for the removal of the product from the reaction zone, a lower partial pressure must exist on the permeate side than on the reaction side. This can be achieved by either a difference in total pressure, or by diluting the permeate side with enough inert to lower the mole fraction of the permeating species. Neither technique is especially attractive as they consume energy or give a highly diluted gas stream that is wasteful of the permeating species. In addition, for porous membranes, the potential for dilution of the reactant exists. This can occur either by back-permeation of inert sweep gas from the permeate side to the reaction side, or by permeation of all reactants to the permeate side, where they are generally lost to further conversion. This is termed the 'reactant loss' problem.

The problem of reactant loss has been addressed in various ways. One approach is the 'hybrid' membrane reactor, which is equivalent to a PFR followed by a PBMR. The PFR brings the reacting mixture to near equilibrium, so that in the porous membrane section the concentration of reactants is reduced and their loss minimized.[34–36] Another approach is to use a PBCMR and attempt to convert reactants as they pass through the membrane, besides the conversion taking place on the reactant side.[37–39] Of course, a dense membrane, such as palladium, for the removal of hydrogen avoids the reactant loss problem. However, early versions of such membranes suffered from low fluxes and poor stability. More recent developments in thin, supported Pd–Ag films may be the answer.

A second motivation for using dense membranes for product removal comes from the fact that mesoporous membranes (*e.g.* Vycor glass, γ-alumina) give poor permselectivity factors. For example, hydrogen has a permselectivity factor over a species of molecular weight equal to 50 of only 5 [$= (50/2)^{\frac{1}{2}}$]. So, in a dehydrogenation, significant amounts of both reactant and dehydrogenated products will pass through the membrane along with the hydrogen, giving a difficult downstream separation problem. In a recent general analysis,[40] it has been concluded that only selective Pd membranes coupled with fast reactions can positively affect conversion.

The general behavior of product-removal membrane reactors has been well studied. More details on porous ceramic membrane reactors can be found in the series of publications by Mohan and Govind.[41–43] An analysis of different flow configurations and the limits of each has been provided for dense Pd membrane reactors by Itoh.[44]

2.2.2 Applications – The earliest product-removal studies were on decomposition reactions (HI, H_2S) and simple alkane dehydrogenations. Tsotsis's group has published extensively on ethane dehydrogenation in a CMR,[45] while Itoh and co-workers studied cyclohexane dehydrogenation to benzene in both porous[46] and dense[47] PBMRs. The industrial importance of the dehydrogenation of ethylbenzene to styrene led to its study by several groups using the PBMR,[37,38] the hybrid PBMR[34] and the PBCMR.[37,38] This reaction is still of considerable interest.[48]

More recent work on product removal has addressed more complicated reaction systems, such as propane dehydrogenation to propylene,[49–52] methane steam reforming and the water-gas shift reaction,[53,54] and dehydrogenation of isobutane[31,55] and *n*-butane.[56] These are all reactions of high industrial interest. Awareness of the need for environmentally-benign processing has also brought forth applications of NH_3 decomposition[57] and has renewed interest in H_2S decomposition.[58]

The PBMR and PBCMR reactors have their counterparts in fluidized bed systems. These have been studied by Adris and co-workers,[59] both theoretically and experimentally. The steam reforming of methane was the system studied, which appeared promising since the FBMR addressed problems of heat transfer and equilibrium limitations simultaneously.

Although most product-removal applications feature H_2 removal, some decomposition reactions in which O_2 is removed are of environmental interest. These use dense mixed-oxide materials (see Section 2.3) to remove O_2 only. Early work involved NO and CO_2 decompositions using zirconias.[60,61] Itoh *et al.*[62] have recently reconsidered the CO_2 decomposition reaction to try to run it with reasonable conversion at lower temperatures using a PBMR. Dixon *et al.*[63] considered the impact of new, high O_2 flux perovskite materials on both CO_2 decomposition and NO decomposition. It appears that further advances in materials will be needed before these applications become viable.

2.3 Membrane Reactors for Distributed Addition of a Reactant – In the second main type of membrane reactor, a membrane is used as a permeable wall, usually in a tube-and-shell configuration. One reactant is fed along the length of the reactor. The membrane can be dense or porous: the separation function is not usually present or is incidental to the main function of the membrane. This configuration is illustrated in Figure 3 where the distributed species B is shown on the shell side with the reaction and catalyst on the tube side, although the opposite configuration is often used.

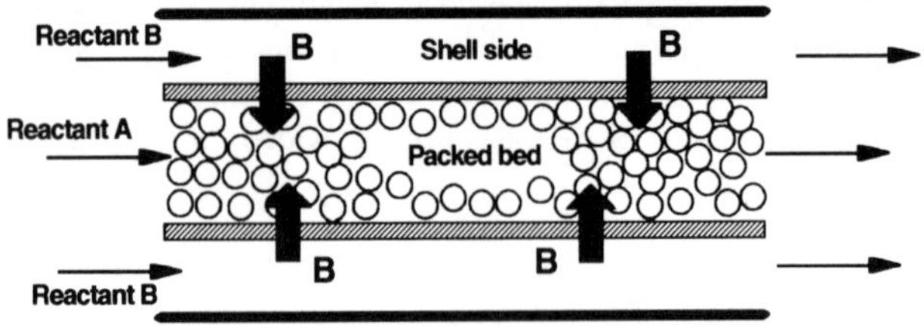

Figure 3 *Schematic of a membrane reactor for reactant feed*

2.3.1 Principles – The idea of a distributed-reactant feed is applied to systems with two competing reactions. A typical example is a partial oxidation of a hydrocarbon. The general reactions have the form,

$$A + \nu_1\, B \rightarrow P \tag{1}$$

$$A + \nu_2\, B \rightarrow S \tag{2}$$

where P is the desired main product, S is the undesired side-product, and A and B are reactants. Let the reaction rates be proportional to the partial pressure of B: $r_1 = k_1\, p_B^{n_1}$; $r_2 = k_2\, p_B^{n_2}$, where any dependence on p_A has been included into k_1 and k_2. If $n_1 < n_2$ the kinetics are termed *favorable*, and a lower p_B will slow reaction (2) more than reaction (1). This gives an enhancement in the selectivity to the desired product. This idea was illustrated in a theoretical study by Reyes *et al.*[64] who also made the point that since the reaction rate decreases with lower p_B, a larger volume of the catalyst will be needed to achieve the same conversion. For some reactions this requirement may increase by a factor of 100, which is clearly impractical. For fast reactions, distributing the feed can be a promising approach.

Both experimental and modelling studies have been carried out. It should be noted that in a membrane reactor, p_B may be much lower than in a conventional co-feed reactor (PFR) of the type usually used for kinetics studies. It is possible that the reaction order could change, since most kinetics reflect a redox mechanism. Thus, kinetics taken from the literature that appear favorable could in fact be unfavorable at the lower partial pressures of B present in a membrane reactor. This may account for the fact that fewer instances of good agreement between theory and experiment have been reported for reactant feed situations than for product removal.

Porous membranes used for reactant feed are typically mesoporous or macroporous. Two problems arise: control of the rate of addition and distribution of reactant B, and back-diffusion of reactant A. To control the uniformity of the distribution of B, we want the membrane to present enough resistance to equalize pressure on the reactant B side, giving a constant transmembrane pressure drop along the tube. This may mean modification of thin γ-alumina mesoporous membranes[65] or use of Vycor glass tubes.[66] A higher pressure drop also helps reduce the driving force for back-diffusion, which was exploited in the 'pneumatic reactor' of Papavassiliou *et al.*[67] The disadvantage of a higher pressure drop, of course, is the decrease in permeation rate of B, which may lead to problems balancing the feed rate to the reaction rate in these reactors.

Dense membranes have been used to feed hydrogen for hydrogenation reactions.[68,69] Improved yields have been observed, due not to the above kinetics reasons, but attributed to the better availability of the active H^+ available on the membrane surface. Most dense membrane use has been to feed oxygen. Early studies considered silver membranes, but cost and low permeation rates did not favor these. More recent work has used solid oxide electrolytes as membranes. Initially, investigators used yittria- and calcia-stabilized zirconias (YSZ or CSZ), which had reasonable oxygen anion conductivity. Their low electron conductivity dictated the use of an external circuit, as shown in Figure 4(a).

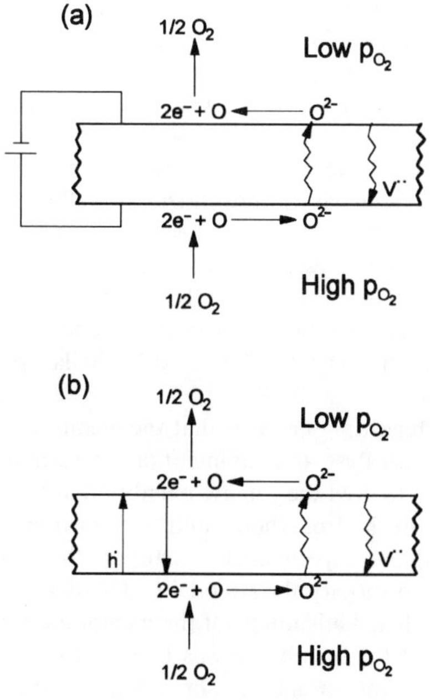

Figure 4 *Dense solid oxide membranes:* (a) *anion conductor with external circuit;* (b) *mixed conductor with no external circuit*

This allowed operation of the membranes as electrochemical oxygen pumps, where an applied voltage could be used to increase the O^{2-} transport across the membrane. This earlier work is described by Eng and Stoukides,[70] who also discuss basic principles in greater depth.

Solid oxide electrolytes with both ionic and electronic conductivities are very attractive as membrane materials as no external circuit is needed [see Figure 4(b)]. The work of Teraoka *et al.*[71] showed the potential of perovskite materials with aliovalent substitutions. A typical perovskite has the formula ABO_3 (*e.g.*

$LaAlO_3$); substitution on the A and B sites of lower valence cations induces anion disorder and increases both conductivities. Although the initial O_2 permeation results appeared to be about one order of magnitude higher than those for YSZ, attempts to reproduce these have not always succeeded. Apparently oxygen out-gassing from order–disorder transitions as the material relaxes toward steady-state may be responsible for the high measured O_2 fluxes, and a long transient may exist for these materials. Surface-exchange rate limitations may also reduce oxygen fluxes under reaction conditions. A thorough review of these questions has been provided by Bouwmeester and Burggraaf.[18]

2.3.2 Applications – The dense, solid oxide membrane materials require high temperatures (> 700 °C) for good O_2 fluxes. This has led to their being restricted to methane activation, and in particular they have been much studied for the oxidative coupling of methane (OCM). Early studies showed that high selectivities could be achieved, but at low conversions and thus low yields, typically less than 1%. More recent work has used the mixed-conducting perovskites in experimental work[72–75] and in simulations.[76] Although yields of > 50% have been predicted theoretically, the experimental results have not shown significant improvement over the best fixed bed results of 25–30%.

Porous membranes do not allow the separation of O_2 from air, but can operate over a wider temperature range. Consequently, more reactions have been studied using porous reactant feed membranes than dense ones. Of course, OCM has been one of the most studied,[66,77–79] but with similar results to the dense membrane studies.

An interesting point regarding OCM is that the membrane material itself may act as a total oxidation catalyst, thus unintentionally turning the PBMR into a PBCMR. Lu *et al.*[75] pre-treated their dense membrane with an OCM catalyst to prevent contact between hydrocarbons and the membrane oxide material. Coronas *et al.*[80] carried out experiments to estimate the contribution from the membrane and used it to modify their model of OCM in a porous membrane. They found that the predicted advantage of the membrane reactor was decreased if the catalytic activity of the membrane was taken into account, and suggested the development of inert membrane reactor materials, and more active OCM catalysts, as possible remedies.

The distributed-reactant membrane reactor has also been studied for several oxidative dehydrogenation reactions: ethane to ethylene,[65] propane to propylene[81] and butane to butene.[82] The results for these reactions have shown more promise, with higher yields for the membrane reactor when compared with a fixed bed, over certain ranges of the operating parameters.

Partial oxidations have not been as frequently studied. Harold and co-workers[83,84] have simulated a partial oxidation reaction scheme in an O_2-fed CMR. Bernstein *et al.*[85] simulated methane partial oxidation, while Dixon *et al.*[63] modelled partial oxidation of *o*-xylene to phthalic anhydride to examine temperature profiles.

2.4 Membrane Reactors for Control of Reactant Contact – The two-sided geometry of a membrane presents an opportunity to contact reactants differently. Two different reactants can be fed from opposite sides, creating an interface within the membrane. Two major application areas of this idea have emerged. In the first, a gas and a liquid phase are on opposite sides of the membrane; in the second, two reactants in the same phase may be kept separate before reaching the catalytic site. For both cases, the membrane again plays little or no separative role.

2.4.1 Multiphase Membrane Reactor – The principle here is to improve the contact between volatile and non-volatile reactants. Reactions studied experimentally are the hydrogenation of α-methylstyrene to cumene over Pd/Al_2O_3,[86] and nitrobenzene hydrogenation to aniline over Pt/Al_2O_3.[87] Modelling studies have also been carried out.[88,89] These reactions are conventionally done in trickle beds, or slurry reactors, where the volatile species must diffuse through the liquid phase to reach the catalyst. This can be the limiting step, since olefin hydrogenation catalysts are often quite active.

The membrane reactor has the liquid on one side and the gas on the other of a catalytically-impregnated porous ceramic tube. Thus the gas can diffuse through the pores directly to the gas–liquid interface inside the membrane, bypassing the diffusion through the liquid. A modelling comparison to a string of suspended pellets with cocurrent gas and liquid flow[89] showed a clear advantage in improved catalyst performance for the membrane reactor for very active catalysts. The experimental results of both groups[86,87] supported these conclusions. The reaction rates of Cini and Harold[86] were 15–20 times higher in a membrane than in a pellet, while Torres *et al.*[87] concluded that CMRs could be efficient for low-temperature three-phase reactions.

2.4.2 Non-permselective CMR – In this type of reactor (sometimes called a CNMR), the membrane plays no separative role, but simply provides a location for the establishment of a reaction zone. This is shown schematically in Figure 5 for the reaction $A + B \rightarrow P$. One reactant (B) is fed on the tube side of the membrane, and the other reactant (A) is fed on the shell side. The partial pressure gradients set up cause them to permeate toward each other inside the membrane, where they react.

For an instantaneous reaction, a reaction plane is established; for slower reactions a reaction zone exists which can be kept completely within the membrane by the right balance of reaction rate and permeability. Keeping reactants completely segregated by this device is possible so that reactant A does not break through to the tube side and reactant B does not break through to the shell side. It is also possible to control the product P to diffuse out of the membrane on one side only, in this case the tube side, which greatly simplifies the downstream separation task.

This type of membrane reactor ensures that the reactants meet in a strict stoichiometric ratio, since at steady-state the permeation flux of a reactant is equal to its rate of consumption in the reaction. A change in reactant flow rates acts to shift the zone to satisfy stoichiometry. It has been investigated experimen-

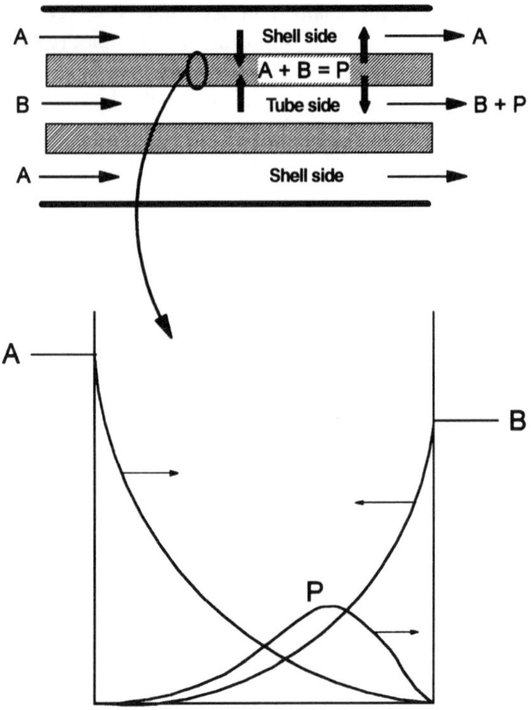

Figure 5 *Catalytic non-permselective membrane reactor with separated feed, showing transmembrane concentration profiles of the species and the direction of permeation*

tally and theoretically for the Claus reaction,[90] and for the selective catalytic reduction (SCR) of NO_x with NH_3.[91,92] In both applications, preventing the slip of one reactant (H_2S, NH_3) to the other side is essential. Also, products can be directed preferentially to one side by applying a pressure difference across the membrane. Veldsink *et al.*[93] studied CO oxidation over Pt and suggested that the CNMR was a good contacting device for fast, exothermic heterogeneously-catalysed reactions, since runaway was prevented due to transport being rate limiting. Keizer *et al.*[94] used a CNMR to regulate the supply of O_2 to the partial oxidation of ethylene to acetylene. Saracco *et al.*[95] deposited a catalyst in the modified pores of a fly-ash filter to combine the SCR of NO with NH_3, and particulate removal in the same unit.

3 Challenges for Catalytic Inorganic Membrane Reactors

With all of the research performed on catalytic inorganic membrane reactors in the last 20–30 years, it is somewhat surprising that there appear to be almost no working catalytic membrane reactors in industry.[13,15] Alternatively, if these reactors find no industrial interest, why is so much research work being done on them? The answer may be that the concepts behind catalytic membrane reactors are appealing, yet technological difficulties and unfavorable economics have so far prevented their realization. In this section, we look at the technical challenges and the comparisons to conventional technology that have so far worked against industrial implementation of catalytic membrane reactors. With this understanding, we will be positioned to identify the driving forces for the innovations to be presented in Section 4.

3.1 Technical and Materials Challenges – These have been considered in several earlier reviews, which presented lists of problems and challenges. It is interesting to compare the earlier concerns with more recent ones. Hsieh[5] highlighted the need for increased thermal and chemical resistance of membranes, problems with module packing (seals), and coking of catalysts. Armor,[9] in an article on challenges in membrane catalysis, specified 200–600 °C as the range of interest, and a need for porous membranes of pore size < 8 Å, to obtain high permselectivities. He saw two configurations for the future: catalyst deposition in porous inorganic membranes of size < 8 Å, and thin metal alloys on a porous support. Several challenges were identified:

- the need for crack-free membranes < 8 Å;
- need for thin, large surface-area layers;
- need for high-temperature seals;
- supported membrane layers must adhere well under temperature cycling;
- the sweep gas used on the permeate side should be eliminated;
- poisoning and fouling of catalytic membranes must be eliminated;
- permselectivity must be increased without reducing permeability.

Recently, the fluidized bed membrane reactor (FBMR) has also been examined from the scale-up and practical points of view.[96] Key factors affecting the performance of a commercial FBMR were analysed and compared to corresponding factors in the PBMR. Challenges to the commercial viability of the FBMR were identified. A very important design parameter was determined to be the distribution of membrane area between the dense bed and the dilute phase. Key areas for commercial viability were mechanical stability of reactor internals, the durability of the membrane material, and the effect of gas withdrawal on fluidization. Thermal uniformity was identified as an advantageous property of the FBMR.

A timely review paper focusing on the hurdles to inorganic membrane use was presented by Saracco *et al.*[13] They listed the main drawbacks as the high cost of membranes, low permeability, defects in permselective layers, instability of membranes and catalysts, and sealing. They were not optimistic for the future of the CNMR or CMR, and considered circumventing equilibrium in a PBMR to be most promising, if membranes with good characteristics become available.

Saracco *et al.*[13] broke their challenges into three groups: materials, catalysis and chemical engineering. This is a useful classification and we will follow it. The main materials issue is permeability *versus* permselectivity *versus* stability. Many attractive materials are metastable and do not stand up to harsh conditions. For example, γ-alumina thin layers open under steam and high temperatures. Stable materials, such as zirconia and α-alumina, have pore sizes that are too big for good permselectivity. Zeolites and porous metals can combine some of these characteristics. This problem can be illustrated by the 'materials box', shown in Figure 6.

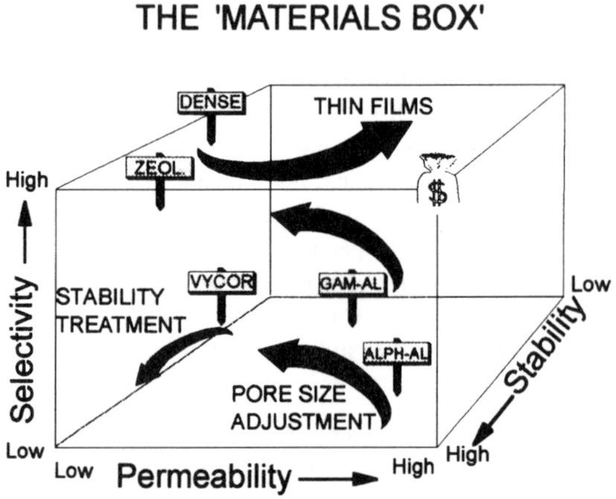

Figure 6 *Schematic representation of the trade-offs between membrane stability, permeability and permselectivity*

The desired material, which will make a bag of money for its developer, would combine high stability, high permselectivity and high permeability. Some materials presently used as inorganic membranes are shown by signposts, where the base of the signpost indicates the approximate values of the three properties. High permselectivity materials often have low permeability and *vice versa*. Materials with reasonable permselectivity and permeability tend not to be stable. Attempts by materials scientists to modify materials are indicated by arrows which show the trends in the properties with modification. For example, heat treating Vycor for stability opens structure and decreases permselectivity; depositing Pd or zeolites in thin films can compromise stability and induce cracks.

The need for thin, high permeability membranes has been confirmed by a technology assessment by Ward *et al.*[97] Their goal was to determine what thickness of membrane is needed for petrochemical applications. They considered only high-selectivity materials, both dense and porous, and calculated the approximate membrane thickness to provide sufficient gas flux to have a substantial impact on the process. Both H_2- and O_2-transporting membranes were considered, and the conditions and typical space velocities of the process. A sample of their results is presented here in Table 2.

Table 2 *Thickness requirements for various membrane and reaction combinations. Summary of findings of Ward* et al.[97]

Reaction	*Material*	*Thickness*
$CH_4 \rightarrow$ Syngas	Perovskite	0.1–1 mm
C_3H_8 Dehydrogenation	PdAg	< 90 μm
Steam reforming of methane	PdAg	< 100 μm
	SiO_2	< 1 μm
Oxidative coupling of methane	Perovskite	0.9 mm
$CH_4 \rightarrow MeOH/CH_2O$	None	—

These numbers show that for C_1-utilization, quite thick membranes (*ca.* 1 mm) can be used, which can be self-supporting. For dehydrogenation reactions very thin films are needed, which will be a difficult challenge for materials scientists to obtain phase and mechanical stability under process conditions. Further evidence of the need for highly selective materials for dehydrogenation is presented by Harold *et al.*[14] and Sheintuch.[40]

Many advances in membrane reactor engineering come from the development of new materials. This subject has such wide scope that we cannot include it here. The innovations of Section 4 are driven more by the challenges discussed below.

Challenges to catalysis scientists set out by Saracco *et al.*[13] include catalyst activation and reproducibility in CMRs, the assessment of catalysts (*i.e.* kinetics and selectivity), avoiding deactivation, and the distribution of the catalyst in the membrane as desired, without changing structure or permeability. The question of location and loading of the catalyst in a membrane is one that continues to be addressed (see Section 4.3.2). It has been determined that loading a membrane with a catalyst does usually reduce permeability,[37] and, unfortunately, does not usually improve permselectivity. Some studies of CMRs have not measured the permeability of the membrane before deposition, after deposition, and after use as a reactor. These measurements are essential to discover if permeability has changed, and to interpret results properly.

The last set of challenges from Saracco *et al.*[13] is for chemical engineers. They suggest the development of models for highly selective transport in micropores (see Section 4.4) and for capillary condensation, the development of models for large-scale modules, increasing area/volume ratios by new reactor configurations, solving problems of heat supply and temperature control (see Section 4.1), and determining the best configurations and flow patterns. Some of these are being addressed and will be discussed below; for modelling issues the reader is referred to the papers previously mentioned.[10,15]

3.2 Competitiveness Challenges – A different set of challenges for catalytic membrane reactors is brought up by the question of whether they provide any real advantage over more conventional reactors. Ultimately, economics must be the deciding factor, although very few studies have included an economic evaluation. One exception is the study of van Veen *et al.*,[98] who investigated the techno-economic feasibility of using porous inorganic membranes in the PBMR configuration. Three applications were investigated: dehydrogenation of propane to propene, dehydrogenation of ethylbenzene to styrene, and the water-gas shift reaction. They also included the downstream separation cost to recover the product. For propane dehydrogenation, they concluded that the selectivity of mesoporous membranes was not high enough, and that the propylene/propane price ratio was too low for a profitable membrane process. For the ethylbenzene to styrene process, it was concluded that prospects were not good due to a high reactant loss in mesoporous membranes, and the fact that the reaction kinetics are limiting for this reaction, and not H_2 removal. This is a similar drawback to that noted for isobutane by Raich and Foley.[33] For the water-gas shift reaction, they concluded that there were good possibilities for Integrated Gasification Combined Cycle (IGCC) power plant CO_2 emission control. The permselectivity of the membrane was adequate, but higher permeability membranes would be needed to decrease the required area and cost.

Most of the comparisons are based on conversion, yield or a similar feature of reactor performance. These have usually been made separately for the two main membrane reactor types.

3.2.1 Product Removal – Early comparisons involved a cocurrent, product-removal membrane reactor for dehydrogenation, with either a sweep gas or lower pressure on the permeate side. The results were typically compared with a PFR fed with the same reactant/inert mixture as the reaction side of the membrane reactor. Many claims were made (and, indeed, are still being made) of exceeding equilibrium conversion. However, the comparison is not fair to the PFR (see Figure 7).

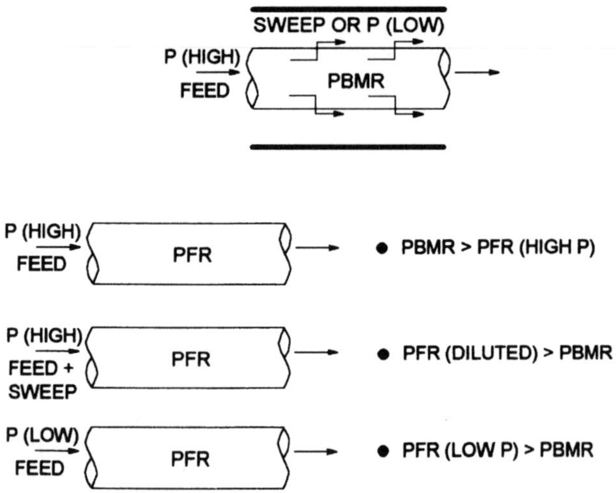

Figure 7 *Basis for comparison between the PFR and the PBMR*

Since at least one species can permeate through the membrane, the reaction side gas mixture does not define a closed system. The proper comparison for equilibrium is a diluted PFR fed with the combined PBMR feed, or run at a lower pressure (if more favorable). If the proper comparison is made, *equilibrium conversion is never exceeded* in a cocurrent membrane reactor. Itoh[44] has shown that exceeding reaction equilibrium in a countercurrent PBMR *is* possible.

The PBMR may still have advantages over the PFR, however. The large gas volume and lowered partial pressures, and thus lowered reaction rates, of the diluted PFR would cause large catalyst volumes to be needed, as noted by Itoh[44] and Tsotsis *et al.*[10] A recent approach by Reo *et al.*[99,100] attempts to make a fair comparison, for the cocurrent PBMR. They argued that the proper comparison is to an optimally-diluted PFR,[99] as long as the PFR used no more dilution gas than the PBMR. The basis of their comparison between isothermal, cocurrent PFR and PBMR was equal catalyst volumes, the same reactant feed, and the same temperature and pressure. A sweep gas was used on the permeate side, at the same pressure as the reaction side. The difficulty of the downstream separation task was not considered. Using a standard form of dehydrogenation kinetics, the PFR model could be solved for a range of *Da* (= reaction rate/reactant feed rate). For a given *Da*, the optimal amount of dilution (inert) gas was then determined to give the best yield of olefin (dehydrogenation product).

The membrane reactor model was also solved for a given *Da*, where the parameter *DaPe* (= maximum reaction rate/maximum permeation rate) could be optimized to give the yield for a given dilution (sweep) gas rate. They found that at low *Da* (kinetic control) the optimal PBMR was the same as an undiluted PFR, whereas for high *Da* (thermodynamic control) the optimal PBMR was the same as a diluted PFR. The membrane could improve the yield over only a narrow range of *Da*, where both kinetics and thermodynamics were important, as shown in Figure 8(a).

This range, of course, depended on the choices of membrane materials and reaction parameters.

To compare the PFR to the PBMR, *Da* was fixed, the PBMR optimized with respect to *DaPe*, then the PFR constrained to the same *Da* and optimized with respect to dilution under the usage constraint mentioned above. Some of their results are shown in Figure 8(b). The range of enhanced yields for the PBMR was small for a low permselectivity of 5 (mesoporous membranes); it was larger and included regions of high yields that were possible industrial operating conditions, only for a high permselectivity of 10 000 (dense membranes).

Their analysis was extended to the porous PBMR with a pressure difference between reaction and permeate sides in a subsequent publication,[100] in which the authors claimed that the appropriate basis for comparison was a PFR network. This allowed one PFR to work at the reaction side pressure, and a second one to work at the permeate side pressure, so that for both PFR and PBMR, two product streams at different pressures were produced. Broadly similar conclusions to the first study were reached.

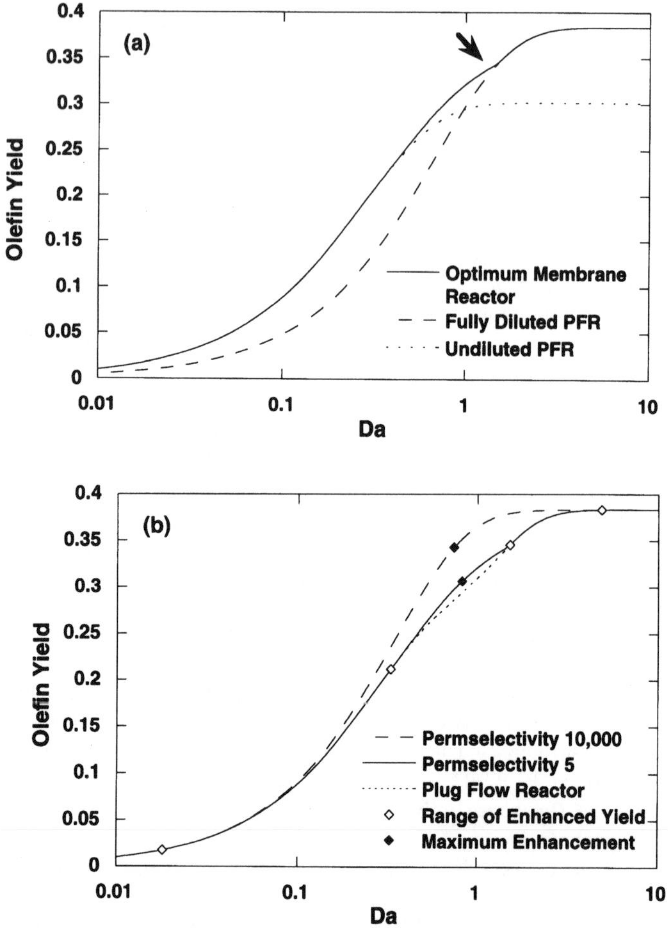

Figure 8 *Comparison between optimal PFR and optimal PBMR:* (a) *range of* Da *over which optimal PBMR is a PFR;* (b) *ranges of* Da *over which PBMR has an advantage over the PFR*
(Reproduced with permission from *AIChE J.*, 1997, **43**, 495. Copyright (1997) American Institute of Chemical Engineers)

3.2.2 Reactant Feed – The theoretical work of Reyes *et al.*[64] leads us to expect that when comparing reactant feed PBMR with the co-fed PFR for partial oxidation or similar reactions, the PBMR can give higher yields, but only in a much larger reactor. When equal size reactors are compared, the decrease in conversion due to the lower reaction rate in the PBMR offsets the increase in selectivity due to the favorable kinetics. This has been shown in general[101] and for OCM specifically.[66,102]

Tonkovich *et al.*[79] investigated the oxidative dehydrogenation of ethane to ethylene in both PBMR and PFR. Typical results are shown in Figure 9.

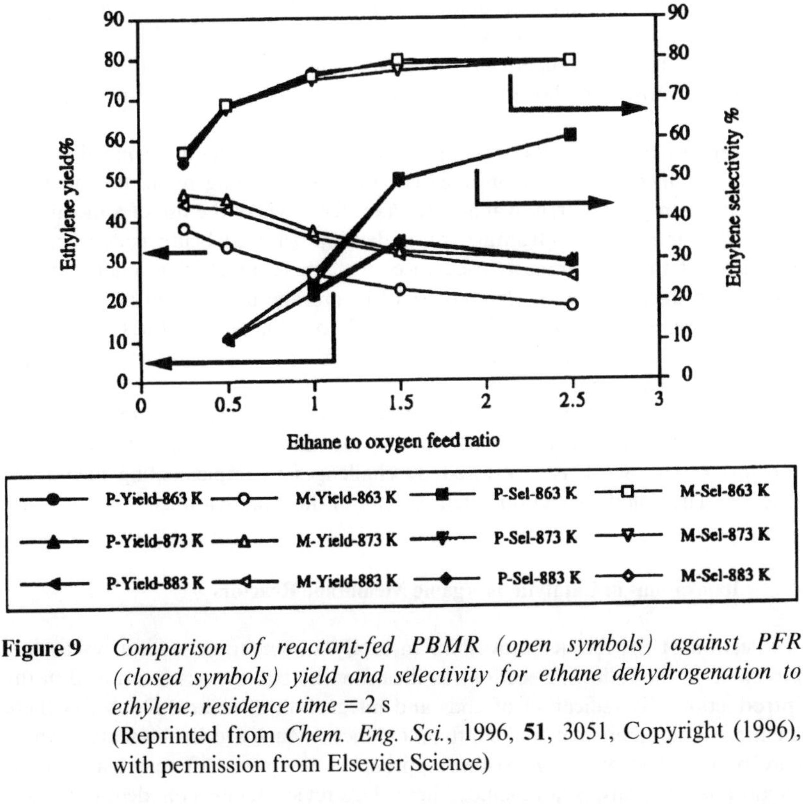

Figure 9 *Comparison of reactant-fed PBMR (open symbols) against PFR (closed symbols) yield and selectivity for ethane dehydrogenation to ethylene, residence time* = 2 s
(Reprinted from *Chem. Eng. Sci.*, 1996, **51**, 3051, Copyright (1996), with permission from Elsevier Science)

They used a porous-wall reactor to obtain high enough O_2 fluxes, and modified it to get better control. Equal catalyst volumes were employed in the two reactors. They found that the PBMR always produced higher yields at low ethane/oxygen feed ratio, but that the PFR did the same or better at high ratios. PBMR selectivity was always higher, as expected. At high feed ratios, PFR selectivity was already high, so the PBMR could not improve the yield by increasing selectivity. At low ratios, the permeating oxygen progressively decreases the ethane contact time as the oxygen/hydrocarbon ratio increases, which favors product. At high ratios, the permeating O_2 has little effect on the higher ethane flow rates. The reaction rate in this study was quite fast, so the PBMR could show some advantage with a reasonable catalyst volume.

3.3 Challenges Lead to Innovations – From Sections 3.1 and 3.2 above, catalytic membrane reactors are clearly no panacea for equilibrium-limited reactions or reactions with selectivity problems. Early optimism has given way to a spirit of healthy scepticism. Several authors have cited the same technical challenges consistently over the years: deposition of the catalyst in the membrane, maintaining high permeability and permselectivity, thermal effects, structural stability of the membrane, sealing, *etc.* High permselectivity is agreed to be key for product removal. Likewise, the earlier, simpler comparative measures that claimed superiority for membrane reactors over PFRs have been replaced by more complex analyses, considering aspects of downstream processing.

In response to these challenges, researchers are broadening the scope of activity in catalytic membrane reactor research. Efforts are being made to find *niche applications* in product removal and reactant feed, where the use of a membrane brings a clear economic advantage. More developments are being made where the *geometry* of the membrane is used to control the contact of reactants and to remove products efficiently. New types of membrane reactor and new ideas for *different configurations* are being considered to address the issues. Also, new objectives are being explored, such as *control of heat release* rather than simple increases in yield or conversion.

When the most recent advances in catalytic membrane reactors are considered, a large fraction of the new work can be identified as falling into the above categories. This work is motivated by challenges to implementing membrane reactors. The following section details several of these innovations.

4 Innovations in Catalytic Inorganic Membrane Reactors

A selection of the recent innovations in catalytic membrane reactor research is presented in the following sections to illustrate the two themes mentioned in the Introduction, diversification of ends and integration of means. Doubtless there are more, and others may make different choices. Much excellent research, which may be classified as using existing approaches but with different reactions or membrane materials, is not included here. This review focuses on identifying new research directions, which illustrate the responses to the above challenges.

4.1 Thermal Effects in Membrane Reactors – The problem of how to supply or remove heat from membrane reactors is an important practical concern. In a shell-and-tube configuration, the catalyst can be put on the shell side, which can then be encased in yet another shell containing a heat-transfer fluid. It is likely that other geometries will, in fact, be used. For example, the burner unit of Ohta *et al.*[103] is shown schematically in Figure 10.

This unit was proposed for the steam reforming of town gas, with removal of product hydrogen through Pd membrane tubes embedded in the annular catalyst ring. The hydrogen could be burned with a fuel to provide the required heat for the endothermic reforming reaction.

A different point of view is that the challenge of managing heat effects provides an opportunity for membrane reactors. The applications described in this section illustrate this approach.

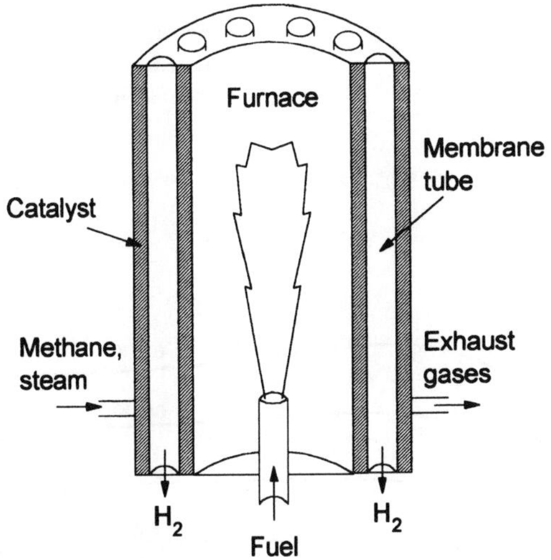

Figure 10 *Heat supply to membrane reactor for the steam reforming of methane town gas to synthesis gas*
(Redrawn based on original presented in ref. 103)

4.1.1 Temperature Control by Reactant Feed – The underlying idea of this innovation is that if the reactant is distributed along the tube in a PBMR configuration, then for exothermic reactions, such as partial oxidations, the release of energy will be more gradual. This could lead to a lower hotspot, and to a wider feasible range of operation.

Dixon *et al.*[63] simulated the partial oxidation of *o*-xylene to phthalic anhydride over a vanadium pentoxide catalyst supported on alumina, in a dense perovskite membrane tube. A non-isothermal model was used, which included the effect of temperature on the permeation rate. The competing reaction, complete oxidation to combustion products, is favored at higher temperatures. Comparisons were made to fixed bed reactors operated under the same conditions. For the fixed bed with inlet temperature 630 K, the usual hotspot near the front of the bed was seen, as shown in Figure 11.

For an inlet temperature of 640 K, the fixed bed exhibited runaway. With the membrane reactor, profiles from inlet temperatures of 630 and 640 K showed no hotspots, and the inlet temperature could be increased to 670 K without runaway. The yield to desired products was lower, however, due to the reduced reaction rates in the O_2-metered membrane reactor. So for the same inlet conditions and catalyst volume, the increased temperature control of the membrane reactor did not improve selectivity enough to overcome the decreased conversion typically seen in the reactant feed configuration.

More recent work[104] has examined whether running the membrane reactor at a higher inlet temperature than the fixed bed is advantageous, at the same catalyst volume. Preliminary results show that a small increase in the inlet temperature of

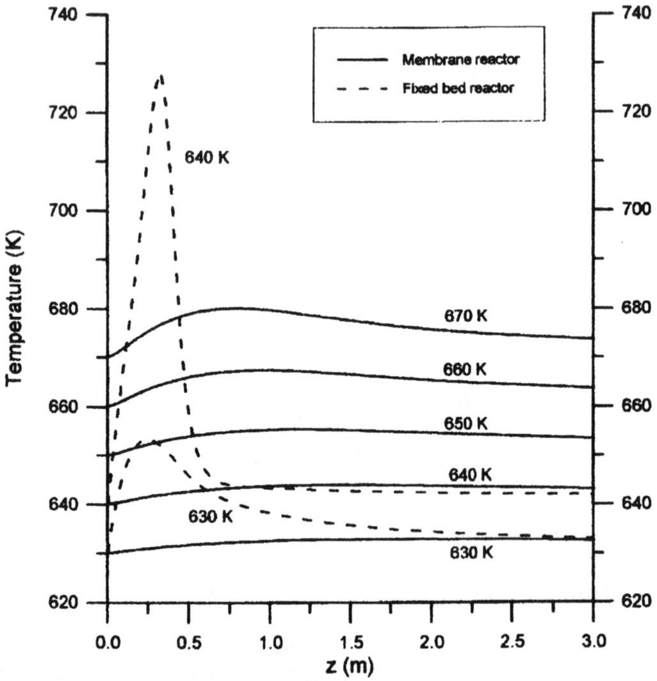

Figure 11 *Comparison of PFR and reactant-fed PBMR axial temperature profiles for* o-*xylene partial oxidation to phthalic anhydride, with different reactant inlet temperatures*

the membrane reactor can give improved yields for a generic partial oxidation reaction.

Similar ideas were explored by Coronas *et al.*[105] for OCM, and by Téllez *et al.*[82] for the oxidative dehydrogenation of butane in a porous ceramic membrane tube reactor. Both used reactant-fed PBMR configurations. The study of Coronas *et al.* claimed the membrane reactor to be an inherently safer contacting device for the gas-phase oxidation of hydrocarbons. They supported this claim by presenting both steady-state and transient experiments, one of the few transient studies of membrane reactors. The aim of the transient experiments was to investigate the response of the membrane reactor to a failure in the reactant feed control system. In steady-state mode, they demonstrated that the shapes of the temperature profiles were very different in the PFR and the PBMR. For the PFR, a sharp peak occurred near the reactor entrance due to the fast consumption of all the available oxygen. For the PBMR, with metered oxygen, a much more symmetrical profile was obtained, since the reaction was more gradual along the bed. They found that the PFR ran away under conditions where stable operation of the PBMR was possible. In the transient experiments, they found that if there was a sudden loss of inert gas to the reactor, the O_2 level rose much more slowly to its final value in the PBMR. The PBMR provided more time for corrective action before the gas mixture entered the explosive range.

In a similar study, Téllez *et al.*[82] compared the measured temperature profiles for a PFR and a PBMR with the catalyst packed on the tube side. Butane was fed to the bed, and an O_2 and He mixture permeated from the shell side. An axial thermowell ran along the centre-line of the bed. Their temperature profiles are shown here in Figure 12.

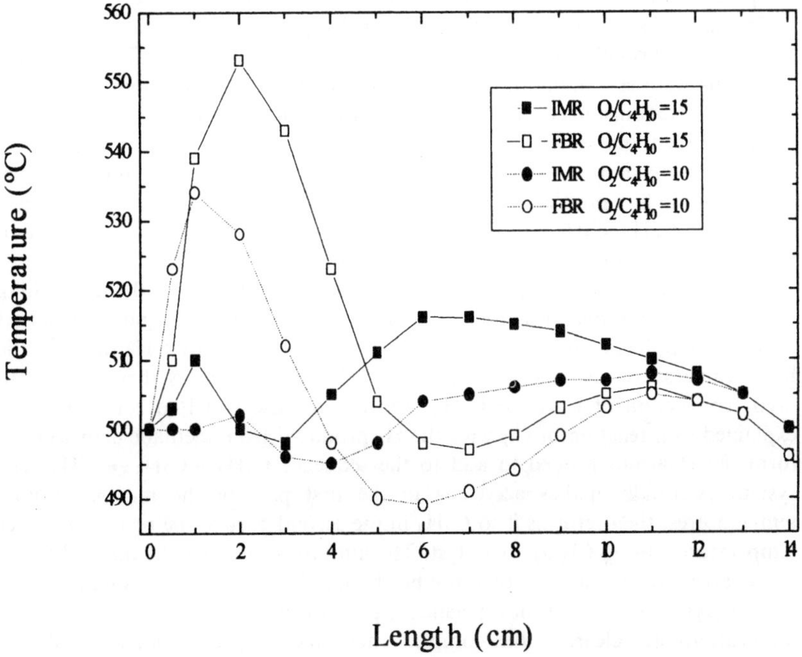

Figure 12 *Measured axial temperature profiles in a PFR (or FBR) and a reactant-fed PBMR (or IMR) for the dehydrogenation of butane to butene*
(Reproduced with permission from *AIChE J.*, 1997, **43**, 777. Copyright (1997) American Institute of Chemical Engineers)

The similarity between these measured profiles, and the simulated ones of Figure 11, albeit for different systems, is striking.

These studies have shown that distributing the O_2 feed *via* a membrane in hydrocarbon oxidations results in both improved selectivities and better temperature control. The challenge will be to adjust reactor operating conditions to maintain these advantages while overcoming the lower catalyst point utilization that accompanies them.

4.1.2 Reaction Control by Thermal and Reactant Fluxes – An ingenious exploitation of the potential of using a membrane reactor to control both the fluxes of energy and reactants has been presented by Rojnuckarin *et al.*[106] They considered the chlorine-catalysed oxidative pyrolysis of methane to C_2 products, such as acetylene and ethylene. The reaction mechanistic pathways involved (i) CH_3Cl and/or CH_2Cl, or (ii) C_2H_6 and/or C_2H_5 as reaction intermediates. A detailed reaction mechanism was considered, involving 74 species and 338 reactions. The richness of this chemistry was essential to the approach as it allowed control of intermediates and reaction pathways by manipulating temperature and species' concentrations through the control of fluxes.

The authors simulated a plug-flow reactor that permitted chemical and heat flux to be added through the sidewall of the reactor, along its length. The governing equations were essentially those of a membrane reactor with a homogeneously-catalysed reaction, and controllable fluxes as functions of position.

Several scenarios were investigated: optimal control of the flux profiles of Cl_2 and heat to maximize C_2H_4 in the output and minimize soot formation; optimal control of the flux profiles of O_2 and heat to maximize C_2H_4 in the output, while minimizing C_6H_6, C_4H_6 and C_4H_4 formation, *etc.* The results corresponding to the first scenario are shown in Figure 13.

This figure shows the optimal fluxes of (a) Cl_2, (b) heat, and the resulting profiles of species' mass fractions (c) and temperature (d). The figure shows that energy is supplied at the beginning of the reactor to generate CH_3 radicals; the high temperature favors C_2H_2 formation. The temperature is then lowered to favor conversion of C_2H_2 to C_2H_4, and an increase in Cl_2 occurs. This was explained as a reaction of Cl_2 with the H_2 produced from methane activation to form the H atoms needed to add to the sequence $C_2H_2 \rightarrow C_2H_3 \rightarrow C_2H_4$. The system essentially makes acetylene in the first part of the reactor at high temperatures, then reforms it to C_2H_4 in the second part of the reactor at lower temperatures, using Cl_2 as a catalyst. The authors argue that the optimal fluxes derived from their analysis would not be obtainable by intuitive reasoning.

Although this work was not specifically performed for a membrane reactor, the implications are clear. If the complex chemistry is understood, a membrane reactor with controlled heating may be applied to favor certain reaction pathways and desired products in other systems of low yields. The practical problems of controlling both energy and species fluxes simultaneously may be formidable; however, with current progress in membrane pore modification techniques, there is reason for optimism.

4.1.3 Thermal Coupling – A different idea for approaching the problem of heat management in a product-removal membrane reactor is that of thermal coupling. The idea is to run an exothermic reaction on one side of the membrane, and an endothermic reaction on the other. The reactant for one reaction is a permeate species of the other. If the rates of heat release can be balanced with the permeation rate and reaction rate, two problems could be solved. The energy needs of one reaction could be largely met, and at the same time the difficulties of efficiently using the often dilute permeate could be overcome.

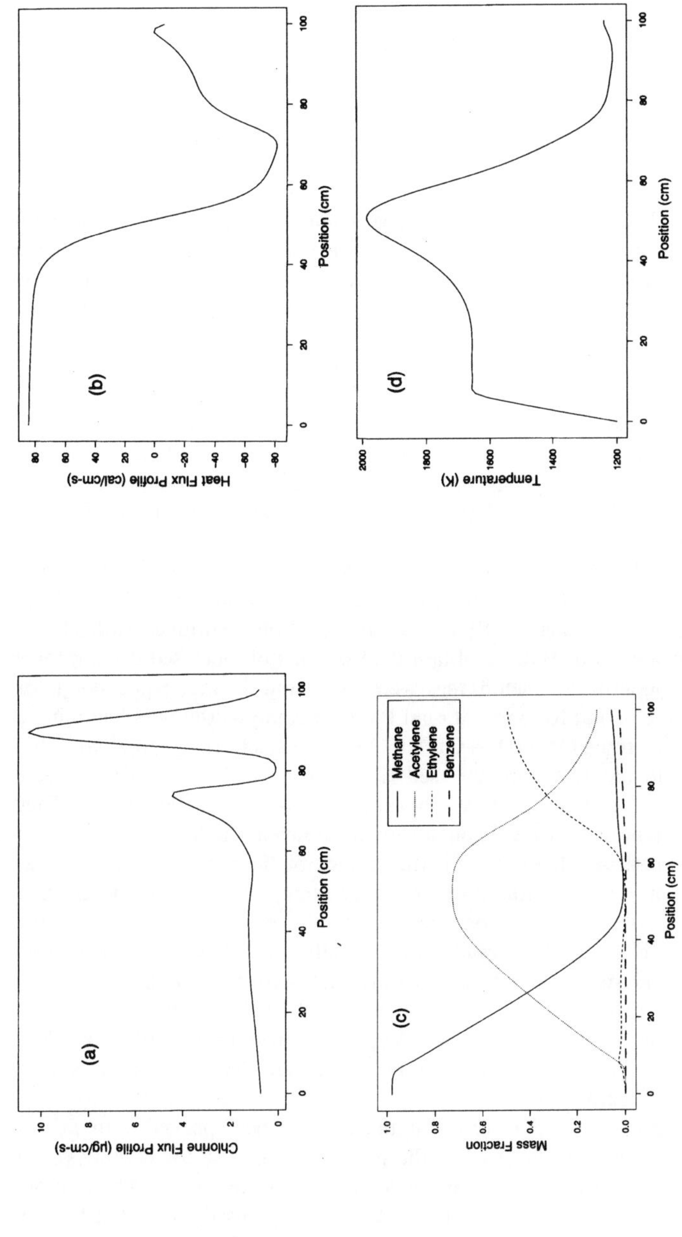

Figure 13 *Optimal control of methane conversion to ethylene and acetylene:* (a) *optimal* Cl_2 *flux profile;* (b) *optimal heat flux profile;* (c) *mass fraction profiles at optimal conditions;* (d) *temperature profile at optimal conditions*
(Reprinted with permission from *Ind. Eng. Chem. Res.* 1996, **35**, 683. Copyright (1996) American Chemical Society)

Early work considered the possibility of coupling dehydrogenation and hydrogenation reactions.[68] The difficulty with this natural pairing is that hydrogenations give up energy at relatively low temperatures, while dehydrogenations require energy at relatively higher temperatures. Thus, the heat-transfer driving force is not favorable. Later work[107,108] has used dehydrogenation coupled with the combustion of H_2 to H_2O. Since combustion releases heat at high temperatures, the driving forces are favorable. This idea may be worth pursuing, especially in combination with the increase in permeation rates obtainable from a reactive sweep gas. Further discussion may be found in Section 4.2.1.

4.2 Synergy between Reaction and Permeation – This section deals with approaches in which the reaction occurring on the permeate side of the membrane was used to increase the permeation rate through the membrane, either by improving the driving force for permeation or by also modifying the membrane characteristics.

4.2.1 Use of the Reaction to Improve Driving Force – We have seen that for product-removal membranes to be viable they must have high permselectivity for one species. This usually means that the permeation rate will be too low. One idea to overcome this problem is to conduct a reaction on the permeate side to decrease the partial pressure of the permeate. This also has the potential to produce a useful product.

Nigara and Cales,[61] in their study of CO_2 decomposition, reacted the permeated O_2 with CO back to CO_2, which increased conversion from 1% with no reaction to 21% with reaction. Several studies by Itoh mentioned earlier[107,108] oxidized permeate H_2 to H_2O to obtain dual benefits of increased driving force and thermal coupling. Pd membranes were used, and an interesting point for kinetic modelling is that for conventional Pd or Pt catalysts the rate-determining step is considered to be $O^* + H^* \rightarrow OH^* + *$, which leads to rate $\propto p_{O_2}^{\frac{1}{2}} p_{H_2}^{\frac{1}{2}}$. In a more recent study,[109] Itoh and Wu used H_2 oxidation to H_2O on the permeate side for the dehydrogenation of cyclohexane to benzene in a Pd PBMR. They achieved a substantial increase in conversion, as shown in Figure 14.

They also noted that hydrogen on the surface of the membrane emerges as dissociated atoms, and they found that the expression, rate $\propto p_{O_2} p_{H_2}$ fitted their data better. This expression corresponds to the limiting step $O_2^* + 2H^* \rightarrow 2OH^* + *$. This result warns us that the kinetics obtained on conventional catalysts may not hold for catalytically-active membranes.

The use of a reactive sweep gas has also been explored by Gobina *et al.*,[48] who studied ethane dehydrogenation using a Vycor glass membrane with a thin Pd–Ag film. Again, air was used on the permeate side. Their set-up is shown schematically in Figure 15.

Kinetics were determined directly on the catalyst bed, packed with Pd on Al_2O_3. Note that the Pd film was on the permeate side and could catalyse the hydrogen oxidation reaction in addition to permeating hydrogen. The authors found a higher conversion when a reactive sweep gas was used, but not as high as their model predicted. They attributed this discrepancy to the formation of a PdO

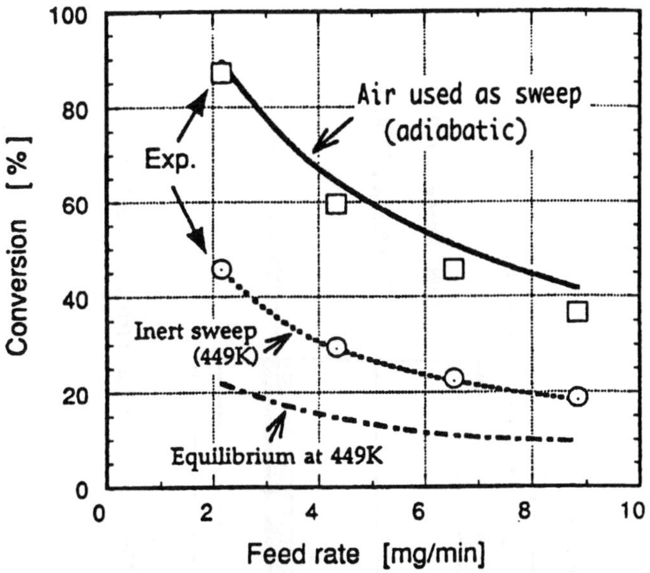

Figure 14 *Comparison of conversion for cyclohexane dehydrogenation for reactive and non-reactive sweep gases*
(Reprinted from *J. Membrane Sci.*, 1997, **124**, 213. Copyright (1997), with permission from Elsevier Science)

oxide layer that prevented the recombination of permeated hydrogen on the surface.

The use of a reactive sweep gas can clearly increase conversion. The challenge will be to find a more productive use of the permeated species, other than combusting it. Heat integration is one possibility, but again combustion may be the only reaction that can supply heat to a dehydrogenation at a high enough temperature. Balancing heat and permeation fluxes will also be a difficult reaction engineering and control task.

4.2.2 Use of the Reaction to Modify the Membrane and Improve Driving Force – The use of dense oxide membranes to feed O_2 to partial oxidation reactions is limited by the need for high temperatures to obtain reasonable permeation rates. The potential exists for selectivity gains and the control of heat effects. One application where dense membranes have a clear role to play is in methane activation. In particular, the partial oxidation of methane to syngas,

$$CH_4 + \tfrac{1}{2} O_2 \rightarrow CO + 2\, H_2$$

is carried out at high temperatures, usually above 750 °C. Air is not used directly in the reactor as it is not economical to separate the nitrogen from synthesis gas downstream. The oxygen separation plant for the feed is the dominant cost of the

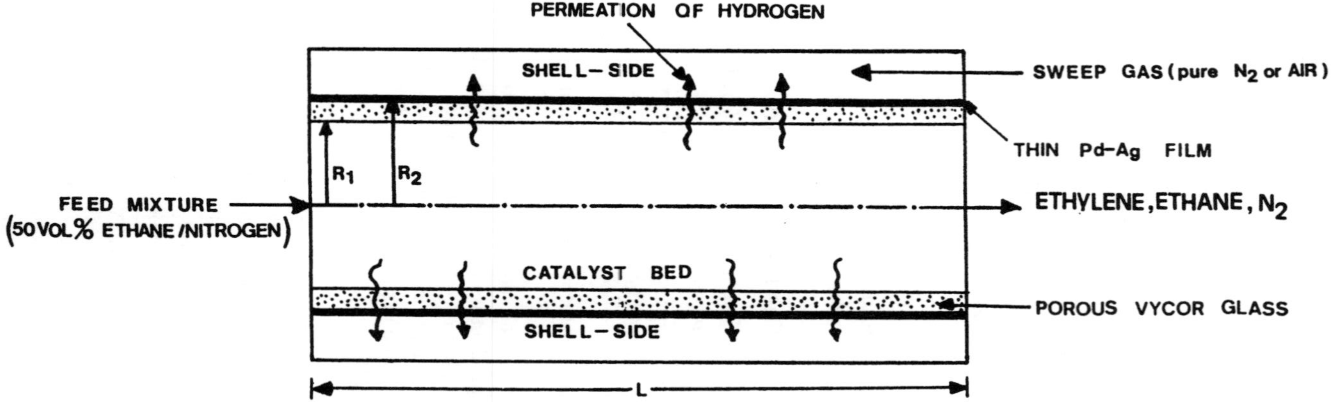

Figure 15 *PBMR configuration for ethane dehydrogenation using a reactive sweep gas*
(Reprinted from *J. Membrane Sci.*, 1995, **105**, 163. Copyright (1995), with permission from Elsevier Science)

process, and its elimination would be desirable. The slightly exothermic partial oxidation reaction is attractive compared with the strongly endothermic steam reforming of methane.

The use of a dense membrane reactor has been studied experimentally by the group at Argonne/Amoco[110–112] and by the group at WPI both theoretically[113,114] and experimentally.[115–117]

The motivation for both was the report of high rates of O_2 permeation through perovskite membranes of the formula $La_{1-x}Sr_xCo_{1-y}Fe_yO_{3-\delta}$ by Teraoka *et al.*,[71] which have been discussed above.

Balachandran and co-workers[111,112] investigated perovskite-type membranes (SFC) of the formula $Sr(Co,Fe)O_x$ in a reactor in which extruded 1 mm thick wall hollow tubes were packed with a Rh-based reforming catalyst, with air on the shell side. The first type they tried broke after a few minutes of operation. An extensive investigation[110] pointed to a lattice mismatch, resulting in strain from the O_2 gradient from the reaction side to the air, which caused a phase change in the membrane material. A second failure type was also observed, long after initiation of the reaction, which was attributed to the expansion of the membrane on the reaction side. This was caused by the decomposition of the membrane material in the presence of the reducing reaction products H_2 and CO. They suggested that the oxygen permeation rate was not high enough to replenish the vacancies, and that control of the operating conditions and a different membrane material was needed to balance O_2 permeation from the air side with vacancies on the reaction side.

A second membrane of non-perovskite SFC-type was investigated. This tube did not fracture, and remained stable over 1000 hours of operation. Results are shown in Figure 16.

The methane conversion was $>98\%$ and the CO selectivity was $>90\%$ with an H_2/CO ratio of 2:1. The properties of this second tube gave stability – no phase transition took place as O_2 was removed. The O_2 flux across the membrane was critical, and was found to depend on the membrane material and the catalytic activity of the surfaces. The authors ran experiments without the reforming catalyst so that all of the oxygen transported through the membrane reacted with methane to form combustion products. This gave a flux of 0.3 scc cm^{-2} min^{-1}. When the reforming catalyst was added, giving products of CO and H_2, a higher O_2 flux of 2–4 scc cm^{-2} min^{-1} was measured. The authors recommended the use of thinner-walled membranes to increase the oxygen flux so that the lattice O_2 lost to the reducing gases could be replaced, thus maintaining stability.

Tsai *et al.*[115,116] also approached the problem of increasing O_2 flux and stability. Their approach was to balance the substitutions on the A and B sites of the ABO_3 perovskite. Stability is strongly influenced by a stable BO_3 skeletal sublattice, and the choice of Fe with a mild Co substitution gives stability with reasonable oxygen anion conductivity. Then different amounts and types of aliovalent cations (Ca^{2+}, Sr^{2+}, Ba^{2+}) were partially substituted for La^{3+} in a $LaFe_{0.8}Co_{0.2}O_{3-\delta}$ perovskite framework to attain higher electron conductivity.

The reactor type was a disk, exposed to air on one side and a CH_4/He mixture on the other, and usually run at 850 °C. Runs were made without the catalyst,

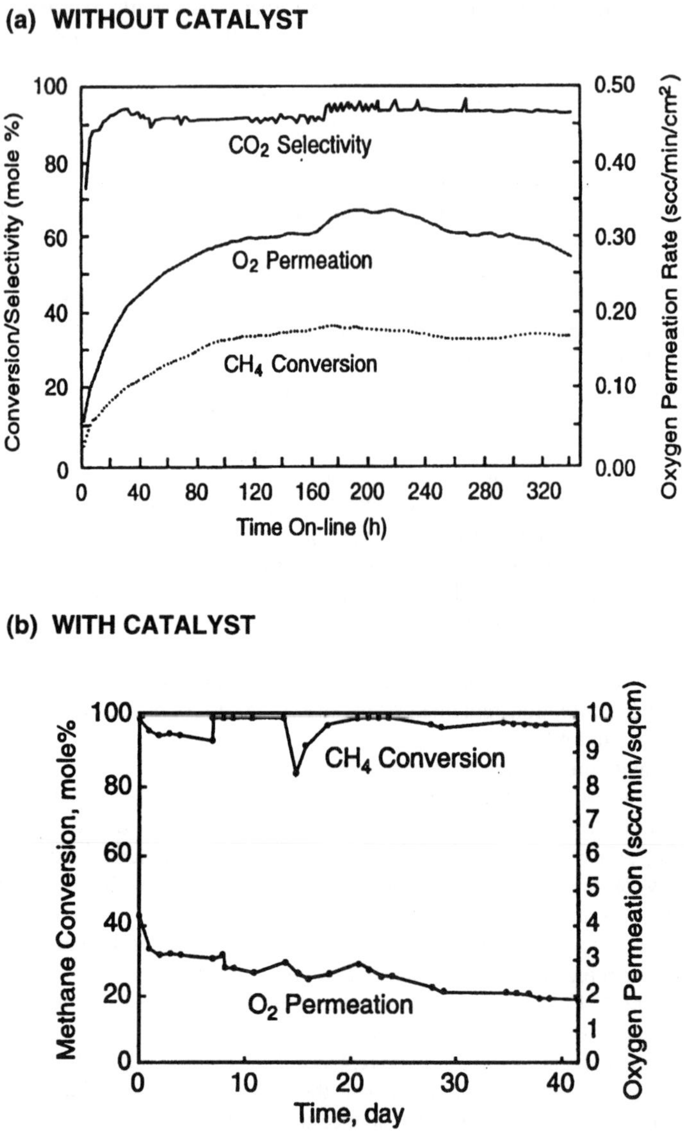

Figure 16 *Experimental results for the conversion of methane to synthesis gas using a dense O_2- permeable membrane for reactant feed*
(Reprinted from *Appl. Catal. A*, 1995, **133**, 19. Copyright (1995), with permission from Elsevier Science)

with the catalyst packed downstream from the membrane and with the catalyst on the membrane surface directly, see Figure 17.

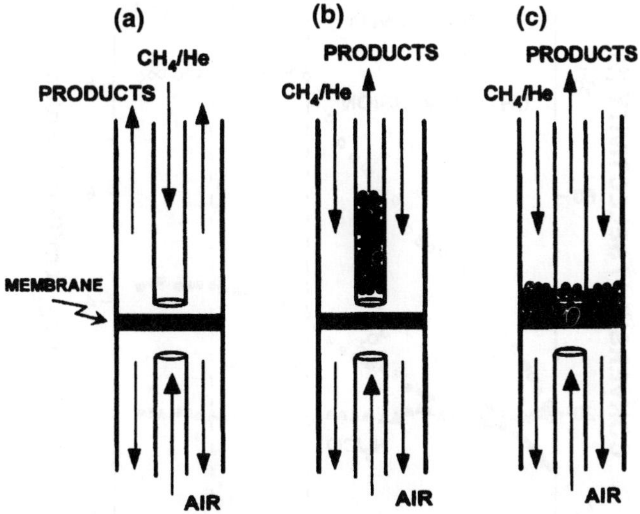

Figure 17 *Reactor configurations for conversion of methane to syngas using a dense oxygen-permeable disk membrane:* (a) *no catalyst;* (b) *catalyst loaded downstream from membrane;* (c) *catalyst loaded directly on to membrane*
(Reproduced with permission from *AIChE J.*, 1997, **43**, 2741. Copyright (1997) American Institute of Chemical Engineers)

Both 5% Ni/Al_2O_3 and 5% Rh/Al_2O_3 catalysts were investigated, although not much difference was found between them, so most experiments used the Ni/Al_2O_3. Oxygen permeation rates were measured and found to decrease with time onstream, in agreement with other results.[18] The highest stable O_2 rate was found for $La_{0.2}Ba_{0.8}Fe_{0.8}Co_{0.2}O_{3-\delta}$ and so this material was used for the reactor experiments.

The reactor runs without the catalyst gave low conversion of CH_4, to CO_2 only. When the catalyst was loaded downstream, a CH_4 conversion of 95% could be attained with CO selectivity of 100%. For both reactor types the oxygen flux was *ca.* 0.7 scc cm^{-2} min^{-1}, and the maximum selectivity occurred when the CH_4/O_2 ratio was close to 2, the stoichiometric ratio.

The results of the configuration with the catalyst placed directly on the membrane are shown in Figure 18.

The membrane was stable for more than 850 hours. From zero to 500 h, the methane conversion increased from 17 to 80%. The oxygen flux J_{O2} increased sharply from zero to 50 h, then continued to increase more slowly until 500 h. During this time the CH_4/O_2 ratio declined from 12 to 2.5, the CO selectivity

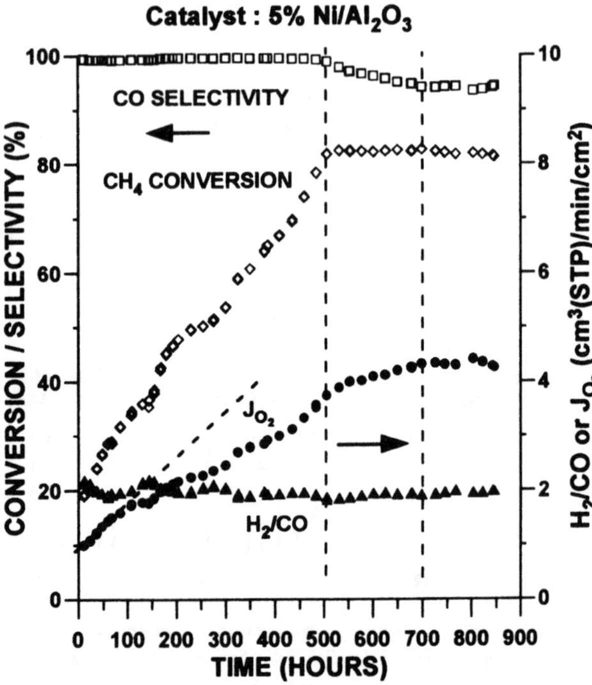

Figure 18 *Experimental results for long-time study of methane to syngas reaction in reactor configuration with catalyst loaded directly on to membrane* (Reproduced with permission from *AIChE J.*, 1997, **43**, 2741. Copyright (1997) American Institute of Chemical Engineers)

remained unchanged. At 500 h the conversion levelled off, while the CO selectivity decreased, and J_{O2} continued to increase but at a much reduced rate. This corresponded to a decrease in CO selectivity due to the formation of CO_2, so the O_2 use increased but the conversion did not. At 700 h, everything levelled out to constant values. The final J_{O2} was *ca.* 4.0 scc cm^{-2} min^{-1}, a five-fold increase under reducing reaction conditions, similar to that of Balachandran *et al.*[111,112]

The observed results were explained as follows: from zero to 50 h lattice oxygen starts being removed, so a fast increase in J_{O2} is observed; the O_2 supply from the air side cannot compensate so the bulk membrane structure adjusts from 50 to 700 h to adapt to the oxygen partial pressure gradient. Surface studies[117] showed that $BaCO_3$ was formed by surface decomposition. Ba segregation took place on the air side due to long-term O^{2-} migration, with Ba^{2+} migration in the opposite direction. Surface segregation and decomposition may have helped slow the increase in the O_2 permeation rate. When the reaction kinetics started to favor CO_2 over CO, a less reducing atmosphere was developed which slowed and ultimately stopped the solid-state transformation and steady-state was reached.

It is clear from these two experimental programs that there can be strong interactions between the catalytic reaction outside the membrane and the permeation rate through the membrane. The membrane structure can be affected by the gas composition that it is contacting. This implies that permeation experiments under non-reacting conditions will not give the complete picture. In addition, the possibility exists that dense membranes can give reasonable oxygen fluxes at lower temperatures if they are coupled to reducing atmospheres. This could widen the range of possible applications of these membranes, beyond methane conversion.

4.3 Control of Contact between Catalyst and Reactants – In this use of catalytic membranes, the membrane is porous and in the CMR configuration, and is either intrinsically active or has had a catalyst deposited within the pores. The membrane geometry allows for a degree of control of the contact time. It is operated in the cross-flow mode, in which all of the reactant is forced to flow through the membrane by feeding it to one side with a closed exit. This is illustrated schematically in Figure 19.

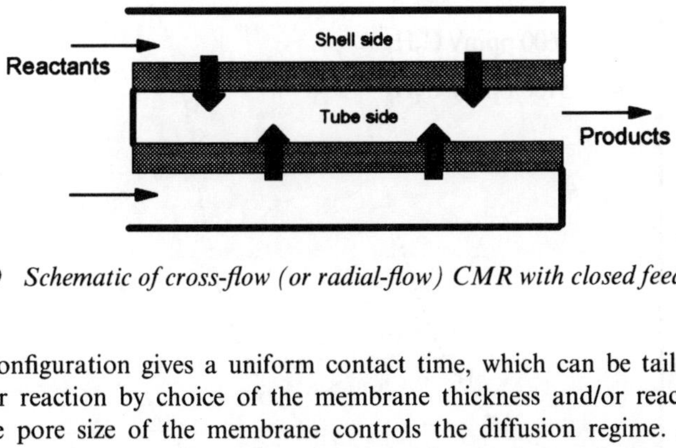

Figure 19 *Schematic of cross-flow (or radial-flow) CMR with closed feed side exit*

This configuration gives a uniform contact time, which can be tailored to a particular reaction by choice of the membrane thickness and/or reactant flow rate. The pore size of the membrane controls the diffusion regime. Also, the membrane geometry can be used to place a catalyst in the membrane optimally, or to control the partial pressure of the reactants in the phase in contact with the catalyst.

4.3.1 Control of Contact Mode – Pina *et al.*[118,119] investigated the use of a CMR as a new type of contactor for the combustion of volatile organic compounds (VOCs). VOC combustion is a problem since there is a wide range of compounds that may have to be treated. Low concentrations (ppm) are usually the case, and a range of combustion temperatures may be involved. Avoiding partial oxidation products is important – these may be more toxic than the original VOCs. The idea is to obtain efficient contact by forcing the gas to contact the catalyst in the Knudsen diffusion regime, where wall–molecule collisions dominate over molecule–molecule collisions. This forces all the VOC molecules into good contact with the catalyst.

A γ-alumina porous ceramic membrane was impregnated with Pt. It was operated in cross-flow mode with a lean VOC-containing stream forced through. The membrane had a 200 nm pore diameter, and γ-Al_2O_3 was deposited in the pores. Pt was then introduced by wet impregnation with chloroplatinic acid, followed by drying and calcination. The Pt loading on the membrane was only 0.13 wt%, but the authors pointed out that the metal was in the γ-Al_2O_3, not in the α-Al_2O_3 that makes up most of the membrane mass. Thus, they felt that they achieved a reasonably high catalyst loading in the thin γ-Al_2O_3 film. The Pt was verified by XPS to be homogeneously distributed, and permeation experiments confirmed that transport was predominantly by Knudsen diffusion.

Results for the conversion of toluene[118] showed that the light-off temperature (defined as 50% conversion) and the total conversion temperature were independent of the flow rate, as expected for the Knudsen flow regime. Typical results are shown in Figure 20 for the effect of concentration on the temperature–conversion curves.

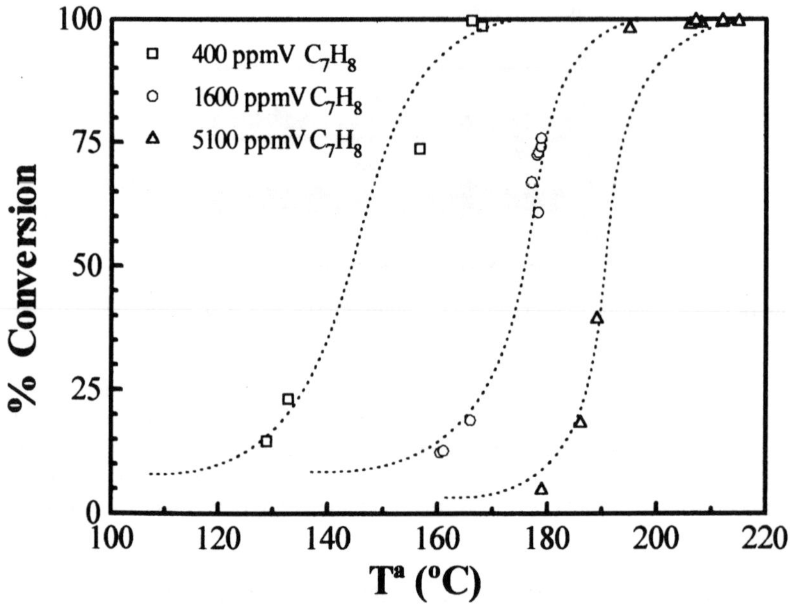

Figure 20 *Conversion* versus *temperature for the combustion of VOCs using a cross-flow CMR*
(Reprinted from *Appl. Catal. B*, 1996, **11**, L19. Copyright (1996), with permission from Elsevier Science)

The light-off temperature increased from 150 °C at 400 ppm to 190 °C at 4700 ppm. It was significant that these were among the lowest reported for VOC combustion. A range of 230–300 °C is more usual. The main drawback to this reactor configuration is the pressure drop across the membrane. The authors pointed out that increased pressure drop means increased operating cost, and that membrane optimization is needed to obtain the right balance between pressure drop and conversion.

In their second paper[119] Pina *et al.* extended their study to the combustion of toluene, methyl ethyl ketone, and mixtures. They also used a different technique, chemical vapor deposition, to deposit the Pt on to the membrane.

A similar configuration was used by Binkerd *et al.*[120] for the OCM reaction. The compound Sm_2O_3 was deposited in an asymmetric porous γ-alumina membrane in the fine pores. The authors used different membranes to look at the effect of pore size on the yield. The pore sizes ranged from 5 μm, giving 50% viscous and 50% Knudsen transport, to 0.02 μm, giving 100% Knudsen transport. The idea behind this was to suppress gas-phase reactions to CO_x products, by contacting reactants with the catalyst in a Knudsen diffusion regime, where molecule–molecule contacts were suppressed, as explained above. Depositing the catalyst in the relatively thin fine membrane layer also served to reduce contact of the desired C_2 products with oxygenated surfaces as they left the reaction zone. Results are shown in Figure 21.

It was found that the smallest pore sizes were best for both conversion and selectivity, confirming the benefits of the Knudsen contacting regime. The yield was better than comparable packed bed reactors with the same catalyst. This suggests that the idea of keeping the contact time short in a radial-flow CMR was beneficial for this reaction.

During the experimentation, three different tubes of pore size 0.02 μm were tested, and they gave different results, although their source and preparation were the same. EDX studies after reaction showed different locations of the deposited catalyst. The best results (shown in Figure 21) were for the catalyst deposited uniformly through the toplayer, as opposed to a large amount of the catalyst deposited at the outer surface of the toplayer. The explanation was considered to be that in the second case, the methane was converted to C_2 products at the outer edge of the toplayer, from which it was fed, then the products passed over enough of the catalyst to convert them to CO_x. This was limited with the uniform catalyst deposition, as the C_2 products were formed throughout the toplayer and not at the inlet to it. This question of catalyst location will be considered further below.

4.3.2 Control of Catalyst Location in a CMR – From the previous section, it can be seen that the optimal location or distribution of a catalyst in a membrane can strongly influence results. This question was studied theoretically by Keller *et al.*[121] for an isothermal, irreversible first-order reaction in a homogeneous membrane with slab geometry. They enumerated some different criteria of optimality: maximize the reaction rate per membrane area; minimize the flux of reactant leaving the membrane; and maximize the purity at the output side, expressed as the ratio (flux of product)/(flux of reactant). The first optimality

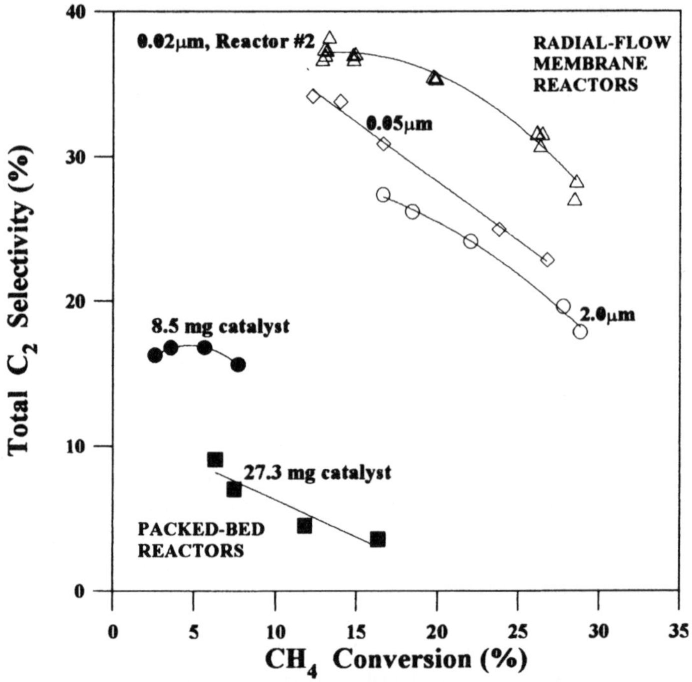

Figure 21 *Comparison of PFR and radial-flow (cross-flow) CMR conversion and selectivity; results for different membrane pore sizes are shown*

criterion, maximize point reaction rate, led to the general result that the optimal distribution was a delta function. That is, the catalyst should be deposited in as thin a film as possible. The other two criteria gave the same solution in the common case that the reaction rate increased with concentration of the reactant; the optimal distribution was a step function in which the membrane could be considered a three-layer sandwich, with the middle layer deposited with the catalyst.

Zaspalis[122] conducted a theoretical study of an asymmetric membrane with a thin, small-pore toplayer on a large-pore support, in both flat and tubular geometries for the simple isothermal reaction $A \rightarrow B$. The best conversion was in the case where all of the catalyst was near the outer surface of the toplayer, on the reactant side. In fact, for this reaction, reactant loss meant that the membrane reactor did worse than the fixed bed reactor. He also claimed that the optimal distribution is a delta function, for both geometries, for all possible kinetics, when diffusion of a reactant occurs from one side only. For segregated reactants, for the reaction $A + B \rightarrow C$ as illustrated in Figure 22, the optimal location of the catalyst was at the toplayer/support interface, assuming that diffusion was the only transport mechanism through the support.

If the support resistance was lower, the optimal location could be within the

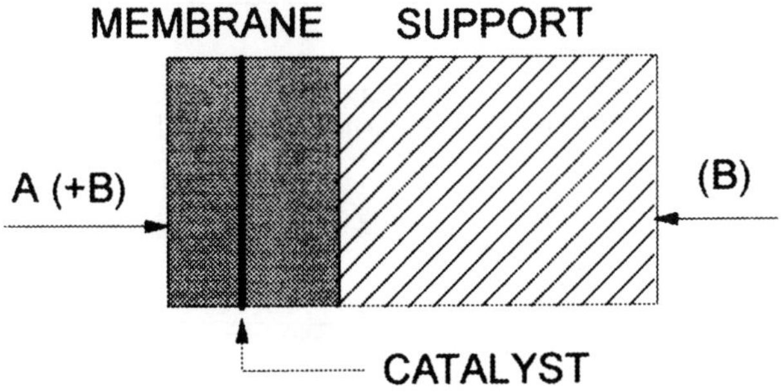

Figure 22 *Schematic showing the δ-function location of catalyst within the toplayer or fine pores of a composite asymmetric porous ceramic membrane*

toplayer. Varma and co-workers[123,124] have also presented a theoretical analysis for optimal catalyst location. In addition, they described a membrane preparation technique, using sequential slip-casting of inert and Pt-containing alumina sols. This gives precise control of catalyst layer thickness and location within a porous alumina membrane.

Varma's group's results build upon their established theory for non-uniform distribution of the catalyst in pellets for a PFR, and they have extended it to the case of a PBMR, obtaining broadly similar results. In Yeung *et al.*[123] they looked at the influence of the location of the catalyst distribution (delta function) for the PFR (PBR in their terminology), the PBMR and the CMR. For the case of interest here, the CMR, taking a first-order reversible reaction with well-mixed (not plug flow) conditions, they also concluded for $A \rightarrow B$ or $A \rightleftharpoons B$ that the optimal location was on the feed side of the membrane. A comparison of the various results with optimal catalyst location is presented in Figure 23, and shows that the CMR with the catalyst at the feed side or the PBMR with the catalyst at the pellet surface do better, at sufficiently high residence times. This maximizes the reactant concentration at the catalyst site, free of diffusion limitations.

Similar results were obtained if other performance indices were used, such as product flow rate or product purity.

The results of these studies are interesting, but still fall somewhat short of real situations, such as in the experiments of Cannon and Hacskaylo[39] or Binkerd *et al.*[120] Studies are needed for realistic kinetics, tube-and-shell geometries with plug flow, non-isothermal effects and external mass transfer. Varma and co-workers did this for a pellet, and found that the optimal catalyst distribution was a multiple step function, which is interesting in the light of the result of Keller *et al.*[121] Experimental verification of theoretical results is needed, which may be forthcoming given the recent advances in membrane synthesis procedures.

4.3.3 Control of Reactant Concentration at the Catalyst – In a series of publications[83,84] Harold and co-workers have attempted to improve the intermediate product yield of consecutive–parallel reaction schemes. They used

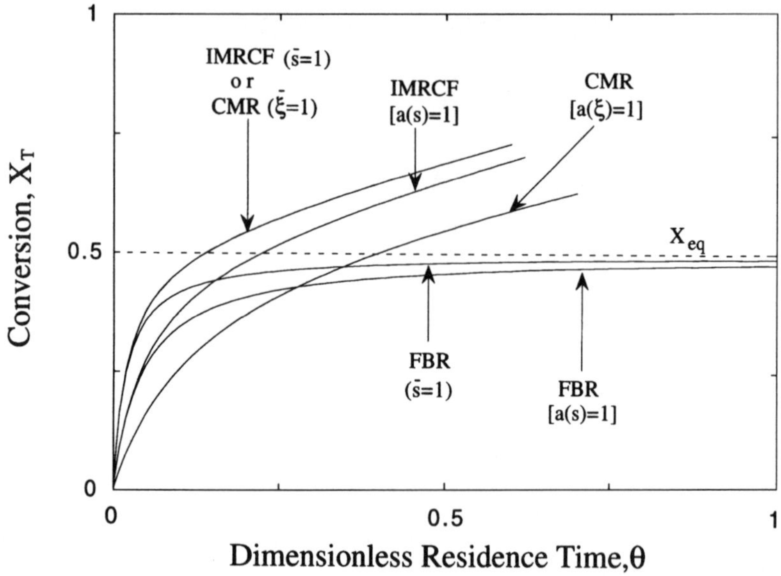

Figure 23 *Comparison of conversion for different reactor configurations [CMR, IMRCF (PBMR) and FBR (PFR)]; uniform catalyst distribution* [a(s) = 1 *or* a(ξ) = 1] *and delta-function catalyst distribution (* s = 1 *or* ξ = 1*)*
(Reprinted from *Chem. Eng. Sci.*, 1994, **49**, 4823. Copyright (1994), with permission from Elsevier Science)

an asymmetric membrane with one reactant fed from the support side, so that owing to diffusional resistance the concentration of that reactant is lowered at the catalyst site in the toplayer. This is similar to the argument for the reactant-fed PBMR, but the use of a CMR allows the product to be withdrawn on the other side. The membrane is mesoporous, so it is essentially non-permselective. The thin reaction layer at one edge of the membrane allows the intermediate product to escape before further reaction can take place. Their results for the isothermal case[83] showed that gains in the yield could be achieved, for sufficiently large reaction rates and provided that the active layer in the membrane was much thinner than the support layer. An extension to the non-isothermal case[84] showed that an improvement in intermediate yield was possible, due to better control of the exothermic reaction heat.

4.4 Zeolite Membrane Reactors – The need for highly permselective membranes, which are stable to high temperatures and resist chemical attack, has led to great interest in microporous materials, especially molecular sieve zeolites. Although a great deal of work is currently underway to develop zeolite membranes, few studies of zeolite membrane reactors are available. This reflects the difficulty of synthesizing defect-free zeolite membranes or thin films, despite a great deal of current work in this area. For a PBMR there will be the need for very thin zeolite films, in order to give permeation rates compatible with the reaction rates, which places very high demands on materials synthesis. If a CMR is considered, thicker membranes may be tolerated, but reactions will need to be chosen carefully or deactivation and/or coking may be an insuperable obstacle.

Experimental studies on zeolite membrane reactors have been reported recently. Casanave *et al.*[125] investigated the dehydrogenation of isobutane to isobutene in a silicalite membrane tube packed with a Pt–Sn/γ-alumina catalyst. Hydrogen was preferentially removed from the reaction, thus enhancing conversion of isobutane. In a second study,[126] a silicalite membrane was used to feed oxygen gradually to the oxidative dehydrogenation of propane to propene. Silicalite has pore dimensions of approximately 5.5 Å, which are too large to separate oxygen from nitrogen, so the use of a microporous membrane does not seem necessary here, as observed by Coronas and Santamaría.[16] However, cation-exchanged zeolite A can do this separation, which would allow air to be used as the feed, similarly to the use of dense membranes for methane partial oxidation to syngas (Section 4.2.2).

A theoretical study of a zeolite membrane CMR was carried out by Harris.[127] The objective was to compare the point performance of a catalytic zeolite membrane with that of a pellet, for two generic reversible reaction schemes, isomerization and dehydrogenation. The simple isomerization, $A \rightleftharpoons B$, was chosen for its simplicity in interpreting diffusion reaction coupling. The analysis of the dehydrogenation reaction $A \rightleftharpoons B + nC$ was a logical extension of the isomerization analysis, in which a decreasing occupancy profile in the zeolite membrane might benefit the rate of a reaction in which the occupancy requirement of the products exceeds that of the reactant. The model of multicomponent diffusion and reaction was based on a continuum formulation that combined a general mass balance for slab geometry with the Generalized Maxwell–Stefan (GMS) theory for micropore single-file diffusion.

The simulation results showed that a rate advantage for the membrane was possible given a favorable occupancy dependence of the diffusivity $Đ_{i0}(\theta)$ and the reaction rate $r(\theta)$. The membrane had an advantage if either $Đ_{i0}(\theta)$ or $r(\theta)$ increased with decreasing occupancy. This result was due to the occupancy gradient set up in the membrane geometry, which decreased from one side of the membrane to the other, as opposed to the pellet, in which the occupancy profile is constrained to be symmetrical. For the reversible reaction $A \rightleftharpoons B$, the membrane's range of advantage for a higher reaction rate and so a higher conversion was limited to high occupancy and low to an intermediate Thiele modulus (reaction rate/diffusion rate). For the dehydrogenation reaction $A \rightleftharpoons B + nC$, in which the products required more vacant adsorption sites than the reactants, the membrane's range of advantage for rate was quite broad, depending upon the value of n.

The other potential advantage for a membrane is the possibility for nearly complete separation of reactant and product. The ideal situation would be to feed reactants to one side of the membrane and withdraw a pure product from the other. A sample of results for $A \rightarrow B$ is shown in Figure 24.

(a)

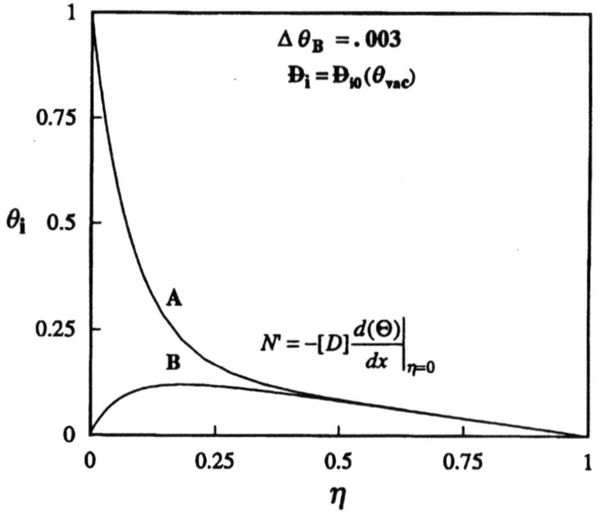

(b)

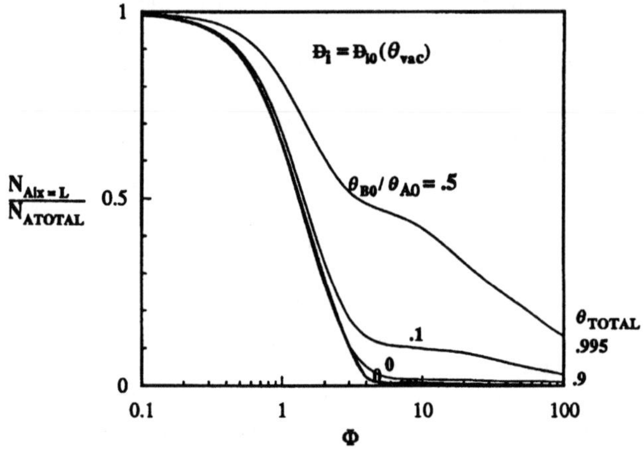

Figure 24 *Simulation results for A → B in a zeolite membrane CMR:* (a) *occupancy profiles across the membrane;* (b) *fraction of reactant A flux that slips to product side* versus *Thiele modulus for different occupancy levels*

In Figure 24(a), a plot is shown of occupancy of species *i* (θ_i) against dimensionless distance into the membrane $\eta = x/L$. The species flux vector N' is proportional to the product of the Fickian diffusivity matrix $[D]$ and the occupancy space derivative vector $d(\Theta)/dx$. In this case the entrainment effect of diffusive coupling between the reintroduction of B and the entering reactant A was significant enough to overcome the opposing gradient presented by the production of B. This is the phenomenon of uphill diffusion. Pure codiffusion of B occurs into the membrane, in spite of the counter-diffusion gradient presented at the feed ($\eta = 0$) side by the occupancy profile of B, as shown in the Figure 24. The slip of reactant A to the permeate side depends on the Thiele modulus, as shown in Figure 24(b). For high enough values of Φ, it is possible to prevent slip of A to the permeate side.

Clearly, phenomena in zeolites are extremely complex, with interactions between diffusion and adsorption effects. This represents both a challenge and an opportunity. The simulations of Harris[127] suggest that by manipulating occupancy effects in zeolites that show the right dependence of diffusion and sorption on occupancy, obtaining complete separation of reactants and products is possible. The use of one species to block another from diffusing through pores is similar to capillary condensation effects, but may be applicable over a wider range of conditions.

4.5 Integration of Functionality through Multiple Membranes – If a single membrane is unable to provide all of the required properties or functionality that the reaction needs, the logical answer is to use more than one membrane in the reactor. In this way, having both a metered feed and a product removal in one operation, or to potentially separate different products from the mixture, is possible *via* this or another combination of any of the functions discussed previously. This idea was first proposed a few years ago, but recently there has been a renewal of interest, although mainly theoretical. The two (or more) membranes can be separate structures or combined into a composite structure. The case where the function of one membrane is as a purely structural support is excluded, as is the usual case in many composite membranes.

4.5.1 Two Separate Membranes – Kim and Datta[128] proposed a supported liquid-phase catalytic membrane reactor–separator (SLPCMRS). The idea is illustrated in Figure 25.

A porous matrix is sandwiched between two membranes. The matrix supports a liquid-phase catalyst. For the reaction A → B, membrane 1 passes A but resists B, and membrane 2 passes both freely; the function of membrane 2 is to encapsulate the catalyst solution. Reactant A is fed external to membrane 1; the concentration of A drops across the catalyst as it is consumed by reaction, due to diffusional resistance. The product diffuses to the right, reactant A does not. The benefits of this reactor are: the liquid phase is encapsulated, the catalyst is separated from the product stream, the product is separated from the reactant, it provides a higher gas–liquid interfacial area, and a product is removed from an equilibrium-limited reaction. The authors suggested that the system be implemented as a shell-and-tube configuration using two different hollow-fiber membranes.

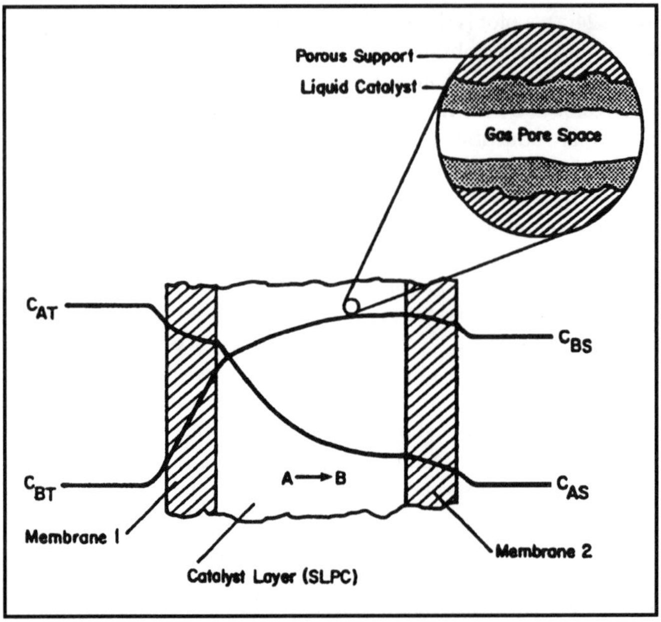

Figure 25 *Schematic of the concept of the supported liquid-phase catalytic membrane reactor*
(Reproduced with permission from *AIChE J.*, 1991, **37**, 1657. Copyright (1991) American Institute of Chemical Engineers)

Kim and Datta[128] tested their proposed system on the hydroformylation of ethylene to propionaldehyde with hydridocarbonyltris(triphenylphosphine)Rh(I) catalyst dissolved in dioctyl phthalate solvent. A differential stirred-cell membrane reactor, consisting of a porous disk between two membranes, was used, with cellulose ester as type 1 (mesoporous 50 Å, favors the reactant slightly) and PTFE as type 2 (2000 Å, permeable to both). The reaction was run at 110 °C. They obtained good agreement between their model of the system and experiment, and recovered 80% of product on side '2', if the liquid loading of the catalyst was adjusted so that the transport resistance of the catalyst layer was lower than that of the membrane. Problems with membrane degradation suggested the use of ceramic membranes rather than polymer membranes. Candidates might be γ-alumina and α-alumina, which have the right pore sizes, but there is really a need for a microporous material with molecular discrimination for membrane 1.

Theoretical studies of a two-membrane system have been made by Tekić *et al.*[129] for the reaction $A \rightleftharpoons bB + cC$. Membrane 1 passes B preferentially and membrane 2 passes C preferentially, so the two-membrane concept does work better for product separation than a single membrane. If the reactant is the slowest permeator, then process conversion is better with two membranes, otherwise reactant loss can affect conversion. Later work by the same group[130] extended the analysis to non-isothermal conditions. They suggested the solar thermal decomposition of water as a candidate reaction, with H_2-permeable and O_2-permeable membranes.

Lu *et al.*[131,132] considered a theoretical two-membrane system for OCM, with O_2 being distributed along the system by one membrane, and C_2 products removed in the other, as shown in Figure 26.

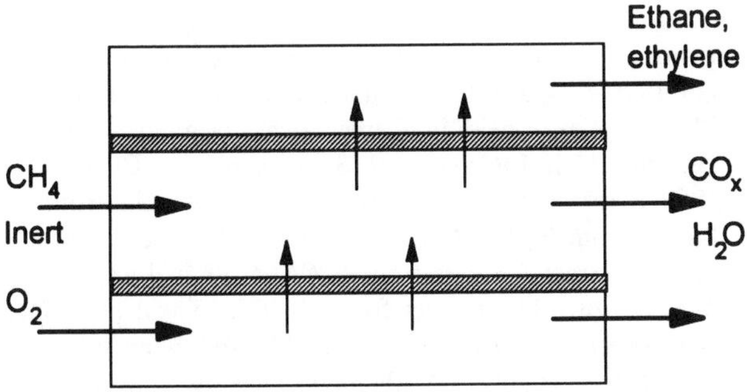

Figure 26 *Schematic of the two-membrane reactor concept for the oxidative coupling of methane*

This led to higher C_2 yields, while the CH_4 permeability was also kept low, and the contact time was high enough. Removal of the desired products helped the distributed feed PBMR reactor more than the conventional co-feed PFR reactor, *i.e.* a synergy between the two membrane functions was observed. Unfortunately a membrane with the desired characteristics of selective removal of C_2 species at high temperature is unknown.

The concept of a two-membrane system is interesting from a theoretical point of view, but more needs to be done to identify selective membranes and candidate reactions.

4.5.2 Multiple Layer Membranes – Two membranes can also be built into different layers in the same membrane. In a theoretical study of methane conversion to syngas, Tsai *et al.*[114] simulated a membrane composed of a dense perovskite layer for oxygen separation from air, and a porous CMR layer for combustion of CH_4. As has already been discussed in Section 4.3.3, Harold and co-workers[83,84] studied a CMR made up of two layers, a diffusion layer next to a catalytically-impregnated layer. The diffusion layer served to reduce the catalytic layer concentration of the oxygen fed at that side of the membrane. As a final example, Coronas and Santamaría[16] report a study in their group, where a porous catalytic membrane contained a thin layer of a dehydrogenation catalyst that also had significant combustion activity. They fed a butane-containing stream from the tube side, and an oxygen-containing stream from the shell side. The idea was that the hydrogen produced from the dehydrogenation would be combusted with the oxygen, thus supplying heat for the endothermic dehydrogenation and also removing a product.

5 Summary of Recent Developments

Following completion of the main body of this review, several major membrane conferences have been held, and some new publications have appeared that are relevant to the subject of catalytic membrane reactors. The 5th International Conference on Inorganic Membranes (ICIM$_5$-98) took place in Nagoya, Japan, from June 22–26, 1998; the 3rd International Conference on Catalytic Membrane Reactors (ICCMR '98) was held in Copenhagen, Denmark from September 8–10, 1998; and the 15th International Symposium on Chemical Reaction Engineering (ISCRE-15) met at Newport Beach, California from September 13–16, 1998. Papers from ICIM$_5$-98 are available in the proceedings volume as four-page extended abstracts, papers from ICCMR '98 will appear in a special issue of *Catalysis Today*, while papers from ISCRE-15 will be published in a special issue of *Chemical Engineering Science*. The following paragraphs summarize a selection of these papers and other very recently published work, with a view towards identifying some of the themes already expounded in this review, and also any emerging new directions.

A lowered emphasis on hydrogen removal membrane reactors was apparent. A main focus of work continuing in this area is on the production of hydrogen for fuel cells,[133] or on the steam-reforming and water-gas shift reactions, again in the IGCC systems.[134] A new reaction that is being studied with hydrogen removal is the dehydrogenation of ethanol to acetaldehyde, as an environmentally-benign alternative to the Wacker process.[135] An interesting use of non-Pd dense membranes for hydrogen removal was described,[136] in which non-oxidative catalytic conversion of methane was predicted to increase to 42% the yield of C_2–C_{10} products by continuous removal of hydrogen through a $SrZr_{0.95}Y_{0.05}O_3$ ceramic membrane. Oxidative dehydrogenation, on the other hand, is thriving, with several new contributions on conversion of *n*-butane to butylene,[137,138] and on the conversion of propane to propylene.[139–141]

The use of oxygen-feed membrane reactors to carry out selective oxidations is increasing, with special interest on the partial oxidation of *n*-butane to maleic anhydride.[142–144] An interesting observation relating to the latter work is that the gains in selectivity are not expected to be high due to favorable kinetics. Rather, it is the control of the heat release afforded by the membrane reactor, which allows higher *n*-butane feed concentrations, that is the driver for the interest. Research also continues on the use of oxygen-permeable dense mixed-conducting membranes for OCM,[145] and for methane conversion to syngas. In the latter area, Balachandran *et al.*[146] have published further work using their SFC membrane that indicates even higher oxygen fluxes than previously, while Sammells and Schwartz[147] have reported a brownmillerite-type of membrane developed by Eltron Research Inc. that has operated continuously at practical rates of natural gas conversion to syngas for one year at 900 °C with no material instability. Methane conversion was also the focus of research by Hamakawa *et al.*[148] who reported on an electrochemical solid oxide membrane reactor with cogeneration of electricity. The group of van Swaaij continues their exploration of the separated-feed CNMR, and reported results obtained on H_2S oxidation[149] and on isobutene oxidation.[150]

Some new work on assessing the limits of membrane reactors, and on comparing them to other reactor configurations, appeared at the ISCRE-15 conference. McGregor *et al.*[151] have extended attainable region theory to include separation processes. This work aims to synthesize the structures of optimal reactor–separator networks, which has implications for the design of membrane reactors. The question of what conversion is achievable in a membrane reactor has been revisited,[152] with identification of operating conditions at which maximum conversion occurs.

Further work on the use of reactive sweep gases has been presented. Schraam *et al.*[153] studied a system in which cyclohexanol dehydrogenation was coupled with cyclohexene hydrogenation using a planar porous membrane, and found increased conversion with the reactive sweep gas. Uemura *et al.*[154] reported on the coupling of $C_6H_{12} \rightarrow C_6H_6 + 3H_2$ with $CO_2 + H_2 \rightarrow CO + H_2O$, thus using permeate hydrogen to convert a greenhouse gas, carbon dioxide, into a useful synthesis gas constituent, carbon monoxide. Their reactor depends on their $LaNi_5$ membrane being both a permeator for H_2 and a catalyst for both reactions. The idea is promising but so far conversions are low, 1–2%, due to low catalyst activity.

Some very interesting results using microporous membranes with cross-flow were obtained by the group of Maier.[155,156] They used very low defect microporous membranes which they prepared using the clean room sol-gel techniques of de Vos and Verweij.[157] For liquid-phase hydrogenation reactions, secondary reactions could be suppressed completely and surprising hydrogenation selectivities were obtained when hydrogen and alkene were both forced to pass through the membrane. This result was attributed to the prevention of the back-mixing of the initial products with the starting reactants. This is a very promising use of a membrane to control the contacting mode in a reactor.

The area of zeolite membranes continues to generate a great deal of interest, especially among Japanese workers. This was especially apparent in the $ICIM_5$-98 conference, where 19 out of 93 oral presentations, and 11 out of 80 poster presentations, were on the subject of zeolite or molecular sieving membranes. The great majority of these, however, dealt with synthesis and/or characterization of such membranes, with several also on zeolite membrane transport. The difficulty of fabricating stable, defect-free zeolite membranes still continues to limit their use in reactors. This is shown by the fact that in the ICCMR '98 conference on membrane reactors, only aproximately 6–8 contributions out of 74 featured zeolite membrane reactors. Most used the zeolite membrane as an inert porous wall. For the separation of hydrogen in the dehydrogenation of isobutane,[158] high separation factors were obtained at high temperatures, whereas at low temperatures the hydrogen flux was blocked by isobutane adsorption. Further work was also reported on the selective oxidation of *n*-butane to maleic anhydride using a tubular zeolite membrane to distribute the oxygen.[159] In possibly the first reported experimental work on a zeolite membrane CMR, Julbe *et al.*[139] looked at the oxidative dehydrogenation of propane using a catalytically-active V-MFI zeolite membrane. The membrane was found to be an efficient contactor between reactant and catalytic sites, but the authors warned about possible side effects due to zeolite acidity.

Several plenary lectures were given at the conferences, including a call from Tsotsis[160] for renewed emphasis on using membrane reactors to reduce or eliminate the separation task. An update to his 1994 review paper was given at the ISCRE-15 conference by van Swaaij,[161] who concluded that the outlook for membrane reactors was perhaps more optimistic than a few years ago. Dalmon presented a timely survey of membrane catalysis for liquid applications,[162] and several recent publications have focused on this, including pervaporation.[163,164] Dalmon also emphasized the area of fine chemicals production, where the use of membranes for three-phase contacting could see a revival of interest. An example of this was presented at the ISCRE-15 conference.[165]

6 Concluding Remarks

Recent years have seen research directions move away from the classical purpose of catalytic membrane reactors, which was simultaneous reaction and separation in equilibrium-limited reactions. The gains are not as easily won, nor are the economics so compelling as was originally anticipated. There has been a move toward the use of membranes to contact reactants differently, or to provide alternative geometries for chemical reactors. It seems clear that applications involving separation and reaction will be done with high-selectivity thin-film dense or molecular-sieving membranes almost exclusively.

The need for careful analysis of catalytic membrane reactor behavior and potential is evident, and it should occur early in any study before too much effort is expended in the research. The theoretical basis for the use of the membrane needs to be established beyond doubt. Experimental verification needs to be made that the membrane will actually perform as expected over a long period under reaction conditions. Practical problems and economic potential should inform all research efforts to develop new catalytic membrane reactors.

This review has attempted to identify trends in innovations and recent research, and to show similar directions in several different research efforts. While new directions and new applications will quickly appear, it is hoped that the principles mentioned here will provide a framework for understanding many future developments.

Acknowledgments

This review is based upon a plenary lecture given by the author at the Engineering Foundation Conference on Chemical Reaction Engineering for Sustainable

Processes and Products, Banff, Alberta, June 8–13, 1997. I would like to thank the Conference Chair Hugo I. de Lasa and the organizers of the conference for the invitation to speak. Thanks are also due to many colleagues who supplied originals of figures from their own publications for this review, and to Theo Tsotsis and Jesus Santamaría for their encouragement and willingness to share their latest thoughts and unpublished work on membrane reactors with me. I would also like to acknowledge my colleagues at Worcester Polytechnic Institute, Yi Hua Ma and William R. Moser, for many stimulating discussions on the subjects of inorganic membranes and membrane reactors, as well as the many graduate students and postdoctoral workers, who taught me more about inorganic membrane reactors than I taught them.

References

1 K.R. Westerterp, *Chem. Eng. Sci.*, 1992, **47**, 2195.
2 A.S. Michaels, *Chem. Eng. Prog.*, 1968, **64**, 31.
3 G. Belfort, *Biotechnol. Bioeng.*, 1989, **33**, 1047.
4 H.N. Chang and S. Furusaki, *Adv. Biochem. Eng. Biotechnol.*, 1991, **44**, 27.
5 H.P. Hsieh, *AIChE Symposium Ser.*, 1989, **85**, 53.
6 J.N. Armor, *Appl. Catal.*, 1989, **49**, 1.
7 H.P. Hsieh, *Catal. Rev. Sci. Eng.*, 1991, **33**, 1
8 J. Shu, B.P.A. Grandjean, A. Van Neste and S. Kaliaguine, *Can. J. Chem. Eng.*, 1991, **69**, 1036.
9 J.N. Armor, *CHEMTECH*, 1992, **Sept.**, 557.
10 T.T. Tsotsis, A.M. Champagnie, R.G. Minet and P.K.T. Liu, in *Computer-aided Design of Catalysts*, eds. E.R. Becker and C.J. Pereira, Dekker, New York, 1993, pp. 471–551.
11 J. Zaman and A. Chakma, *J. Membr. Sci.*, 1994, **92**, 1.
12 G. Saracco and V. Specchia, *Catal. Rev. Sci. Eng.*, 1994, **36**, 305.
13 G. Saracco, G.F. Versteeg and W.P.M. van Swaaij, *J. Membr. Sci.*, 1994, **95**, 105.
14 M.P. Harold, C. Lee, A.J. Burggraaf, K. Keizer, V.T. Zaspalis and R.S.A. de Lange, *MRS Bulletin*, 1994, **April**, 34.
15 J. Sanchez and T.T. Tsotsis, in *Fundamentals of Inorganic Membrane Science and Technology*, eds. A.J. Burggraaf and L. Cot, Elsevier Science B. V., Amsterdam, 1996, pp. 529–568.
16 J. Coronas and J. Santamaría, *Catal. Today*, 1999, in press.
17 P.J. Gellings and H.J.M. Bouwmeester, *Catal. Today*, 1992, **12**, 1.
18 H.J.M. Bouwmeester and A.J. Burggraaf, in *Fundamentals of Inorganic Membrane Science and Technology*, eds. A.J. Burggraaf and L. Cot, Elsevier Science B. V., Amsterdam, 1996, pp. 435–528.
19 J.E. Koresh and A. Soffer, *Sep. Sci. Technol.*, 1987, **22**, 973.
20 J.J. Hammel, PPG Co., U.S. Patent 4 853 001, 1989.
21 J.G. Tsikoyiannis and W.O. Haag, *ZEOLITES*, 1992, **12**, 126.
22 N. Nourbakhsh, T.T. Tsotsis and I.A. Webster, *Appl. Catal.*, 1989, **50**, 65.
23 A.B. Shelekhin, A.G. Dixon and Y.H. Ma, *AIChE J.*, 1995, **41**, 58.
24 S. Uemiya, T. Matsuda and E. Kikuchi, *J. Membr. Sci.*, 1991, **56**, 325.
25 P.P. Mardilovich, Y. She, Y.H. Ma and M.-H. Rei, *AIChE J.*, 1998, **44**, 310.
26 Y.H. Ma and S. Xiang, WPI, PCT/US Patent 9 302 384, 1993.

27 E.R. Geus, H. van Bekkum, W.J.W. Bakker and J.A. Moulijn, *Microporous Mater.*, 1993, **1**, 131.
28 M.D. Jia, B. Chen, R.D. Noble and J.L. Falconer, *J. Membr. Sci.*, 1994, **90**, 1.
29 Y. Yan, M.E. Davis and G.R. Gavalas, *Ind. Eng. Chem. Res.*, 1995, **34**, 1652.
30 Z.A.E.P. Vroon, K. Keizer, M.J. Gilde, H. Verweij and A.J. Burggraaf, *J. Membr. Sci.*, 1996, **113**, 293.
31 T. Ioannides and G.R. Gavalas, *J. Membr. Sci.*, 1993, **77**, 207.
32 L.A. Bernstein and C.R.F. Lund, *J. Membr. Sci.*, 1993, **77**, 155.
33 B.A. Raich and H.C. Foley, *Appl. Catal. A*, 1995, **129**, 167.
34 J.C.S. Wu and P.K.T. Liu, *Ind. Eng. Chem. Res.*, 1992, **31**, 322.
35 Y.V. Gokhale, R.D. Noble and J.L. Falconer, *J. Membr. Sci.*, 1993, **77**, 197.
36 F. Tiscareño-Lechuga, C.G. Hill, Jr. and M.A. Anderson, *J. Membr. Sci.*, 1996, **118**, 85.
37 Y.L. Becker, A.G. Dixon, W.R. Moser and Y.H. Ma, *J. Membr. Sci.*, 1993, **77**, 233.
38 A.G. Dixon, Y.L. Becker, Y.H. Ma and W.R. Moser, *Preprints – Separations Topical Conference*, AIChE Annual Meeting, Miami Beach, 1992.
39 K.C. Cannon and J.J. Hacskaylo, *J. Membr. Sci.*, 1992, **65**, 259.
40 M. Sheintuch, *Ind. Eng. Chem. Res.*, 1998, **37**, 807.
41 K. Mohan and R. Govind, *AIChE J.*, 1986, **32**, 2083.
42 K. Mohan and R. Govind, *AIChE J.*, 1988, **34**, 1493.
43 K. Mohan and R. Govind, *Ind. Eng. Chem. Res.*, 1988, **27**, 2064.
44 N. Itoh, *Catal. Today*, 1995, **25**, 351.
45 A.M. Champagnie, T.T. Tsotsis, R.G. Minet and E. Wagner, *J. Catal.*, 1992, **134**, 713.
46 N.Itoh, Y. Shindo, K. Haraya, K. Obata, T. Hakuta and H. Yoshitome, *Int. Chem. Eng.*, 1985, **125**, 138.
47 N. Itoh, *AIChE J.*, 1987, **33**, 1576.
48 E. Gobina, K. Hou and R. Hughes, *J. Membr. Sci.*, 1995, **105**, 163.
49 Z.D. Ziaka, R.G. Minet and T.T. Tsotsis, *AIChE J.*, 1993, **39**, 526.
50 J.P. Collins, R.W. Schwartz, R. Sehgal, T.L. Ward, C.J. Brinker, G.P. Hagen and C.A. Udovich, *Ind. Eng. Chem. Res.*, 1996, **35**, 4398.
51 Y. Yildirim, E. Gobina and R. Hughes, *J. Membr. Sci.*, 1997, **135**, 107.
52 H. Weyten, K. Keizer, A. Kinoo, J. Luyten and R. Leysen, *AIChE J.*, 1997, **43**, 1819.
53 T.T. Tsotsis, A.M. Champagnie, S.P. Vasileiadis, Z.D. Ziaka and R.G. Minet, *Chem. Eng. Sci.*, 1992, **47**, 2903.
54 G. Barbieri and F.P. Di Maio, *Ind. Eng. Chem. Res.*, 1997, **36**, 2121.
55 Y.V. Gokhale, R.D. Noble and J.L. Falconer, *J. Membr. Sci.*, 1995, **105**, 63.
56 M.E. Rezac, W.J. Koros and S.J. Miller, *Ind. Eng. Chem. Res.*, 1995, **34**, 862.
57 J.P. Collins and J.D. Way, *J. Membr. Sci.*, 1994, **96**, 259.
58 Y.H. Ma, W.R. Moser, S. Pien and A.B. Shelekhin, in *Proc. 3rd Int. Conf. on Inorganic Membranes*, ed. Y.H. Ma, Worcester Polytechnic Institute, MA, 1994, p. 281.
59 A.M. Adris, C.J. Lim and J.R. Grace, *Chem. Eng. Sci.*, 1997, **52**, 1609.
60 S. Pancharatnam, R.A. Huggins and D.M. Mason, *J. Electrochem. Soc.*, 1975, **122**, 869.
61 Y. Nigara and B. Cales, *Bull. Chem. Soc. Jpn.*, 1986, **59**, 1997.
62 N. Itoh, M.A. Sanchez C., W.-C. Xu, K. Haraya and M. Hongo, *J. Membr. Sci.*, 1993, **77**, 245.
63 A.G. Dixon, W.R. Moser and Y.H. Ma, *Ind. Eng. Chem. Res.*, 1994, **33**, 3015.
64 S.C. Reyes, E. Iglesia and C.P. Kelkar, *Chem. Eng. Sci.*, 1993, **48**, 2643.

65 A.L.Y. Tonkovich, R. Secker, E. Reed, E. Roberts and J. Cox, *Sep. Sci. Technol.*, 1995, **30**, 397.
66 A.M. Ramachandra, Y. Lu, Y.H. Ma, W.R. Moser and A.G. Dixon, *J. Membr. Sci.*, 1996, **116**, 253.
67 V. Papavassiliou, C. Lee, J. Nestlerode and M.P. Harold, *Ind. Eng. Chem. Res.*, 1997, **36**, 4954.
68 V.M. Gryaznov, *Platinum Metals Rev.*, 1986, **30**, 68.
69 H. Nagamoto and H. Inoue, *Chem. Eng. Commun.*, 1985, **34**, 315.
70 D. Eng and M. Stoukides, *Catal. Rev.-Sci. Eng.*, 1991, **33**, 375.
71 Y. Teraoka, H.-M. Zhang, S. Furukawa and N. Yamazoe, *Chem. Lett.*, 1985, 1743.
72 S.J. Xu and W.J. Thomson, *AIChE J.*, 1997, **43**, 2731.
73 T. Hibino, T. Sato, K. Ushiki and Y. Kuwahara, *J. Chem. Soc., Faraday Trans.*, 1995, **91**, 4419.
74 J.E. ten Elshof, H.J.M. Bouwmeester and H. Verweij, *Appl. Catal. A*, 1995, **130**, 195.
75 Y. Lu, A.G. Dixon, W.R. Moser and Y.H. Ma, submitted to *Catal. Today,* 1998.
76 W. Wang and Y.S. Lin, *J. Membr. Sci.*, 1995, **103**, 219.
77 D. Lafarga, J. Santamaría and M. Menéndez, *Chem. Eng. Sci.*, 1994, **49**, 2005.
78 J. Coronas, M. Menéndez and J. Santamaría, *Chem. Eng. Sci.*, 1994, **49**, 2015.
79 A.L.Y. Tonkovich, D.M. Jimenez, J.L. Zilka and G.L. Roberts, *Chem. Eng. Sci.*, 1996, **51**, 3051.
80 J. Coronas, A. Gonzalo, D. Lafarga and M. Menéndez, *AIChE J.*, 1997, **43**, 3095.
81 G. Capinelli, E. Carosini, F. Cavani, O. Monticelli and F. Trifiro, *Chem. Eng. Sci.*, 1996, **51**, 1817.
82 C. Téllez, M. Menéndez and J. Santamaría, *AIChE J.*, 1997, **43**, 777.
83 M.P. Harold, V.T. Zaspalis, K. Keizer and A.J. Burggraaf, *Chem. Eng. Sci.*, 1993, **48**, 2705.
84 M.P. Harold and C. Lee, *Chem. Eng. Sci.*, 1997, **52**, 1923.
85 L.A. Bernstein, S. Agarwalla and C.R.F. Lund, *ACS Symp. Ser.*, 1993, **523**, 427.
86 P. Cini and M.P. Harold, *AIChE J.*, 1991, **37**, 997.
87 M. Torres, J. Sanchez, J.A. Dalmon, B. Bernauer and J. Lieto, *Ind. Eng. Chem. Res.*, 1994, **33**, 2421.
88 J.F. Akyurtlu, A. Akyurtlu and C.E. Hamrin, Jr., *Chem. Eng. Commun.*, 1988, **66**, 169.
89 M.P. Harold, P. Cini, B. Patenaude and K. Venkataraman, *AIChE Symp. Ser.*, 1989, **85**, 26.
90 H.J. Sloot, G.F. Versteeg and W.P.M. van Swaaij, *Chem. Eng. Sci.*, 1990, **45**, 2415.
91 H.J. Sloot, G.F. Versteeg, C.A. Smolders and W.P.M. van Swaaij, *Key Eng. Mater.*, 1991, **61/2**, 261.
92 V.T. Zaspalis, W. van Praag, K. Keizer, J.G. van Ommen, J.R.H. Ross and A.J. Burggraaf, *Appl. Catal.*, 1991, **74**, 249.
93 J.W. Veldsink, R.M.J. van Damme, G.F. Versteeg and W.P.M. van Swaaij, *Chem. Eng. Sci.*, 1992, **47**, 2939.
94 K. Keizer, V.T. Zaspalis, R.S.A. de Lange, M.P. Harold and A.J. Burggraaf, in *Membrane Processes in Separation and Purification*, ed. J.G. Crespo and K.W. Böddeker, Kluwer Academic Publishers, The Netherlands, 1994, pp. 415–429.
95 G. Saracco, S. Specchia and V. Specchia, *Chem. Eng. Sci.*, 1996, **51**, 5289.
96 A.M. Adris and J.R. Grace, *Ind. Eng. Chem. Res.*, 1997, **36**, 4549.
97 T.L. Ward, G.P. Hagen and C.A. Udovich, in *Proc. 3rd Int. Conf. on Inorganic Membranes*, ed. Y.H. Ma, Worcester Polytechnic Institute, MA, 1994, p. 335.
98 H.M. van Veen, M. Bracht, E. Hamoen and P.T. Alderliesten, in *Fundamentals of*

Inorganic Membrane Science and Technology, eds. A.J. Burggraaf and L. Cot, Elsevier Science B. V., Amsterdam, 1996, pp. 641–680.
99 C.M. Reo, L.A. Bernstein and C.R.F. Lund, *AIChE J.*, 1997, **43**, 495.
100 C.M. Reo, L.A. Bernstein and C.R.F. Lund, *Chem. Eng. Sci.*, 1997, **52**, 3075.
101 Y. Lu, A.G. Dixon, W.R. Moser and Y.H. Ma, *Chem. Eng. Sci.*, 1997, **52**, 1349; *ibid.*, 1997, **52**, 2443.
102 Y. Lu, A.G. Dixon, W.R. Moser and Y.H. Ma, *Ind. Eng. Chem. Res.*, 1997, **36**, 559.
103 Y. Ohta, M. Gondaira, K. Kobayashi, Y. Fujimoto and K. Kuroda, Paper presented at the AIChE Annual Meeting, November 1994, San Francisco, CA, USA.
104 A.G. Dixon, Paper presented at the AIChE Spring National Meeting, March 1998, New Orleans, LA, USA.
105 J. Coronas, M. Menéndez and J. Santamaría, *J. Loss Prev. Process Ind.*, 1995, **8**, 97.
106 A. Rojnuckarin, C.A. Floudas, H. Rabitz and R.A. Yetter, *Ind. Eng. Chem. Res.*, 1996, **35**, 683.
107 N. Itoh and R. Govind, *Ind. Eng. Chem. Res.*, 1989, **28**, 1557.
108 N. Itoh, *J. Chem. Eng. Japan*, 1990, **23**, 81.
109 N. Itoh and T.-W. Wu, *J. Membr. Sci.*, 1997, **124**, 213.
110 S.Pei, MS Kleefisch, T.P. Kobylinski, J. Faber, C.A. Udovich, V. Zhang-McCoy, B. Dabrowski, U. Balachandran, R.L. Mieville and R.B. Poeppel, *Catal. Lett.*, 1995, **30**, 201.
111 U. Balachandran, J.T. Dusek, S.M. Sweeney, R.B. Poeppel, R.L. Mieville, P.S. Maiya, MS Kleefisch, S.Pei, T.P. Kobylinski, C.A. Udovich and A.C. Bose, *Am. Ceram. Soc. Bull.*, 1995, **74**, 71.
112 U. Balachandran, J.T. Dusek, R.L. Mieville, R.B. Poeppel, MS Kleefisch, S.Pei, T.P. Kobylinski, C.A. Udovich and A.C. Bose, *Appl. Catal. A*, 1995, **133**, 19.
113 C.-Y. Tsai, Y.H. Ma, W.R. Moser and A.G. Dixon, in *Proc. 3rd Int. Conf. on Inorganic Membranes*, ed. Y.H. Ma, Worcester Polytechnic Institute, MA, 1994, p. 271.
114 C.-Y. Tsai, Y.H. Ma, W.R. Moser and A.G. Dixon, *Chem. Eng. Commun.*, 1995, **134**, 107.
115 C.-Y. Tsai, Y.H. Ma, W.R. Moser and A.G. Dixon, Paper presented at 4th Int. Conf. on Inorganic Membranes, Gatlinberg, TN, 1996.
116 C.-Y. Tsai, A.G. Dixon, W.R. Moser and Y.H. Ma, *AIChE J.*, 1997, **43**, 2741.
117 C.-Y. Tsai, A.G. Dixon, W.R. Moser, Y.H. Ma and M.R. Pascucci, *J. Am. Ceram. Soc.*, 1998, **81**, 1437.
118 M.P. Pina, M. Menéndez and J. Santamaría, *Appl. Catal. B*, 1996, **11**, L19.
119 M.P. Pina, S. Irusta, M. Menéndez, J. Santamaría, R Hughes and N. Boag, *Ind. Eng. Chem. Res.*, 1997, **36**, 4557.
120 C. R. Binkerd, Y.H. Ma, W.R. Moser and A.G. Dixon, Paper presented at 4th Int. Conf. on Inorganic Membranes, Gatlinberg, TN, 1996.
121 J.B. Keller, M.S. Falkovitz and H. Frisch, *Chem. Eng. Sci.*, 1984, **39**, 601.
122 V.T. Zaspalis, PhD Thesis, Twente University, 1990.
123 K.L. Yeung, R. Aravind, R.J.X. Zawada, J. Szegner, G. Cao and A. Varma, *Chem. Eng. Sci.*, 1994, **49**, 4823.
124 K.L. Yeung, J.M. Sebastian and A. Varma, *J. Membr. Sci.*, 1997, **131**, 9.
125 D. Casanave, A. Giroir-Fendler, J. Sánchez, R. Loutaty and J.A. Dalmon, *Catal. Today*, 1995, **25**, 309.
126 A. Pantazidis, J.A. Dalmon and C. Mirodatos, *Catal. Today*, 1995, **25**, 403.
127 J. Harris, personal communication, 1996.
128 J.S. Kim and R. Datta, *AIChE J.*, 1991, **37**, 1657.

129 M.N. Tekić, R.N. Paunović and G.M. Ćirić, *J. Membr. Sci.*, 1994, **96**, 213.
130 R.P. Omorjan, R.N. Paunović and M.N. Tekić, *J. Membr. Sci.*, 1998, **138**, 57.
131 Y. Lu, A.G. Dixon, W.R. Moser and Y.H. Ma, Paper presented at 4th Int. Conf. on Inorganic Membranes, Gatlinberg, TN, 1996.
132 Y. Lu, A.G. Dixon, W.R. Moser and Y.H. Ma, *Catal. Today*, 1997, **35**, 443.
133 E. Kikuchi, Paper PL1, III Int. Conf. Catalysis in Membrane Reactors, Copenhagen, Denmark, Sept. 8–10, 1998.
134 P.P.A.C. Pex, Paper EU1, III Int. Conf. Catalysis in Membrane Reactors, Copenhagen, Denmark, Sept. 8–10, 1998.
135 B.A. Raich and H.C. Foley, *Ind. Eng. Chem. Res.*, 1998, **37**, 3888.
136 R.W. Borry and E. Iglesia, Paper P-20, 15th Int. Symp. Chemical Reaction Engineering, Newport Beach, CA, Sept. 13–16, 1998.
137 T.M. Raybold and M.C. Huff, Paper O6, III Int. Conf. Catalysis in Membrane Reactors, Copenhagen, Denmark, Sept. 8–10, 1998.
138 C. Téllez, M. Menéndez and J. Santamaría, Paper 12-a, 15th Int. Symp. Chemical Reaction Engineering, Newport Beach, CA, Sept. 13–16, 1998.
139 A. Julbe, D. Farrusseng, C. Guizard, J.C. Jalibert and C. Mirodatos, Paper O20, III Int. Conf. Catalysis in Membrane Reactors, Copenhagen, Denmark, Sept. 8–10, 1998.
140 J.C. Volta and C. Mirodatos, Paper EU4, III Int. Conf. Catalysis in Membrane Reactors, Copenhagen, Denmark, Sept. 8–10, 1998.
141 R. Ramos, M. Menéndez and J. Santamaría, Paper P11, III Int. Conf. Catalysis in Membrane Reactors, Copenhagen, Denmark, Sept. 8–10, 1998.
142 J. Perregaard and P.E. Højlund Nielsen, Paper EU5, III Int. Conf. Catalysis in Membrane Reactors, Copenhagen, Denmark, Sept. 8–10, 1998.
143 E. Xue and J. Ross, Paper O19, III Int. Conf. Catalysis in Membrane Reactors, Copenhagen, Denmark, Sept. 8–10, 1998.
144 R. Mallada, M. Menendez and J. Santamaria, Paper O18, III Int. Conf. Catalysis in Membrane Reactors, Copenhagen, Denmark, Sept. 8–10, 1998.
145 S.J. Xu and W.J. Thomson, *Ind. Eng. Chem. Res.*, 1998, **37**, 1290.
146 U. Balachandran, J.T. Dusek, P.S. Maiya, B. Ma, R.L. Mieville, MS Kleefisch and C.A. Udovich, *Catal. Today*, 1997, **36**, 265.
147 A.F. Sammells and M. Schwartz, Paper PL4, III Int. Conf. Catalysis in Membrane Reactors, Copenhagen, Denmark, Sept. 8–10, 1998.
148 S. Hamakawa, T. Hayakawa, K. Suzuki, K. Murata and K. Takehira, in 'Inorganic Membranes $ICIM_5$-98', Proc. 5th Int. Conf. on Inorganic Membranes, Nagoya, Japan, June 22–26, 1998, pp. 350–353.
149 H.W.J.P. Neomagus, W.P.M. van Swaaij and G.F. Versteeg, *J. Membr. Sci.*, 1998, **148**, 147.
150 H.W.J.P. Neomagus, W.P.M. van Swaaij and G.F. Versteeg, Paper O17, III Int. Conf. Catalysis in Membrane Reactors, Copenhagen, Denmark, Sept. 8–10, 1998.
151 C. McGregor, D. Hildebrandt and D. Glasser, Paper 23-b, 15th Int. Symp. Chemical Reaction Engineering, Newport Beach, CA, Sept. 13–16, 1998.
152 D. Chmielewski, Z. Ziaka and V. Manousiouthakis, Paper P-123, 15th Int. Symp. Chemical Reaction Engineering, Newport Beach, CA, Sept. 13–16, 1998.
153 O. Schramm, A. Tuchlenski and A. Seidel-Morgenstern, Paper P15, III Int. Conf. Catalysis in Membrane Reactors, Copenhagen, Denmark, Sept. 8–10, 1998.
154 Y. Uemura, Y. Ohzuno and Y. Hatate, in 'Inorganic Membranes $ICIM_5$-98', Proc. 5th Int. Conf. on Inorganic Membranes, Nagoya, Japan, June 22–26, 1998, pp. 620–623.

155 C. Lange, B. Tesche, S. Stork and W.F. Maier, *J. Catal.*, 1998, **175**, 280.
156 W.F. Maier, C. Lange, I. Tilgner and B. Tesche, Paper O22, III Int. Conf. Catalysis in Membrane Reactors, Copenhagen, Denmark, Sept. 8–10, 1998.
157 R.M. de Vos and H. Verweij, *Science*, 1998, **279**, 1710.
158 P. Ciavarella, H. Moueddeb, S. Miachon, K. Fiaty and J.A. Dalmon, Paper P16, III Int. Conf. Catalysis in Membrane Reactors, Copenhagen, Denmark, Sept. 8–10, 1998.
159 S. Mota, M. Abon, J.-C. Volta and J.A. Dalmon, Paper P14, III Int. Conf. Catalysis in Membrane Reactors, Copenhagen, Denmark, Sept. 8–10, 1998.
160 T.T. Tsotsis, Paper PL5, III Int. Conf. Catalysis in Membrane Reactors, Copenhagen, Denmark, Sept. 8–10, 1998.
161 G. Saracco, H.W.J.P. Neomagus, G.F. Versteeg and W.P.M. van Swaaij, Paper Plenary VII, 15th Int. Symp. Chemical Reaction Engineering, Newport Beach, CA, Sept. 13–16, 1998.
162 J.-A. Dalmon, Paper PL3, III Int. Conf. Catalysis in Membrane Reactors, Copenhagen, Denmark, Sept. 8–10, 1998.
163 B. Park, V.S. Ravi-Kumar and T.T. Tsotsis, *Ind. Eng. Chem. Res.*, 1998, **37**, 1276.
164 B. Park, V.S. Ravi-Kumar, L. Yang and T.T. Tsotsis, in 'Inorganic Membranes $ICIM_5$-98', Proc. 5th Int. Conf. on Inorganic Membranes, Nagoya, Japan, June 22–26, 1998, pp. 382–385.
165 K. Daub, R. Dittmeyer, R. Gavagnin, G. Strukul, G. Centi, M. Tomaselli, M. Marella, M.-J. Chollier and P. Ruiz, Paper 3-a, 15th Int. Symp. Chemical Reaction Engineering, Newport Beach, CA, Sept. 13–16, 1998.

3

Kinetic Coupling in and between Catalytic Cycles

BY MICHEL BOUDART

1 Introduction

The fascination of catalysis is due to its apparent improbability as well as to its economic impact on our modern world. The latter aspect is well illustrated in a recent book.[1] The former is related to the seeming improbability of discovering a new catalyst for a new reaction. Even the development and optimization of a catalytic process are awesome tasks in the hands of large multidisciplinary teams of industrial technologists who labor many years before the startup of a chemical plant. But the discovery of a new catalytic cycle remains an improbable event. The cycle turns over successive elementary steps at a sufficient frequency and for an acceptable lifetime. At the time of discovery, little is known about the composition or structure of the catalyst. The nature of the elementary steps remains the object of wild speculations. Their thermodynamic and kinetic feasibility are even more inscrutable. Practical success then seems to be the product of luck rewarding a prepared mind, according to Pasteur's dictum. While the role of luck may frequently be important, there are other critical elements responsible for industrial success: optimizing chaos, fundamental thinking, perseverance, moonlighting, some arrogance, daring, and timing.[2]

Here we are concerned about the fascination of discovery preceding eventual industrial success. It will be shown and illustrated how kinetic coupling in and between catalytic cycles can assist discovery in the face of ignorance about how they turn over. Examples will be borrowed from both homogeneous and heterogeneous catalysis as well as from free-radical chain reactions. Some of the examples will be borrowed from a book chapter dealing with catalyst design.[3]

2 The Chain or Catalytic Cycles

As first clarified by Christiansen, chain and catalytic reactions consist of a closed sequence of elementary steps involving stable reactants, intermediates, and products reacting with reactive intermediates. The first such reactive intermediate in the first step of the sequence is regenerated in the last step of the sequence, closing the chain or catalytic cycle.[4]

Catalysis, Volume 14

It is generally assumed that the kinetics of each elementary step can be treated by transition state theory, as described in all textbooks.[5] The stoichiometric number σ_i of an elementary step in the cycle is the number of times that the step must occur for the cycle to turn over once, as corresponding to its stoichiometric equation.

3 The De Donder Relation: A Thermodynamic Measure of the Kinetic Irreversibility of an Elementary Step

For any elementary step that can be treated by transition state theory, it is easy to show that:

$$\nu_+/\nu_- = \exp\,(A/RT) \tag{3.1}$$

where ν_+ and ν_- are the rates of the forward and reverse reactions, and A is the affinity of the reaction. The affinity is defined as,

$$\left(\frac{\delta(-\Delta G)}{\delta\xi}\right)_{T,p} \tag{3.2}$$

where G is the Gibbs free energy and ξ is the extent of reaction. Clearly, if A is positive, the reaction may move forward; if A is negative it will move in the reverse direction; if $A = 0$, equilibrium is reached with $\nu_+ = \nu_-$. Kinetically, all reactions must be reversible. It is convenient to consider a forward reaction as a one-way reaction when $\nu_+ >> \nu_-$.[6] Otherwise, the reaction is called a two-way reaction. To the best of the author's knowledge, eqn. (3.1) was first written by De Donder, the founder of the Brussels school of irreversible thermodynamics.[7] It is useful in extending the principle of microscopic reversibility away from equilibrium.[8] It will provide the quantitative basis for kinetic coupling in chain and catalytic reactions, a concept introduced by the author in 1983.[9]

4 The Bodenstein Kinetic Steady State

When the concentration of reactive intermediates such as free radicals or adsorbed species is sufficiently smaller than that of stable reactants, intermediates of products, there exists a kinetic steady state expressed by the relations,

$$\sigma_i\,\nu = \nu_{+i} - \nu_{-i} \tag{4.1}$$

where i denotes the ith elementary step in the chain or catalytic cycle proceeding at a net rate ν. The stoichiometric number σ_i of the ith step was introduced in Section 2. Because of the conditional restriction concerning the relative amounts of reactive intermediates and stable species, the Bodenstein steady state is frequently called the steady state approximation or the quasi-steady state approximation. There is no approximation if the reaction takes place in a continuous completely back-mixed reactor. The steady state must be reached after the start of a reaction in a batch reactor. Toward the completion of the reaction, the validity of the approximation may fail because the ratio of concentrations of reactive and stable species becomes large enough. Such a situation may occur in reactors for the control of pollutants in ppm amounts.[10] Generally, the statement of eqn. (3.1) is the clear valid expression of kinetic coupling in chain and catalytic cycles. As shown in Figure 1 for two elementary steps, their net rate at the steady state is the same and is equal to the net turnover rate of the cycle duly weighted by the appropriate stoichiometric number. This equality at the steady state is established by kinetic coupling. To denote a step as a slow step is inappropriate, confusing, and unacceptable. Quantitatively, in Figure 1, the affinity driving the step above is smaller than that driving the step below, according to the De Donder relation: the step above is nearer to equilibrium than the step below. Quantitative examples are treated in the following sections.

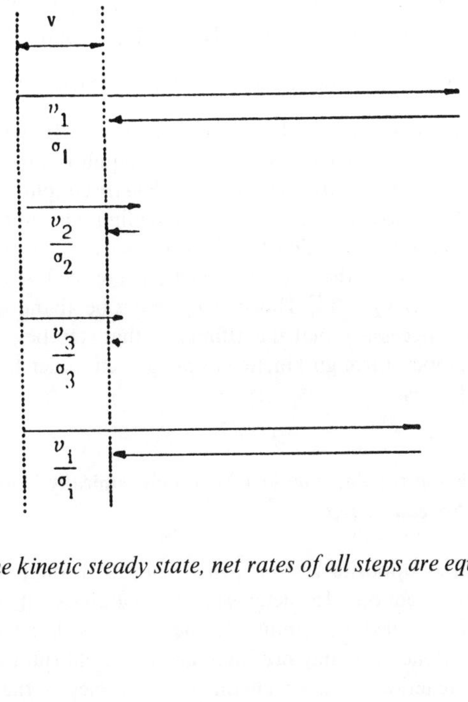

Figure 1 *At the kinetic steady state, net rates of all steps are equal*

5 Depletion of the Product of a Thermodynamically Unfavorable Step by a Following Step

The classical gas phase between H_2 and Br_2 to form HBr between 200 and 300 °C is not limited by equilibrium. Yet Bodenstein and Lind found that its rate is inhibited by its product.[11] Although the authors obtained the correct rate expression, they were unable to explain their kinetic observations. The explanation came 13 years later, independently and almost simultaneously.[12-14] Without knowing it, Bodenstein and Lind had discovered the first free-radical chain reaction with two chain propagation steps:

$$Br + H_2 \underset{k-1}{\overset{k_1}{\rightleftharpoons}} HBr + H \tag{5.1}$$

$$H + Br_2 \overset{k_2}{\rightarrow} HBr + Br \tag{5.2}$$

The first step is a two-way process with a very unfavorable equilibrium: $A_1^{\ominus} = -\Delta G_1^{\ominus} = -15.7$ kcal mol^{-1}, at 298 K, with the superscript $\ominus$ standing for standard state. The second one is a one-way process, with a rate constant k_2 that is 10 times larger than k_{-1} over an extended temperature range. The kinetic coupling between both steps at the kinetic steady state has the effect of pumping away the hydrogen atom concentration from its equilibrium value $[H]_e$ in the absence of the second step below, down to its steady state value $[H]_s$. In fact, application of the De Donder equation to an equimolar H_2–Br_2 mixture at half conversion to HBr at the steady state yields with $[H_2] = [Br_2] = [HBr]$,

$$v_+/v_- = 1 + (k_2/k_{-1}) = [H]_e/[H]_s = \exp(A_1/RT) \tag{5.3}$$

as shown in detail elsewhere.[9] With $k_2 = 10\ k_{-1}$, eqn. (5.3) yields $A_1 = +1.9$ kcal mol^{-1} at 400 K. Thus, in spite of its unfavorable equilibrium ($A_1^{\ominus} < 0$), the first step can move forward ($A_1 > 0$) because of the *kinetic* coupling between steps (1) and (2). It must be stressed that kinetic coupling has nothing to do with thermodynamic coupling according to which the succession of the coupled first reaction with $A_1 v_{+1} < 0$ by the coupling reaction $A_2 v_2 > 0$ would move forward provided that $A_1 v_1 + A_2 v_2 > 0$.[15] This concept must be abandoned: for a step to move forward, it is necessary that the affinity of that step be positive. As shown above, this can happen through kinetic coupling even under thermodynamically unfavorable conditions.

6 Accumulation of the Reactant in a Thermodynamically Unfavorable Step through a Preceding Step

This is the situation opposite to that just discussed through the example of a free-radical chain reaction. In heterogeneous catalysis, it is traditional to consider classical so-called Langmuir–Hinshelwood mechanisms. According to such mechanisms, reactants and products are in equilibrium with the catalyst surface, and the reaction between chemisorbed species is then the rate-determining step by definition. In fact, there are many well-investigated cases in heterogeneous catalysis that do not proceed through Langmuir–Hinshelwood mechanisms. Rather, entry into the surface and/or exit from the surface are not equilibrated steps but in fact are one-way, or close to one-way steps. In particular, the exit step is frequently found to be non-equilibrated as witnessed by the fact that the exiting product is not an inhibitor of the rate of the overall reaction. This situation can be explained by kinetic coupling in the catalytic cycle.[16]

Let us consider a desorption step (d) and ask how far from equilibrium it takes place in a catalytic cycle, irrespective of its detailed mechanism. All we want to know is how we can measure the affinity A that drives desorption of a molecule M adsorbed (a) on a site*:

$$*M \underset{v_a}{\overset{v_d}{\rightleftharpoons}} * + M \tag{6.1}$$

At the steady state of the catalytic reaction, the affinity A for desorption is by definition,

$$A = \mu_{M*} - \mu_M - \mu^* \tag{6.2}$$

where the μ values are the chemical potentials of the species M$*$, M and $*$ at the *steady state.* Let us define a virtual fugacity of M, denoted by $M_{,v}$ that corresponds to *equilibrium* between M on the one hand, and M$*$ and $*$ on the other hand, at their steady state value of fugacity. Since this is a situation at equilibrium, the corresponding affinity is zero and (6.2) becomes:

$$0 = \mu_{M,*} - \mu_{M,v} - \mu^* \tag{6.3}$$

Subtraction of (6.3) from (6.2) side by side yields:

$$A = \mu_{M,v} - \mu_M \tag{6.4}$$

But, with the superscript $\ominus$ standing for standard state, we have,

$$\mu_M = \mu^{\ominus} + RT \ln [M] \tag{6.5}$$

$$\mu_{M,v} = \mu^{\ominus} + RT \ln [M_{,v}] \tag{6.6}$$

where the quantities between brackets are values of fugacity. Substitution of (6.5) and (6.6) into (6.4) gives finally, with the help of the De Donder relation (3.1):

$$[M_{,v}]/[M] = \exp (A/RT) = v_d/v_a \tag{6.7}$$

Thus, if values of virtual and real fugacity in the gas phase can be *measured at the same value of surface coverage*, the affinity of the desorption step can be obtained, and with it a measure of the one-way behavior of the adsorption step. In particular, with a large value of A, the desorbed product will not appear as an inhibitor of the turnover freqency of the catalytic cycle, since in that case $v_d >> v_a$.

The decomposition of ammonia was studied at low pressures and high temperatures on foils of W and Mo in the laboratory of Tamaru.[17,18] Because of the low pressure, the surface coverage by adsorbed N could be determined by Auger electron spectroscopy (AES) during the reaction and in a series of equilibrium adsorption isotherms of nitrogen. From the latter, the heat of adsorption was obtained: it decreases with increasing surface coverage. As an example of catalytic coupling for Mo, for desorption of N_2 at 1000 K and a number of N_2 molecules per cm^3 equal to 8.2×10^8, the value of A was found to be 41 kJ mol^{-1} by using eqn. (6.7) with values of $[M_{,v}]$ and [M] determined at the same value of surface coverage during reaction and from the equilibrium isotherms. This corresponds to $v_d/v_a = 142$. Thus desorption is assisted appreciably by the high coverage of the surface by nitrogen as a result of steps preceding desorption. This effect of kinetic coupling is enhanced by the decrease in the heat of adsorption of N with increasing surface coverage.

Unfortunately, to measure surface coverage during reaction at pressures that are too high for electron spectroscopies to be usable remains a challenge. The development of sum frequency generation, a new method to use surface infrared spectroscopy at normal pressures, looks promising.[19] In the meantime, reactions for which the value of the affinity of a given step can be obtained readily are those with a rate-determining step, *rds*, in the true sense of Horiuti, *i.e.* when all other steps in the catalytic cycle are quasi-equilibrated.[6] One such reaction is ammonia synthesis or decomposition at normal pressures. Under these conditions the *rds* is the two-way chemisorption of dinitrogen. Clearly then, the value of the affinity for the *rds* is that of the overall reaction, as obtained from thermodynamics, since the affinity value of all other steps is closer to zero. The celebrated case of ammonia synthesis and decomposition was the first one for which a Langmuir–Hinshelwood mechanism was replaced by the kinetic coupling mechanism of Temkin and Pyshev.[20,21] In this mechanism, adsorption and desorption of dinitrogen are coupled to the surface-equilibrated step with the added simplification that chemisorbed nitrogen is the most abundant surface intermediate, the *masi*.[22] Later, this species was called the most abundant reactive intermediate, *mari*,[23] because a *masi* may be just a spectator species that is not a participant in the catalytic cycle.[23]

Even when the affinity of a desorption step cannot be determined, qualitative observations can often lead to clear evidence of kinetic coupling. A much investigated example by Tamaru and co-workers is the catalytic decomposition of GeH_4 on a growing Ge film to surface Ge and gaseous H_2.[24] The rate of decomposition is zero order. A Langmuir–Hinshelwood mechanism might suggest that the measured rate is that of GeH_4 adsorbed in a monolayer equilibrated with gaseous GeH_4. But dynamic experiments suggest that the monolayer consists of chemically bound hydrogen. A kinetic coupling situation is suggested by the following observations. First, H_2–D_2 equilibration to HD takes place readily on a Ge surface only partially covered with adsorbed hydrogen under conditions similar to those used in the decomposition of GeH_4. Second, when mixtures of GeH_4 and GeD_4 are decomposed on Ge, equilibrated mixtures of H_2, HD, and D_2 are produced. Third, when mixtures of GeH_4 and D_2 react with Ge, no HD is observed. Thus, kinetic coupling during reaction leads to a surface coverage by hydrogen that is considerably higher than its equilibrium value, blocking access to the surface by D_2 during the decomposition of GeH_4, and facilitating the desorption of H_2.

Another example is the dehydrogenation of methylcyclohexane to toluene and H_2 under conditions such that the reaction is not limited by equilibrium.[25] Still, the rate is not inhibited by toluene in spite of the intuitively evident stronger chemisorption of toluene than that of methylcyclohexane. Again, a Langmuir–Hinshelwood mechanism seems improbable. This conclusion receives strong support from the critical observation that benzene added to the feed does not inhibit the rate of toluene formation. Qualitatively, as a result of kinetic coupling, the virtual fugacity of adsorbed toluene is higher than its real fugacity. Thus, as shown above in the case of nitrogen desorption during ammonia decomposition, the affinity for toluene desorption has a high positive value. Thus, benzene is denied access to the surface during reaction and toluene itself is not an inhibitor of the rate of reaction. Even though these observations remain qualitative at the moment, they illuminate the power of kinetic coupling in catalytic cycles and help us to understand how catalytic cycles turn over. In particular, the examples treated in this section show how desorption, the exit process normally expected to be limited by equilibrium, can be assisted by adsorption steps preceding the exit step.[16,26]

7 Kinetic Coupling between Chain Cycles

7.1 A Case for Negative Catalysis – In biocatalysis, many complex reactions in the living cell are taking place through successive coupled enzymatic reactions.[27] In industrial catalysis, multifunctional catalysts consist of two or more catalytic sites working in sequence though separated over distances that can be overcome by diffusion of intermediates between them.[28] In such cases, two or more catalytic cycles are linked and kinetic coupling operates between them. In this section, this situation is illustrated by a free-radical gas phase reaction of the type used on a very large industrial scale in steam cracking, *i.e.* the pyrolysis of hydrocarbons in the presence of steam used as a heat transfer agent.[29] Kinetic coupling between two chain cycles sharing common elementary steps is demonstrated quantitatively in a study of the pyrolysis of neopentane $C(CH_3)_4$ denoted as μH to an alkene, methyl propene $CH_2=C(CH_3)_2$ called m, and a smaller alkane, methane, or βH, as shown in the upper cycle (a) of Figure 2.

In the nomenclature of the Nancy school of kinetics, μ is a free radical that *decomposes mono*molecularly, while β is a free radical that *reacts bi*molecularly.[30] What happens to the reaction $\mu H \rightarrow \beta H + m$ when a hydrogen donor YH is added to μH? As shown on the new coupled cycle on Figure 2(b), YH reacts with β to form βH and Y with the latter being regenerated to YH by reaction with μH. Thus, YH is a true catalyst as it enters into the upper cycle but is regenerated at the end of that cycle. The question is: what kind of a catalyst is YH? The experimental answer is shown in Figure 3.[31,32]

As can be seen, H_2S is a positive catalyst in the pyrolysis of neopentane but a negative catalyst in the pyrolysis of ethane. The experimental data are fitted nicely by the theoretical curves obtained by the kinetic treatment of the cycles taken from refs. 31 and 32. This kinetic treatment is reproduced in ref. 3, pp. 155–158. In brief, YH is a positive catalyst when the coupling rate constant k_{21} is larger than the coupling rate constant k_{11} of the coupled main cycle. Conversely, YH is a negative catalyst when k_{21} is smaller than k_{11}. This is readily understood: with the introduction of YH, some of the most abundant reactive intermediates, namely β free radicals, are side-tracked in the coupling cycle in the form of Y radicals. If the Y radicals react with μH faster than β, *i.e.* $k_{21} > k_{11}$,

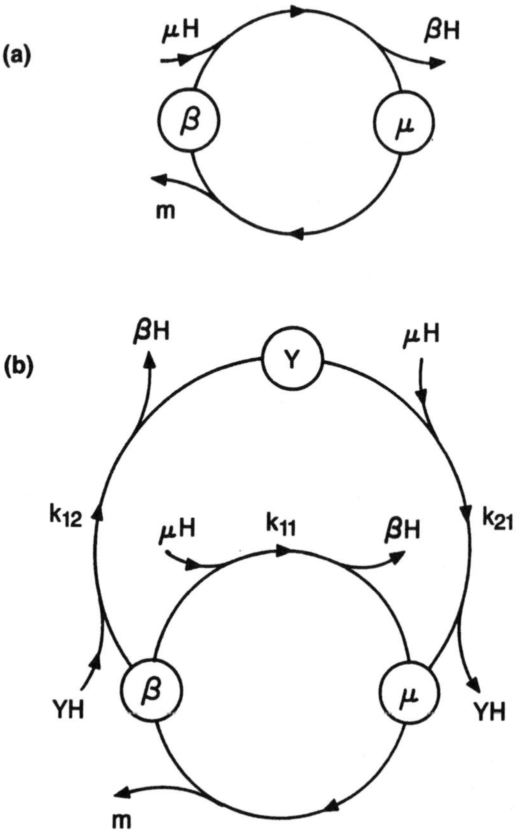

Figure 2 (a) *Chain propagation cycle for the gas phase pyrolysis reaction* μH→βH + m, *where* μH *is an alkane,* βH *an alkane of lower molecular weight, and* m *an alkene;* (b) *coupled cycles: same cycle as above coupled with a catalytic cycle involving the catalyst* YH

YH is a positive catalyst. If $k_{21} < k_{11}$, the reverse is true. Details are given in ref. 3.

It is generally admitted that a negative catalyst is a contradiction in terms since the traditional function of a catalyst is to accelerate the rate of a reaction. The above example demonstrates that a substance may be a negative catalyst through the kinetic coupling between two cycles. The author recalls a discussion at a meeting of a IUPAC committee in 1972 at Palm Beach, Florida. In this discussion, G.K. Boreskov defended the notion that a catalyst could be either positive or negative. But in the absence of a convincing example, like the one above that was unknown to the members of the committee, Boreskov's opinion was not retained. Yet, he was right.

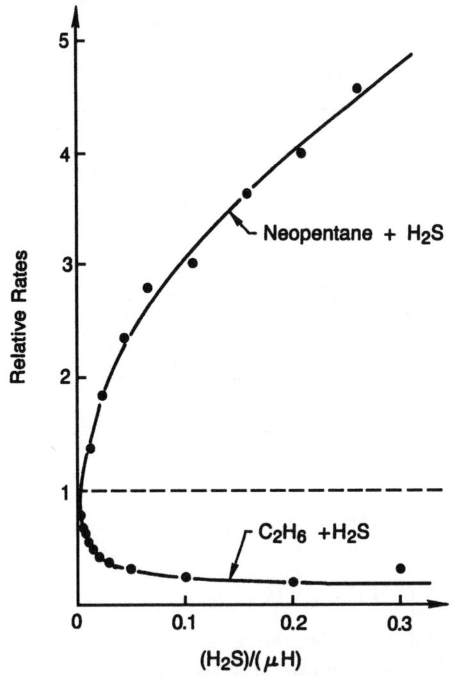

Figure 3 *The additive H_2S as a positive catalyst in the pyrolysis of neopentane and a negative catalyst in the pyrolysis of ethane. Experimental data are shown together with full lines calculated as indicated in the text.* (Adapted from refs. 31 and 32)

7.2 Changes in Selectivity through Kinetic Coupling between Chain Cycles – Kinetic coupling at the steady state does not only affect the rate of chain reactions in parallel or in series (ref. 5, Chapter 5), but more importantly, the coupling affects selectivity. As an example, consider the partial oxidation of methane to formaldehyde followed by the oxidation of formaldehyde to carbon monoxide. Both radical chain cycles are kinetically coupled as shown in Figure 4. At the steady state, the coupling steps between the chain cycles above and below must proceed at the same rate so that:

$$k_{12}\,(OH)\,(CH_2O) = k_{21}\,(HO_2)\,(CH_4) \tag{7.1}$$

Both upper and lower cycles normally proceed at different rates. But since they represent two reactions in series, the maximum amount of intermediate product is reached when the rates of both successive reactions become equal. Thus, the optimum selectivity, *i.e.* the maximum amount of formaldehyde will be reached when:

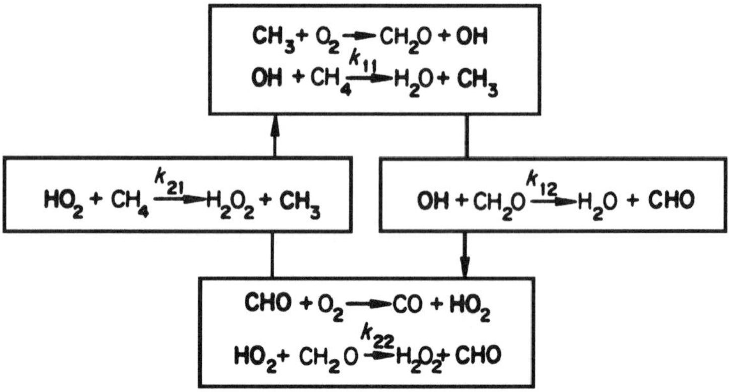

Figure 4

$$k_{11}\,(\text{OH})\,(\text{CH}_4) = k_{22}\,(\text{HO}_2)\,(\text{CH}_2\text{O}) \tag{7.2}$$

Dividing (7.1) and (7.2) side by side, the maximum concentration of formaldehyde is found to be,

$$(\text{CH}_2\text{O})_{\text{max}} = (k_{11}k_{21}/k_{21}k_{12})^{\frac{1}{2}}(\text{CH}_4)_0 \tag{7.3}$$

where $(CH_4)_0$ is the initial concentration of methane, since, unfortunately, it is reached very early in the reaction. This result was first obtained and verified experimentally by the school of Semenov.[33] The use of kinetic coupling to obtain eqn. (7.3) was first published in ref. 5, Chapter 5.

The treatment of copolymerization is very similar to the above example. It is another example of selectivity regulated by kinetic coupling at the steady state between chain cycles with the added advantage that the selectivity is materialized in the alternation of monomers in the copolymer chain.[34]

7.3 Changes in Selectivity through Kinetic Coupling between Cycles in Homogeneous Catalysis – There are many examples of multiple catalytic cycles grafted on each other by kinetic coupling. In homogeneous catalysis, a good illustration is the carbonylation of methanol.[35] But kinetic details are few and far between. A notable exception is the thorough kinetic investigation by the Halpern group of the stereoselective reaction with dihydrogen B of an alkene A with a chiral soluble rhodium catalyst designated as * in Figure 5.

The six kinetically significant rate constants shown in Figure 5 were determined.[36] The work has been reviewed from the viewpoint of kinetic coupling between catalytic cycles.[37] Both branches, left and right, are kinetically coupled through the free catalyst site. As emphasized by Landis and Halpern, the oddity of the situation is that the *major* active intermediate, on the left, yields the *minor* product, and *vice versa* for the right branch.

The reason, as demonstrated quantitatively by Landis and Halpern, is that the rate constant k_2 on the right side is much larger than k_2 on the left side. Kinetic

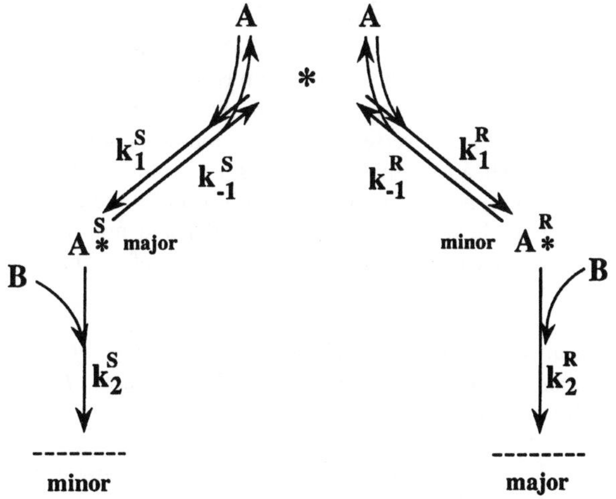

Figure 5

coupling then upsets the thermodynamic expectation: the kinetic pull is winning and the major product comes out of the right branch. The additional beauty of the situation is that the kinetic coupling can be, and was, manipulated to control selectivity between the two branches. Indeed, the pull of kinetic coupling can be enhanced by varying the pressure of dihydrogen: the higher it is the larger is the pull and thus the selectivity toward the right branch.

Besides, from the available kinetic data, selectivity values can be extrapolated to zero and infinite pressure of dihydrogen. As infinite pressure is approached, the binding steps of the substrate to the catalyst become *one-way*. By contrast, as zero pressure is approached, the same steps become *equilibrated*. At intermediate pressures accessible to experimentation, none of these two common simplifications is applicable: the entering steps are neither equilibrated nor one-way. The most important result is that, through kinetic coupling between catalytic cycles, selectivity can be controlled by varying the pressure of H_2.

7.4 Control of Selectivity through Kinetic Coupling between Heterogeneous Catalytic Cycles – Kinetic parameters in heterogeneous catalytic cycles are even less numerous than in homogeneous catalysis. Even the chemical identity of elementary steps is less securely known in heterogeneous catalysis than in homogeneous catalysis. But useful hints on how to control selectivity can be made available through mere qualitative knowledge of kinetic coupling between cycles. A classical example comes from work of Wauquier and Jungers.[38] They studied the liquid phase hydrogenation of non-polar cyclohexene, C, and polar acetone, A, on a nickel catalyst. The reaction is zero order when A and C are studied separately with zero-order rate constants k_A and k_C that are unchanged in various polar and non-polar solvents. But when A and C are hydrogenated together, the ratio of rates, *i.e.* the selectivity in favor of C, is,

$$S = k_C K_C / k_A K_A \tag{7.4}$$

because both reactants partition the available surface between themselves according to the magnitude of their adsorption equilibrium constants K_C and K_A. If the solvent is polar, C will be driven off the surface and the selectivity as defined will go down. The reverse is true in a non-polar solvent. Similarly, if a non-reactive non-polar adsorptive is added as a modifier to the solution, it will drive the polar compound off the surface and S will also increase. The reverse is true with a polar modifier. By using various polar and non-polar solvents and additives, the selectivity could be changed by a factor of 2500 as summarized in Table 1.

Table 1 *Selectivity in the liquid phase co-hydrogenation of cyclohexene and acetone, in favor of cyclohexene, at* 25 °C, 1 atm *of* H_2, *on a Raney nickel catalyst*

Solvent	Polar dioxane	Non-polar cyclohexane
Modifier	Non-polar stearic acid	Polar piperidine bicarbonate
Selectivity to cyclohexene	50	0.02

This example of kinetic coupling is particularly interesting because it deals with selectivity: the effect on selectivity is not due to a change in kinetic parameters but to a change in thermodynamic parameters.

A similar situation has been documented quantitatively in the classical case of isomerization of *n*-pentane to *i*-pentane on naphtha-reforming catalysts consisting of platinum supported on acidic alumina.[39] Under normal conditions of use, *n*-pentane is dehydrogenated to *n*-pentene on the platinum clusters and is in equilibrium with *n*-pentene and hydrogen. Then *n*-pentene diffuses to neighboring acid sites on the alumina and is isomerized there on *i*-pentene. The latter diffuses back to the metallic clusters where it is re-hydrogenated to the product *i*-pentane. The two catalytic cycles are kinetically coupled through an intermediate species that is in thermodynamic equilibrium with the original reactant as a result of reaction in the first cycle. This intermediate species then reacts in a second kinetically-coupled cycle to yield a product by reaching back within the first inverse cycle (Figure 6).

The proof that *n*-pentene is in equilibrium with *n*-pentane and dihydrogen during the isomerization reaction on Pt/Al_2O_3 was obtained as shown in Figure 7.

The data represented by circles correspond to calculated partial pressures of *n*-pentene in equilibrium with *n*-pentane and H_2 *vs.* the measured rate of isomerization of *n*-pentane on Pt/Al_2O_3. The slope of the straight line in the doubly logarithmic plot is one-half as observed separately. The experimental rate represented by a square is that of isomerization of *n*-pentene alone on the same alumina as used for the Pt/Al_2O_3 catalyst with the same acidity. The

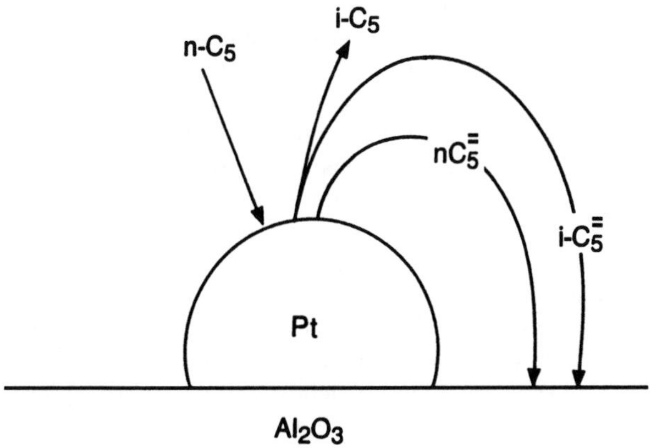

Figure 6 *Schematic representation of two coupled catalytic cycles in the isomerization of* n-*pentane to* i-*pentane on a bifunctional catalyst consisting of platinum supported on acidic alumina*

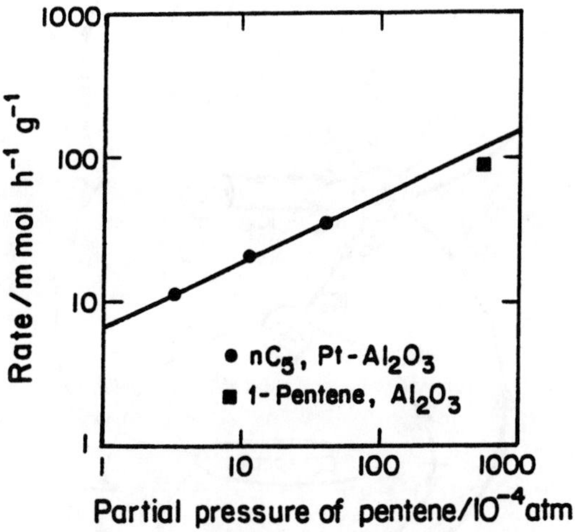

Figure 7 *Rate of isomerization* versus *partial pressure of pentene. The slope gives the observed reaction order* a = 0.5 *for the isomerization of pentene*

location of the square point very near the straight line extrapolated over a 200-fold pressure of *n*-pentene is an elegant proof of the *kinetic* coupling between both bifunctional cycles linked by a *thermodynamic* equilibrium partial pressure of the species diffusing between the two functions. It must be stressed again that this is not an example of thermodynamic coupling as defined in ref. 15.

7.5 Kinetic Coupling between Multiple Catalytic Cycles: Cracking of Alkanes on Zeolites – Great progress has been made recently in the kinetic analysis of complex reaction networks in heterogeneous catalysis, paralleling similar advances in combustion systems involving free radicals. In the latter case, extensive microkinetic data banks exist for many rate constants of elementary steps.[16] In the case of catalytic reactions, rate constants are estimated, and the results of the computation, *or microkinetic analysis*, are validated by comparison with experimental data.[40] Catalytic cracking of alkanes provides a good example of the power of the method developed independently by the groups of Froment, Dumesic, and others.[41–44]

As an example, consider the cracking of isobutane over Y-zeolites.[43] Twenty-one *elementary steps* (or *single events*, as called by the Froment group) were retained. The pre-exponential factors and activation energy of each one were estimated. Catalytic cracking proceeds through eight kinetically-coupled cycles involving the Brønsted acid sites of the zeolite and seven most abundant reactive carbenium ions paired with surface basic sites (Figure 8).

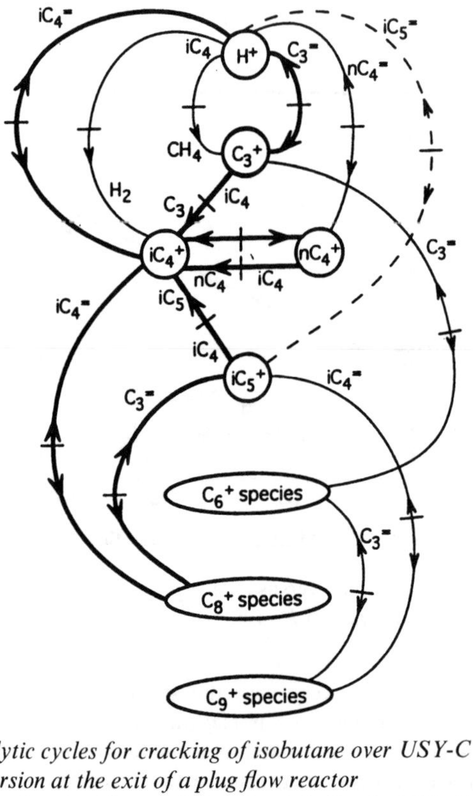

Figure 8 *Catalytic cycles for cracking of isobutane over USY-C at* 773 K *at* 11% *conversion at the exit of a plug flow reactor*
(Reproduced with permission from *J. Catal.*, 1995, **153**, 65. Copyright (1995) Academic Press, Inc.)

Forward and net rates of the elementary steps were used to draw the catalytic cycles of Figure 8 by starting at a surface species and following the reaction line connecting various surface species until returning to the original one. Gaseous reactants and/or products, if any, are specified on the reaction segment. Relative rates of individual steps are symbolized by the aspect of the reaction line, from dashed to continuous to fat. The arrows on the reaction lines indicate their direction and relative rates: a single arrow indicates a one-way process, opposite arrows of equal length stand for an equilibrated process, and opposite arrows of unequal length signify a two-way step with a faster rate in one direction. A given elementary step may participate in more than one cycle. But since the kinetic steady state remains valid, the net rate of production and consumption of a given surface species by all reaction lines connected to that species must be equal. Catalytic cycles such as those of Figure 6 account for product selectivity and the effect on selectivity of variables such as reactor length, degree of conversion, temperature, origin of the catalyst, acidity, and de-alumination by steaming. Selectivity means nature and abundance of products, alkenes or alkanes. The microkinetic analysis is validated by comparison of its predictions with experimental data. In a more recent paper, the authors of the paper on isobutane discuss selectivity exhibited by more complex catalytic cycles in the cracking of 2-methylhexane on various zeolites.[44] The analysis included consideration of 33 elementary steps. Similar work has also been reported by the Froment group.[45] With elaborate types of microkinetic analysis, a graphic summary, such as that on Figure 6, is the best way to convey the operation of coupled catalytic cycles and the factors that determine selectivity.

8 Summary and Conclusions

Radical chain reactions, as well as homogeneous and heterogeneous catalytic cycles, may be pictured as kinetically self-assembled systems. Their improbable practical success can be perceived as the effect of kinetic coupling between a large number of elementary steps that help each other in turning over single or linked catalytic cycles, thanks to the kinetic steady state.

The mutual assistance between elementary steps in a cycle at the kinetic steady state relies on the principle of Le Chatelier that helps a thermodynamically unfavored reaction by accumulating reactants or withdrawing products. In this essay, examples have been provided to illustrate kinetic coupling in and between cycles that turn over at the steady state.

With growing data banks of rate constants of elementary steps, it appears that kinetics assisted design of new catalytic processes will accelerate the development of improved catalytic processes. With powerful computational methods of thermodynamic and kinetic parameters, microkinetic analysis has emerged as a mature method to describe existing processes in combustion, polymerization, homogeneous and heterogeneous catalysis. In catalysis, the method is now poised to become one of the tools of catalyst development with the help of surface science, organometallic chemistry, materials science, computer modelling, combinatorial chemistry, quantum chemistry, and catalytic reaction engineering.

To conclude, I offer first an encouraging note and finally a word of caution. If during an on-going microkinetic analysis, one of the prospective elementary steps is severely limited by thermodynamics, kinetic coupling may well overcome this apparent obstacle. Yet, it still appears that the discovery of *new* catalytic reactions and of *new* catalysts will continue to be the creation of a lone investigator, with unfettered vision and the luck that Pasteur promised to the prepared mind.

References

1 Chemicals and Long-term Economic Growth, ed. A. Arora, R. Landau and N. Rosenberg, John Wiley & Sons, Inc., New York, 1998.
2 M. Boudart, in *Technology and Economics*, National Academy Press, Washington, D.C., 1991, pp. 77–83.
3 M. Boudart, in *Catalyst Design: Progress and Perspectives*, ed. L. Louis Hegedus, John Wiley & Sons, New York, 1987, ch. 5, pp. 141–162.
4 J.A. Christiansen, *Advan. Catal.*, 1953, **5**, 311.
5 M. Boudart, *Kinetics of Chemical Processes*, Butterworth-Heinemann, London, 1991, ch. 2, pp. 32–59.
6 M. Boudart and K. Tamaru, *Catal. Lett.*, 1991, **9**, 15.
7 Th. De Donder, in *L'Affinité*, Gauthier-Villars, Paris, 1927, p. 43.
8 W.L. Holstein and M. Boudart, *J. Phys. Chem. B*, 1997, **101**, 9991.
9 M. Boudart, *J. Phys. Chem.*, 1983, **87**, 2786.
10 M. Boudart, *Ind. Eng. Chem. Res.*, 1994, **33**, 2900.
11 M. Bodenstein and S.C. Lind, *Z. Phys. Chem.*, 1907, **57**, 168.
12 J.A. Christiansen and K. Dan, *Vidensk. Selsk., Mat. Phys. Medd.*, 1919, **1**, 14.
13 K.F. Herzfeld, *Ann. Phys.*, 1919, **59**, 635.
14 M. Polanyi, *Z. Elektrochem.*, 1920, **26**, 49.
15 I. Prigogine and R. Defay, in *Chemical Thermodynamics*, translated by D.H. Everett, Longmans, London, 1954, p. 38.
16 M. Boudart, *Ind. Eng. Chem. Fundam.*, 1986, **25**, 70.
17 H. Shindo, C. Egawa, T. Onishi and K. Tamaru, *J. Chem. Soc., Faraday Trans. 1*, 1980, **76**, 280.
18 M. Boudart, C. Egawa, S.T. Oyama and K. Tamaru, *J. Chimie Phys.* 1981, **78**, 987.
19 G.A. Somorjai and M.X. Yang, *J. Mol. Cat. A*, 1997, **115**, 389.
20 M.I. Temkin and V. Pyshev, *Zhur. Fiz. Khim.*, 1939, **13**, 851.
21 M. Boudart, *Topics Catal.* 1994, **1**, 405.
22 Ref. 5, p. 67.
23 M. Boudart, in *Perspectives in Catalysis*, ed. J.M. Thomas and K.I. Zamaraev, IUPAC Monograph, Blackwell, Oxford, 1991, p. 187.
24 K. Tamaru, in *Dynamic Heterogeneous Catalysis*, Academic Press, London, 1978, pp. 81–88.
25 J.H. Sinfelt, H. Hurwitz and R.A. Shulman, *J. Phys. Chem.*, 1960, **64**, 1559.
26 M. Boudart, in special issue in honor of K. Tamaru, *J. Mol. Cat.*, in press.
27 Y. Termonia and J. Ross, *Proc. Nat. Acad. Sci. USA*, 1981, **78**, 2952.
28 J.H. Sinfelt, H. Hurwitz and J.C. Rohrer, *J. Phys. Chem.*, 1960, **64**, 892.

29 A.G. Buckens and G.F. Froment, in *Chemical Reaction, Proceedings of the 4th European Symposium*, Pergamon Press, Oxford 1971.

30 P. Goldfinger, M. Letort and M. Niclause, in *Volume Commémoratif Victor Henri: Contribution à l'Etude de la Structure Moléculaire*, Desoer, Liège, 1947, p. 283.

31 G. Scacchi, F. Baronnet, R. Martin and M. Niclause, *J. Chimie Phys.*, 1968, **65**, 1671.

32 G. Scacchi, M. Dzierzynski, R. Martin and M. Niclause, *Int. J. Chem. Kinet.*, 1970, **2**, 1970.

33 N.N. Semenov, in *Some Problems in Chemical Kinetics and Reactivity*, translated by M. Boudart, Princeton University Press, Princeton, NJ, 1959.

34 P.J. Flory, *Principles of Polymer Chemistry*, Cornell University Press, Ithaca, NY, 1953, Chapter V.

35 D. Forster, *Adv. Organometal. Chem.*, 1979, **17**, 255.

36 C.R. Landis and J. Halpern, *J. Am. Chem. Soc.*, 1987, **109**, 1746 and 6217.

37 M. Boudart and G. Djéga-Mariadassou, *Catal. Lett.*, 1994, **23**, 7.

38 J.P. Wauquier and J.C. Jungers, *Bull. Soc. Chim.*, 1957, 1280.

39 J.H. Sinfelt, H. Hurwitz and J.C. Rohrer, *J. Phys. Chem.*, 1960, **64**, 892.

40 J.A. Dumesic, D.F. Rudd, L.M. Aparicio, J.E. Rekoske and A.A. Treviño, *The Microkinetics of Heterogeneous Catalysis*, American Chemical Society, Washington, D.C., 1993.

41 P.A. Willems and G.F. Froment, *Ind. Eng. Chem. Res.*, 1988, **27**, 1959.

42 W. Feng, E. Vynckier and G.F. Froment, *Ind. Eng. Chem. Res.*, 1993, **32**, 2997.

43 G. Yaluris, J.E. Rekoske, L.M. Aparicio, R.J. Madon and J.A. Dumesic, *J. Catal.*, 1995, **153**, 65.

44 R.D. Cortright, J.A. Dumesic and R.J. Madon, *Topics Catal.*, 1997, **4**, 15.

45 G.F. Froment, I. Vasalos, N. Markatos and N. Skandalis, *Belg. Appl. Therm. Eng.*, 1997, **17**, 837.

4
Catalytic Resource Recovery from Waste Polymers

BY LINDA J. BROADBELT

1 Introduction

Environmental problems have become global concerns. Much of the world's collective attention has recently focused on pollutants in the air such as carbon dioxide, nitrogen oxides and other greenhouse gases, as climate change has been definitively labelled as a global problem.[1,2] However, water pollution and soil contamination have no small influence on the health of the global environment. Indeed, all of these media are linked through a complex set of transport processes, and a 'systems' viewpoint has emerged as the dominant approach to the world's pollution prevention efforts. The Pollution Prevention Act of 1990 in the United States, for example, declares that it is the 'national policy of the United States that pollution should be prevented or reduced at the source whenever feasible' and that 'pollution that cannot be prevented should be recycled in an environmentally safe manner whenever feasible' without an emphasis on any one particular medium.[3] One pollution prevention effort with which the general public is familiar is plastics recycling, as individuals in many communities are critical components in the collection, sortation and processing system that enables successful resource recovery from waste plastics. However, empirical evidence reveals that average municipal recycling rates are around 18% across America.[4] Market forces and technological limitations have not yet made plastics recycling attractive enough to result in its widespread practice. Therefore, both environmental concerns and economic limitations are driving the exploration of new strategies for viable plastics resource recovery.

Waste polymers represent a tremendous source of untapped energy. In the United States alone, over 70 billion pounds of plastics are manufactured annually, while only 10% of this amount is recycled or incinerated.[5] Therefore, by simply increasing the total amount of material processed in material recovery facilities, the amount of energy gained is quite significant. The ultimate challenge, however, will be to increase the potential for significant energy savings through improvement of resource recovery processes. A study conducted by the Polymer Processing Institute[5] indicates that a reasonable projection of the maximum gross energy opportunity in the year 2010 from plastics resource recovery is 5.2×10^{14} Btu year^{-1}, an amount equivalent to approximately 225 000 barrels per day of

Catalysis, Volume 14

crude oil. These figures allow for the most optimistic scenario, one in which high-valued chemicals are obtained directly from the resource recovery process. In order to realize the great energy savings potential, a new generation of resource recovery technologies must be developed.

Currently, material recovery facilities operate using one of four categories of recycling processes which are summarized in Table 1.[6–8] Primary and secondary recycling methods are limited in application since, ultimately, the properties of the polymer have degraded to the point that only tertiary or quaternary methods can be used. Tertiary and quaternary are attractive since they do not require costly sorting of the typical mixed plastic waste. However, incineration or combustion is an inefficient use of resources and has a negative public image as an environmental liability.[8] Tertiary recycling is a radical departure from the traditional processes and is projected to grow 10-fold in the next 10 years.

Table 1 *Classifications of resource recovery strategies*

Category	*Product*
Primary	Similar to starting material
Secondary	Lower-valued polymeric material
Tertiary	Fuels, petrochemical feedstocks, chemicals
Quaternary	Energy

Tertiary recycling may employ both thermal and chemical methods, depending on the polymer feedstock. Condensation polymers, such as polyester, nylon and polyurethanes, are decomposed using glycolysis, methanolysis and hydrolysis[9,10] and are the most tractable components of plastic waste because of the reversible nature of their polymerization. However, addition polymers such as vinyls, acrylics and polyolefins, which comprise 60% of the plastic waste stream,[9] are significantly more intractable. Energy must be added to break the bonds of the polymer backbone, often requiring extremely high temperatures or harsh environments to promote decomposition. Under these harsh conditions, a wide range of reactions occur, and therefore, high selectivity to high-valued products, *e.g.* monomers, is not achieved.

Clearly, a strategy is required that directs the chemistry of addition polymer decomposition along favorable reaction pathways. Catalytic resource recovery is a promising technology for the conversion of polymeric waste materials to valuable products with high selectivity that has received considerable attention in only the last 10 years. Through analogies to the petrochemical industry, it has been proposed that catalysts will be able to direct the complex chemistry of high molecular weight materials, and the distribution of products can be tailored by manipulation of the catalyst properties. The majority of the investigations examining catalytic resource recovery from waste polymers has focused on obtaining high-valued fuels, but attention has also been paid to obtaining particular feedstock chemicals with high selectivity. The unifying theme of these efforts is to obtain useful and valuable products from waste polymers with maximum economic benefit and minimal environmental liability.

This review concentrates on three commodity addition polymers, polyethylene, polypropylene and polystyrene, and is organized into five sections. Since resource recovery from waste polymers through thermal decomposition has been more extensively studied and provides the baseline to which catalytic degradation may be compared, Section 2 summarizes the product distributions observed from reacting the three polymers of interest at high temperatures in the absence of catalysts. Although the term 'catalytic degradation of polymers' has been widely used in the literature, there is not a consistent definition of the contacting pattern between the catalyst and the polymer that this phrase implies. Therefore, Section 3 outlines the different catalyst–polymer contacting patterns used and their potential influence on the products. In Section 4, work that has been performed using acid catalysts is reviewed. With their known ability to cleave carbon–carbon bonds and their widespread use in the hydrocarbon processing industry, acid catalysts have been the most extensively studied. Although the number of studies is more limited, base catalysts have also been examined for decomposing polymeric materials, and this work is summarized in Section 5. To exploit existing process equipment and the extensive knowledge base that has been already assembled, much attention has focused on the catalytic conversion of polymers in the presence of co-reactants, primarily petroleum feedstocks and coal. Co-processing studies are the focus of Section 6. A summary of studies investigating the catalytic degradation of polymers is provided in Table 2. This summary designates the reactants that were studied, both polymeric and co-reactants, the particular catalysts used, the way in which the polymer and the catalyst were contacted and whether thermal baseline information is clearly provided in the reference cited. Finally, a summary of the current status of catalytic resource recovery from waste polymers is presented in Section 7.

2 Thermal Degradation

The thermal degradation of polyolefins such as polyethylene, polypropylene and polystyrene as an independent process became a topic of great interest at the end of the 1940s when researchers aimed to achieve the selective extraction of valuable gaseous hydrocarbons from polymeric feedstocks.[11] However, with rare exceptions, the selectivity to monomer is very low; the formation of a wide range of low molecular weight products is typical. Although the chemistry is very complex, the majority of the products can be rationalized in terms of typical free radical reactions including bond fission, hydrogen abstraction, β-scission, radical addition, isomerization and termination. The thermal decomposition process is typically reported by monitoring the changes in the molecular weight of the polymer, the amount and identity of volatile and non-volatile products and the rate of decomposition and its dependence on temperature. Representative studies and the products of thermal degradation of polyethylene, polypropylene and polystyrene are summarized below.

2.1 Polyethylene – The thermal degradation of polyethylene has been studied by numerous researchers (see Table 2). Thermal degradation forms predominantly *n*-paraffins and 1-olefins; aromatic products are formed in trace amounts. Ohkita *et al.*[12] examined the thermal decomposition of low density polyethylene using 15 g of polymer in a semibatch reactor operating at 673 K with a nitrogen sweep of 50 mL min^{-1}. Gases (C_1–C_4) and oils (C_5–C_{20}) were quantified, and the products consisted of 20 wt% gases and 44 wt% oils. The balance of the products was identified as waxes or residue remaining in the reactor. In preparing a feedstock for subsequent catalytic degradation, Ng *et al.*[13] reacted polyethylene in a batch reactor in a nitrogen atmosphere. The reaction temperature was 723–773 K, and the reaction was carried out for 10 min. The final pressure at the reaction temperature ranged from 1.38 to 16.13 MPa. The gases were analysed by gas chromatography, and the other products were separated into naphtha, gas oil and residue by distillation. The amount of residue decreased with temperature, and the amount of gas increased with temperature. The weight yield of gas oil was maximized at 743 K, and it consisted primarily of normal saturates and lesser amounts of 1-olefins. Only small amounts of branched hydrocarbons were detected, which is consistent with results obtained in our own laboratory.[14]

2.2 Polypropylene – Although the reactor configuration and the reaction conditions affect the product distribution observed, thermal decomposition of polypropylene results in a complex mixture of different hydrocarbons as revealed by the numerous studies investigating its reaction (*e.g.* refs. 15 and 16). Tsuchiya *et al.*[16] decomposed isotactic polypropylene under vacuum at 633, 653 and 673 K and analysed the volatile products from C_1 to C_{12} hydrocarbons. The weight percentages of products are summarized for a reaction temperature of 633 K in Table 3. Pentane was clearly the most abundant product formed, and propylene and 2-methyl-1-pentene were also formed in high yields relative to the other products. These compounds were also the major products observed at higher temperatures, but 2,4-dimethyl-1-heptene became more significant as temperature increased, in part due to the increased volatility of this heavier product at higher temperature. The formation of the main products was rationalized based on a free radical mechanism in which intramolecular radical transfer was assumed to play an important role. Similar thermal degradation studies were carried out by Audisio *et al.*[15] to compare to their catalytic degradation experiments. At both 673 and 873 K, C_9 and C_{12} products were observed in the highest yields, with smaller yields of products reported in the C_5–C_8 and C_{10}–C_{11} ranges. The relative amounts of the products in each carbon number fraction were essentially insensitive to temperature.

2.3 Polystyrene – Thermal degradation is the simplest of the current techniques used to recover feedstock chemicals from styrene-based polymers and has therefore been studied extensively.[17–27] Investigations of the product distributions from thermal degradation of polystyrene have mainly focused on liquid products.[17] It has been observed that the yield and the composition of liquid products vary strongly with temperature[17,22,27–30] and the reactor configuration,

Table 2 *Summary of studies investigating the catalytic degradation of polyethylene, polypropylene or polystyrene*

First author	*Citation*	*Reactant(s)*[a]	*Catalyst(s)*[b]	*Contacting pattern*[c]	*Thermal baseline*
Aguado	*Chemical Communications*, 1996, **6**, 725–726.	PE	MCM-41, SA, HZSM-5	COMS	No
Anderson	*Preprints of the Division of Fuel Chemistry of the American Chemical Society*, 1993, **38**, 816–822.	PE, PS, coal	Ammonium tetrathiomolybdate	COMB	Yes
Arandes	*Ind Eng Chem Res*, 1997, **36**, 4523–4529.	PP, PS, LCO	REY	BIRR	Yes
Audisio	*J Anal Appl Pyrol*, 1985, **8**, 87–93.	PP	$(BiO)_2CO_3$, Sb_2O_3	COMS	Yes
Audisio	*J Anal Appl Pyrol*, 1984, **7**, 83–90.	PP	Al_2O_3, SiO_2, NaY, HY, REY, SA	COMS	Yes
Audisio	*Polymer Degradation and Stability*, 1990, **29**, 191–200.	PS	HY, REY, SA, SiO_2, Al_2O_3	COMS	Yes
Ayame	*J Japan Petrol Inst*, 1979, **22**, 280–287.	PE	CaX	FB	No
Azhar Uddin	*Poly Deg Stab*, 1997, **56**, 37–44.	PE	SA	COMB	Yes
Baramboim	*Polymer Science USSR*, 1979, **21**, 487–494.	PE	$AlCl_3$, $SnCl_4$, $FeCl_3$, $ZnCl_2$		Yes
Beltrame	*Polymer Degradation and Stability*, 1989, **26**, 209–220.	PE	Al_2O_3, SiO_2, HY, REY, SA	COMS	Yes
Butcher, Jr.	US Patent 5,315,055, 1994.	PE, PP, PS	$NaOH/KOH/Cu_2O$	COMS	
Carniti	*Applied Catalysis A: General*, 1995, **127**, 139–155.	PS	HY, REY, SA, SiO_2	COMB	Yes
Carniti	*Polym Deg Stab*, 1997, **57**, 301–306.	PS	SiO_2/MgO, SiO_2/TiO_2, SA, MOR	COMB	Yes
Carrick	*Journal of Polymer Science: Part A-1*, 1970, **8**, 215–223.	PE, PS, PP	$AlCl_3$	COMB	No
Day	*Polymer Degradation and Stability*, 1995, **48**, 341–349.	PP	Cu, Fe_2O_3	COMS	Yes
Deng	*Preprints of the Division of Fuel Chemistry of the American Chemical Society*, 1997, **42**, 987–992.	PE	$Pt/ZrO_2/SO_4$, ZrO_2/SO_4	COMS	Yes
Ding	*Fuel Processing Technology*, 1997, **51**, 47–62.	PE	$TiCl_3$, HZSM-5	COMB	Yes
Drago	*New J Chem*, 1994, **18**, 937–939.	PE, PP, PS	$(SG)_n$–$AlCl_2$	COMB	No
Fernandes, Jr.	*Journal of Thermal Analysis*, 1997, **49**, 225–260.	PE	HZSM-5	COMS	Yes
Fernandes, Jr.	*Progress in Zeolite and Microporous Materials*, 1997, **105**, 941–947.	PE	HZSM-5	COMS	Yes
Frank	*Monatschefte Chem*, 1963, **94**, 393–399.	PS	$AlCl_3$		
Garforth	*Thermochimica Acta*, 1997, **294**, 65–69.	PE	SA, HY, USY, HZSM-5, MCM-41	COMS	Yes
Human	*Preprints of the Division of Fuel Chemistry of the American Chemical Society*, 1995, **40**, 34–37.	PE, PP, coal	HZSM-5, ultrafine ferrihydrite, Al/Si/ferrihydrite SA, MOR, HZSM-5	COMB	Yes
Ibrahim	*Fuel Processing Technology*, 1996, **49**, 65–73.	PE, PP	Al_2O_3	COMS	Yes

Ide	*J Anal Appl Pyrol*, 1984, **6**, 69–80.	PS	Active charcoal	FB	Yes
Ide	*J Appl Polym Sci*, 1984, **29**, 2565–2571.	PS	SA	COMS	Yes
Ide	*J Chem Soc Japan*, 1983, 1657–1663.	PS	SA		
Ishihara	*Bull Chem Soc Jpn*, 1991, **64**, 3585–3592.	PE	NaY	COMS	Yes
Ishihara	*Polymer*, 1992, **33**, 3482–3486.	PE	SA	COMS	No
Ishihara	*J Appl Polym Sci*, 1989, **38**, 1491–1501.	PE, PP	SA	COMS	Yes
Ishihara	*Fuel*, 1990, **69**, 978–984.	PE	SA	FB, COMS	Yes
Ishihara	*Fuel*, 1993, **72**, 1115–1119.	PP	SA	COMS	No
Ishihara	*Bull Chem Soc Jpn*, 1989, **62**, 2981–2988.	PP	SA	COMS	No
Ivanova	*Polymer Science*, 1991, **33**, 268–275.	PE, PP	$LiAlCl_4$, $NaAlCl_4$, $KAlCl_4$		No
Ivanova	*Russian Chemical Reviews*, 1991, **60**, 225–234.	PE	$AlCl_3$, $NaAlCl_4$, $KAlCl_4$, $LiAlCl_4$, $MgCl_2{\cdot}AlCl_2$, $MgCl_2{\cdot}H_2O$, $MgCl_2{\cdot}2H_2O$, $MgCl_2{\cdot}4H_2O$, $BaCl_2{\cdot}2H_2O$		Yes
Ivanova	*Prog Polym Sci*, 1990, **15**, 193–215.	PE	$AlCl_3$, $NaAlCl_4$, $KAlCl_4$, $LiAlCl_4$, $MgCl_2{\cdot}AlCl_3$, $MgCl_2{\cdot}H_2O$, $MgCl_2{\cdot}2H_2O$, $MgCl_2{\cdot}4H_2O$, $BaCl_2{\cdot}2H_2O$		Yes
Joo	*Energy & Fuels*, 1997, **11**, 801–812.	PE, coal, resid	$NiMo/Al_2O_3$	COMB	Yes
Joo	*Energy & Fuels*, 1996, **10**, 603–611.	PE, PS, coal, resid	$NiMo/Al_2O_3$	COMB	Yes
Lin	*Energy & Fuels*, 1998, **12**, 767-774.	PE, PP	HZSM-5, SA, USY, MOR, MCM-41	COMS	No
Lin	*J Appl Polym Sci*, 1997, **63**, 1287–1298.	PS	HZSM-5, SA, ZrO_2/SO_4	COMS	Yes
Lin	*Preprints of the Division of Fuel Chemistry of the American Chemical Society*, 1997, **42**, 982–986.	PS	HZSM-5, SA, ZrO_2/SO_4	COMS	Yes
Lin	*J Appl Polym Sci*, 1995, **58**, 1151–1159.	PE	HZSM-5, SA, ZrO_2/SO_4	COMS	Yes
Lin	*ICHME Research Event/Second European Conference for Young Researchers*, 1996, **2**, 576–578.	PE	USY	COMS	Yes
Liu	*Preprints of the Division of Fuel Chemistry of the American Chemical Society*, 1995, **40**, 9–16.	Mixed plastics waste, coal	Fe_2O_3/SO_4^{2-}, ZrO_2/SO_4^{2-}, Al_2O_3/SO_4^{2-}, $Ni–Mo/Al_2O_3$	COMS	Yes
Luo	*Fuel Processing Technology*, 1996, **49**, 177–196.	PE, PS, PP, coal	HZSM-5, Low Alumina, Super Nova-D	COMB	Yes
Luo	*Fuel Processing Technology*, 1996, **49**, 91–117.	PE, PS, coal	Sulfided molybdenum naphthenate, sulfided iron naphthenate, Low Alumina, Super Nova-D, Octacat-50G, HZSM-5	COMB	Yes

Continued

Table 2 *Continued*

First author	*Citation*	*Reactant(s)*[a]	*Catalyst(s)*[b]	*Contacting pattern*[c]	*Thermal baseline*
Luo	*Preprints of the Division of Fuel Chemistry of the American Chemical Society*, 1997, **42**, 1062–1066.	Mixed plastics waste, coal	Sulfided molybdenum naphthenate, sulfided iron naphthenate, Low Alumina, HZSM-5	TD/CD	Yes
Mordi	*Poly Deg Stab*, 1994, **46**, 57–62.	PE, PP	HZSM-5	COMB	No
Mordi	*J Anal Appl Pyrol*, 1994, **29**, 45–55.	PE	MOR, HZSM-5, H-theta-1	COMB	Yes
Mordi	*J Chem Soc, Chem Commun*, 1992, **4**, 374–375.	PP	HZSM-5, MOR, H-theta	COMB	No
Nakamura	*Catalysis Today*, 1996, **27**, 175–179.	PP	Carbon	COMB	Yes
Nanbu	*Polymer Degradation and Stability*, 1987, **19**, 61–76.	PS	$AlCl_3$, SA	COMB	No
Ng	*Energy & Fuels*, 1995, **9**, 735–742.	PE	USY	TD/CD	Yes
Ng	*Energy & Fuels*, 1995, **9**, 216–224.	PE, VGO	Fluidized cracking catalyst (KOB-627, Akzo)	FB	Yes
Ochoa	*Fuel Processing Technology*, 1996, **49**, 119–136.	PE	SA, SiO_2, Al_2O_3, HY, HZSM-5	COMB	Yes
Ogawa	*J Appl Polym Sci*, 1982, **27**, 859–869.	PS	SA	FB	Yes
Ohkita	*Ind Eng Chem Res*, 1993, **32**, 3112–3116.	PE	SA, SiO_2, Al_2O_3, ZnO, MgO, TiO_2, carbon, HZSM-5	TD/CD	Yes
Pei	*Fuel*, 1997, **76**, 363–366.	PE	$RhCl_3{\cdot}3H_2O/Al_2O_3$		No
Pukánszky	*Polymer Bulletin*, 1981, **5**, 469–476.	PS	Bu_2SnCl_2, $AlCl_3$	COMB	No
Robbins	*Preprints of the Division of Fuel Chemistry of the American Chemical Society*, 1995, **40**, 92–96.	PE, PS, coal	Ni/Mo		No
Rothenberger	*Preprints of the Division of Fuel Chemistry of the American Chemical Society*, 1994, **39**, 688–694.	PE, PS, coal	Ni-Mo/Al_2O_3, Fe-S	COMB	
Rothenberger	*Energy & Fuels*, 1997, **11**, 849–855.	PE, PP, PS, coal	Molecular sieve, Ni–Mo/Al_2O_3	COMB, COMS	Yes
Sakata	*J Anal Appl Pyrol*, 1997, **43**, 15–25.	PE	SA, HZSM-5, KFS-16B	COMS	Yes
Simionescu	*Thermochimica Acta*, 1988, **134**, 301–305.	PE, PP, PS	MnO_2/K_2O, Cr_2O_3/K_2O, CuO, SiO_2, SA, MOR, HZSM-5	TD/CD	Yes
Songip	*Applied Catalysis B: Environmental*, 1993, **2**, 153–164.	PE	HY, REY, HZSM-5, SA	TD/CD	No

Songip	*Energy & Fuels*, 1994, **8**, 136–140.	PE	REY	TD/CD	No
Songip	*Energy & Fuels*, 1994, **8**, 131–135.	PE	REY	TD/CD	No
Taghiei	*Preprints of the Division of Fuel Chemistry of the American Chemical Society*, 1993, **38**, 810–815.	PE, coal	HZSM-5, highly dispersed iron-based	COMB	Yes
Taghiei	*Energy & Fuels*, 1994, **8**, 1228–1232.	PE, PP, coal	Fe, HZSM-5	COMB	Yes
Tochihara	*Trans Mat Res Soc Jpn*, 1994, **18A**, 713–716.	PE	SA, SiO_2, Al_2O_3, ZnO, MgO, TiO_2, carbon, HZSM-5	TD/CD	Yes
Uemichi	*Kobunshi Ronbunshu*, 1993, **50**, 887–892.	PS	SA, MOR, HY, HZSM-5	FB	
Uemichi	*Ind Eng Chem Res*, 1997, **37**, 867–872.	PE	SA, HZSM-5, HY, MOR	FB	No
Uemichi	*J Japan Petrol Inst*, 1980, **23**, 35–43.	PE	NaX, SA	FB	No
Uemichi	*Journal of Chromatography*, 1983, **259**, 69–77.	PE	SA	FB	No
Uemichi	*Chemistry Letters*, 1984, **1**, 41–44.	PE	Carbon, Pt/SA, Pt/Al_2O_3	FB	Yes
Uemichi	*J Anal Appl Pyrol*, 1989, **14**, 331–344.	PE	Carbon, carbon with Pt, Fe, Mo, Zn, Co, Ni, Cu	FB	No
Uemichi	*J Anal Appl Pyrol*, 1989, **16**, 229–238.	PP	Pt/carbon, Fe/carbon, carbon	FB	No
Uemichi	*Bull Chem Soc Jpn*, 1983, **56**, 2768–2773.	PP	SA, CaX	FB	Yes
Vasile	*Revue Roumaine de Chimie*, 1995, **40**, 679–691.	PE, PP	HZSM-5, SA, Cr_2O_3/Al_2O_3	TD/CD	No
Vasile	*Acta Polymerica*, 1988, **39**, 306–310.	PE, PP	SA, HZSM-5	TD/CD	No
Wingfield	Ford Motor Co., US Patent 4,515,659, 1985.	ASR	Na_2CO_3, $CaCO_3$, K_2CO_3, $MgCO_3$, $ZnCO_3$	COMB	Yes
Xiao	*Preprints of the Division of Fuel Chemistry of the American Chemical Society*, 1995, **40**, 4–8.	PE, PP	Fe_2O_3/SO_4, ZrO_2/SO_4, Al_2O_3/SO_4	COMB	No
Yamaguchi	*Chemosphere*, 1973, **1**, 7–10.	PE, PP, PS	$AlCl_3$		Yes
Yamamoto	*Nippon Kagaku Kaishi*, 1976, **5**, 802–807.	PS	SA	COMB	Yes
Yoshida	*Bull Jpn Petrol Inst*, **17**, 218–225.	PE	NaA, CaA, CaX, NaX, NaY, SA	TD/CD	Yes
Zhang	*Ind Eng Chem Res*, 1995, **34**, 4514–4519.	PS	HZSM-5, SA, MgO, CaO, BaO, K_2O, carbon	TD/CD	Yes
Zhao	*Poly Deg Stab*, 1996, **53**, 199–206.	PP	Ferrierite	COMS	Yes
Zhibo	*Catalysis Today*, 1996, **29**, 303–308.	PE, PS	HZSM-5, SA, SiO_2, Al_2O_3, ZnO, MgO, TiO_2, carbon	TD/CD	Yes

[a]PE = polyethylene; PP = polypropylene; PS = polystyrene; ASR = automobile shredder residue; VGO = vacuum gas oil; LCO = light cycle oil. [b]SA = SiO_2/Al_2O_3; MOR = mordenite. [c]TD/CD = thermal degradation followed by catalytic degradation; COMB = mixed polymer and catalyst in a batch reactor; COMS = mixed polymer and catalyst in a semibatch reactor; FB = fixed bed flow reactor; BIRR = Berty internal recycle reactor.

Table 3 *Weight percentage of volatile products evolved from thermal decomposition of isotactic polypropylene under vacuum at* 633 K [16]

Product	*Weight percentage of polymer*
Methane	0.0055
Ethane	0.0057
Propane	0.0021
Propylene	0.0296
Isobutylene	0.0083
Pentane	0.102
2-Methylpentane	0.0066
2-Methyl-1-pentene	0.079
2,4-Dimethyl-1-pentene	0.0019
4-Methylheptane	0.013
2,4-Dimethyl-1-heptene	0.040
4,6-Dimethylnonane	0.0085
2,4,6-Trimethyl-1-nonene	0.0264

ranging from high selectivity to short oligomers, to high yields of saturated products such as ethylbenzene and toluene.[17,21,22,28,31,32] The mechanism of thermal degradation proceeds through free radicals, and the selectivity to the major products is dictated by the relative contributions of the propagation steps, depolymerization resulting in monomer, and hydrogen abstraction. Owing to the presence of the easily abstractable hydrogens along the backbone, the ability of thermal degradation to recover high yields of styrene monomer from polystyrene is limited. However, in comparison to other addition polymers such as polyethylene, the recovery of styrene monomer from polystyrene *via* thermal degradation is relatively high. Representative product distributions of low molecular weight compounds from thermal degradation of polystyrene in a flow reactor and a batch reactor are summarized in Tables 4 and 5, respectively.

Table 4 *Weight percentage of low molecular weight products evolved from thermal degradation of polystyrene in a flow reactor with a residence time of 4 s and at two different temperatures*[22]

Product	*Temperature* (°C)	
	350	400
Styrene	35.4	50.1
2,4-Diphenyl-1-butene	12.3	9.5
2,4,6-Triphenyl-1-hexene	24.9	18.2
Residue	11.6	9.3

3 Catalyst–Polymer Contacting Patterns

Although the concept of the ‘catalytic degradation of polymers’ has been significantly advanced in the last decade, the terminology has been used to

Table 5 *Selected product selectivities (wt%) for the thermal and acid-catalysed degradation of polystyrene at 350 °C after 60 min of batch reaction*[72]

Product	*Thermal*	*HY*	*REY*	*SiO_2/Al_2O_3*
Benzene	0	45.6	41.2	47.4
Toluene	7.6	6.2	4.7	0.7
Ethylbenzene	2.7	23.3	22.2	15.9
Styrene	25.3	0	0	0
Cumene	0.9	5.8	6.9	8.8
Indan	0	2.5	2.4	4.7
Methylindans	0	4.6	4.4	9.2
Naphthalene	0	1.9	1.1	0.5
2-Methylnaphthalene	0	2.3	1.5	1.0
1,3-Diphenylpropane	20.5	2.0	4.3	3.3
Diphenylpropenes	11.5	0	0	0
Diphenylbutanes	4.2	0	0.8	0.2
Diphenylbutenes	15.7	1.5	4.7	2.9

characterize a broad range of catalyst–polymer contacting patterns and reactor configurations. Depending on the reactor configuration employed, the catalyst may be in direct contact with the polymer, or the catalyst may be used only to transform volatiles that are evolved from thermal degradation of polymers. The diversity of the reactor configurations used makes it difficult to compare product distributions among different studies that are all defined as catalytic degradation of polymers. The same catalyst–polymer system may affect the product distribution differently depending on the way in which the polymer and its product are contacted with the catalyst. Three common reactor configurations reported in the literature that are all defined as catalytic degradation of polymers are depicted in Figure 1 and summarized below.

3.1 Thermal Degradation Followed by Catalytic Decomposition – One strategy that has been widely applied is to convert waste plastics to useful chemicals using a pyrolysis–reforming reactor.[12,31,33–38] The reactor consists of two integrated zones: one for thermal degradation, *i.e.* with no catalyst present, and the other to convert products of thermal degradation in the presence of a catalyst. The catalytic zone may be placed directly below the pyrolysis zone, in which case heavy oils are the reactants for the catalytic transformation. The catalytic zone may also be placed above the pyrolysis zone, such that volatile compounds swept from the pyrolysis zone using an inert carrier gas are converted *via* catalysis. Both of these configurations do not result in the true catalytic degradation of *polymers*, but rather the degradation of the thermal decomposition products of the polymer. However, they have been widely labelled as such and have been extensively used.

The configuration of the combined pyrolysis/catalysis reactor can pose serious engineering problems as pointed out by Songip *et al.*[38] Because the catalytic zone relies on the products of thermal degradation as its feedstock, it is difficult to

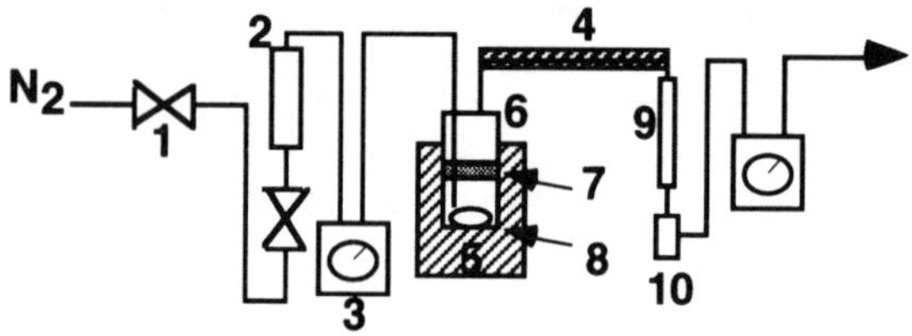

(a) Thermal degradation followed by catalytic degradation of volatile products: 1, regulator; 2, flow meter; 3, gas meter; 4, flexible heater; 5, furnace; 6, reactor; 7, catalyst; 8, polymer sample; 9, condenser; 10, trap

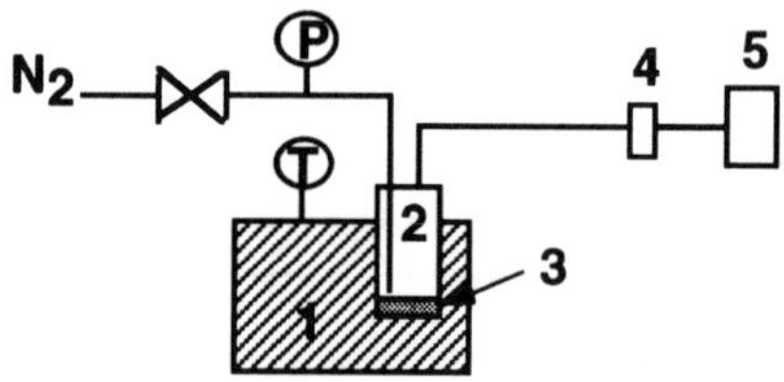

(b) Co-milling of polymer and catalyst in semibatch mode: 1, furnace; 2, reactor; 3, co-milled polymer and catalyst; 4, trap for liquid fraction; 5, trap for gas fraction

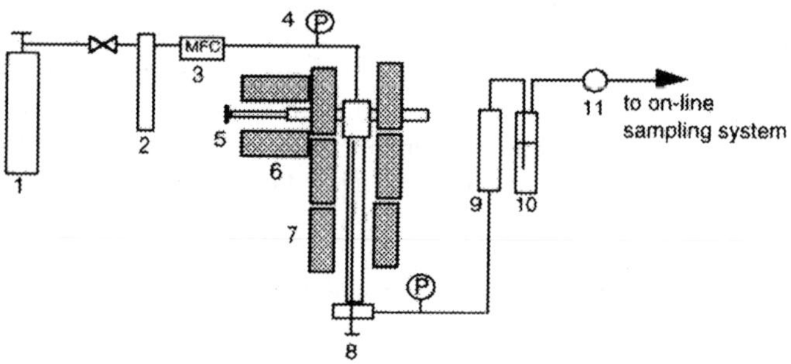

(c) Fixed bed flow reactor system: 1, He cylinder; 2, trap; 3, mass flow controller; 4, pressure transducer; 5, syringe pump; 6, preheat/melt zone; 7, three zone furnace; 8, thermocouple; 9, trap; 10, trap; 11, back pressure regulator

Figure 1 *Different catalyst–polymer contacting patterns and reactor configurations, each of which has been used to study the catalytic degradation of polymers*

measure the exact mass flow rate of the reactant contacting the catalyst. Furthermore, it is an unsteady state process, even though a fixed bed of catalyst is typically used, since the identity and quantity of the thermal degradation products, and therefore the reactant for the catalytic zone, are a changing function of time. Developing a kinetic model for this type of reactor configuration that could ultimately be used in process design and optimization would be a challenging task. However, this configuration does prevent a viscous polymer melt from contacting the catalyst which would lead to severe diffusion limitations and potentially facilitate catalyst deactivation.[39]

A typical approach for carrying out reactions in which the vapors from thermal degradation of polymers are subsequently converted catalytically is that of Ohkita *et al.*[12] In their experiments, 15 g of polymer powder were packed at the bottom of a stainless steel reactor. At the middle of the reactor, 1.5 g of the appropriate catalyst were placed between layers of stainless steel gauze. The volatile products from thermal degradation were swept over the catalyst by flowing nitrogen at a flow rate of 50 mL min^{-1}. The products were then collected in a series of traps, and an extraction protocol was used to quantify the gases, liquids and higher molecular weight compounds evolved. The products were not sampled as a function of reaction time but rather represented the composite of the products evolved over the particular reaction time chosen for the thermal degradation. As a result, although the operation is continuous with respect to catalytic conversion of the volatiles, the analysis protocol is practiced batchwise.

In some cases, the two stages of pyrolysing the polymers and catalysing the products of thermal degradation have been decoupled and carried out in two separate steps. This has the advantage over an integrated system that the feedstock to the catalytic reaction can be well characterized. For example, Songip *et al.* carried out a series of studies in which they examined the catalytic cracking over Y-type zeolites of heavy oil derived from waste plastics.[33,34] The feed oil was prepared by pyrolysing polyethylene at 723 K and distilling at 473 K to remove lighter hydrocarbons to obtain a residual oil having a heavy oil content of 95%. The carbon/hydrogen ratio was equal to 2, suggesting that the fuel oil was composed of linear paraffins. However, experiments in our own laboratory show that the predominant products of decomposition of polyethylene are linear α-olefins, with paraffins present in lower yields by a factor of 10.[14] The evolution of α-olefins as a function of carbon number and parametric in reaction time is shown in Figure 2 for polyethylene ($M_w = 125\,000$) reacted at 693 K.

3.2 Co-milling of Polymer and Catalyst

3.2.1 Semibatch Mode – Intimate mixing of the polymer and the catalyst followed by heating with release of the volatiles has been used extensively to study the decomposition of polymers in the presence of catalysts.[22,29,40–58] The catalyst and the polymer of interest are usually blended in a ball mill or similar device, often with heating, to ensure that intimate contact between the two materials is achieved. Catalyst to polymer ratios are high, typically ranging from 0.1 to 1.0 by weight. The catalyst–polymer mixture is loaded into the reactor, the temperature is raised to the degradation temperature desired, and the products exit the reactor, *i.e.* the reactor is operated in semibatch mode. The products are usually classified into three groups: liquid products which are condensable at room temperature, gaseous products, and residues which include carbonaceous species and waxy compounds.[55] Ishihara *et al.* have used this mode of operation

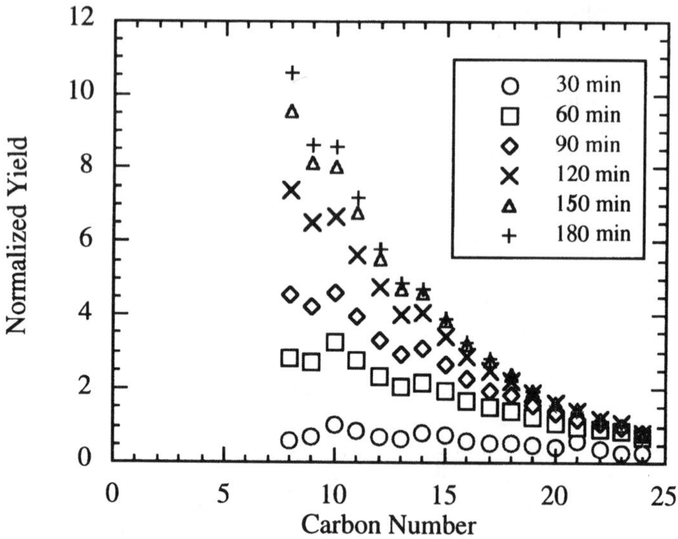

Figure 2 *Comparison of normalized yields of α-olefins as a function of time and carbon number for pyrolysis of HDPE at* 693 K

extensively.[47–52] Specifically, they immerse a Pyrex tube reactor with the catalyst–polymer mixture in the bottom into a molten metal salt bath, sweep away the products with nitrogen flow, and collect the products in a series of traps.

Thermal gravimetry (TG) is a second example of semibatch mode of operation that has been used extensively and allows for the simultaneous monitoring of the weight loss of the reactant.[41,43–45,54,59,60] The products are swept away by flowing nitrogen and can be quantified and identified by gas chromatography and mass spectrometry. By varying the heating rate and using standard techniques for analysis of TG results, information about global kinetics can be obtained, including activation energies. The results are typically reported in terms of the effect of the catalyst on T_i, T_m and T_f, the temperatures corresponding to initial weight loss, maximum rate of weight loss and final constant weight, respectively. The presence of the catalyst most often leads to a lowering of these values.

3.2.2 Batch Mode – Batch reactions have also been carried out to examine the decomposition of polymers in the presence of catalysts.[61–73] The polymer and the catalyst are both placed in the reactor, and the reactor is immersed in a molten metal bath or a fluidized sand bath at a given temperature for a specified period of time. In some cases, a shaking apparatus is used to promote contact of the catalyst, the polymeric reactant and the decomposition products.[66] After the reaction time is complete, the reactor is opened, and the reaction products are typically analysed by chromatographic techniques. For example, in order to probe the full range of decomposition products, we developed an analytical protocol which allowed the molecular weight distribution as a function of time to be monitored by gel permeation chromatography (GPC). The effluent from the GPC analysis is passed to a fraction collector, where low molecular weight products are isolated and subsequently analysed by gas chromatography and mass spectrometry.[74]

In contrast to operation in semibatch mode, the reaction products evolved during a batch reaction can interact with the catalyst and the reactant. This may have a beneficial effect on the ultimate product distribution, as the higher molecular weight products may themselves be broken down further. However, the complex interactions resulting from significant secondary reactions may make reaction pathways and kinetics more difficult to unravel.

3.3 Flow Systems – To probe the feasibility of carrying out catalytic decomposition of polymers as a continuous process, some investigators have evaluated reaction of polymeric melts over a fixed bed of catalyst.[13,75–85] For example, Uemichi *et al.* used a flow reactor system similar to the one depicted in Figure 1(c).[84–87] The polymer was melted at 583 K under a helium atmosphere. After melting, the pressure of helium in the melter was set to 0.14 MPa, by which the polyethylene melt was pressed through a capillary into the reactor packed with catalyst. Reaction was carried out at space times of 7–23 g catalyst min/g polymer. Because of the deactivation of the catalyst observed by Uemichi and co-workers, the reaction was only carried out for 15 min, and the kinetics of deactivation and the feasibility of catalyst regeneration were probed. To analyse the products, they were trapped at liquid nitrogen temperatures to collect both gases and liquids, and then the gaseous and liquid products were analysed by gas chromatography. In addition, wax was observed adhering to the wall at the outlet of the reactor.

To combat the problems associated with deactivation of the catalyst in fixed bed continuous systems, direct conversion of plastic wastes in fluidized catalytic cracking units has been carried out.[18,39] However, the polymers are not processed neat, but rather are dissolved in traditional refinery feedstocks. Although this limits control of the product distribution derived from the polymer itself, this strategy does have the advantage that existing processing units may be used, and the catalyst may be regenerated as is typically carried out in fluidized catalytic cracking. By analogy with the petroleum industry, developing a 'refinery' for polymeric waste has tremendous promise as a viable processing strategy. However, the other reactor configurations discussed above are valuable means of unravelling the complex kinetics and reaction pathways underlying catalytic degradation of polymers.

4 Acid-catalysed Degradation

For the addition polymers of interest in this review, the predominant focus in the literature has been on using acid catalysts to degrade them. With their known ability to cleave carbon–carbon bonds and their widespread use in the hydrocarbon processing industry, using these materials to catalyse the degradation of polyethylene, polypropylene and polystyrene was a natural extension. The two classes of acidic materials that have received the most attention are amorphous silica–aluminas and zeolites. Additionally, early work was carried out using metal chlorides as typical Lewis acids, and recently, mesoporous materials such as MCM-41 and materials such as sulfated zirconias have been investigated. The effect of these acid catalysts on the degradation of polyethylene, polypropylene and polystyrene and the product distributions are discussed. Mechanistic interpretation put forth in the literature to account for the product distributions observed is then summarized.

4.1 Product Distributions

4.1.1 Silica–Alumina

4.1.1.1 Polyethylene – The degradation of polyethylene has been most widely studied in the presence of amorphous silica–alumina (SA) catalysts.[12,35,40,45,49,50,53,55,61,68,70,76–79,85,88] With the larger pores available in these materials, the polymer and its decomposition products have greater access to the internal structure of the catalyst than in the case of narrow pore zeolites. In all studies, the presence of silica–alumina enhances the rate of conversion of polyethylene as compared to thermal reaction alone. For example, Lin and White showed that the initial weight loss temperature as measured by thermal gravimetric analysis decreased by 100 K when silica–alumina was co-milled with high density polyethylene.[53] Furthermore, the product distribution is altered; in general, the distribution of products shifts to lower carbon numbers. To probe the impact of SA catalysts on polyethylene degradation in more detail, various silica to alumina ratios have been used to monitor the effect of catalyst acidity, and the amount of catalyst has been varied. Studies have also measured the deactivation behavior of these catalysts. As pointed out above, the reactor configuration used also has a strong influence on the catalytic behavior observed.

Using the reactor configuration in which only the thermal degradation products passed over the catalyst, Ohkita *et al.* probed the relationship between the acid strength and catalyst amount and the compositions of products formed at 673 K.[12] The products were distinguished as gases, oils or waxes. As the alumina content of the SA catalyst was varied from 0 to 100%, the amount of gases passed through a maximum, and the amounts of waxes and oils achieved minima. As revealed in Figure 3, the dependence of the product yields as a function of alumina content corresponded to the amount of Brønsted acid sites. Furthermore, the amount of aromatics in the oils tracked the quantity of Lewis acid sites as measured by titration with *n*-butylamine and infrared spectroscopy of pyridine adsorption.

These results are in contrast to the results of Beltrame *et al.* who carried out the degradation of co-milled polyethylene and two different silica–alumina catalysts with alumina contents of 24.2 and 13.2 wt%.[40] At the same temperature used by Ohkita *et al.*, *i.e.* 673 K, the degradation resulted in the evolution of over 90 wt% oils, less than 8 wt% gas and negligible amounts of residue.[12] However, it is important to note the differences in the reactor configurations used in these two

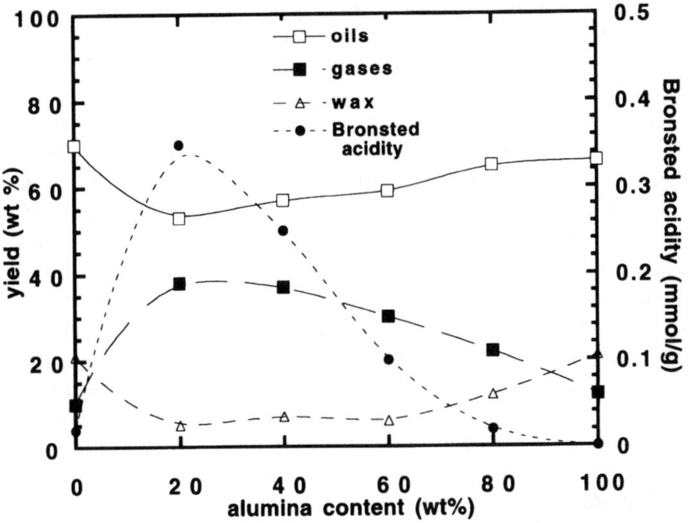

Figure 3 *Changes in the fractions of gases, oils and wax produced from degradation of polyethylene thermal decomposition products over a series of silica–alumina catalysts correlate with the Brønsted acidity*[12]

studies. Beltrame and co-workers used a semibatch reactor and swept away the volatiles as they were formed. Thus, the volatile products in the oil range would simply be evolved and collected, whereas in the work of Ohkita *et al.*, the volatiles would contact the catalyst and react to increase the gas yield.

The work of Beltrame *et al.* can be compared to that of Sakata *et al.* to probe the effect of temperature on the product distribution since the reactor configurations used by the two groups were similar.[55] Using a silica–alumina catalyst with 16.7 wt% alumina, 67.8 wt% liquid products and 23.7 wt% gaseous products were observed at 703 K, revealing that the higher temperature leads to increased gas formation. Sakata *et al.* provided additional insight into the reaction of polyethylene over SA catalysts by reporting the carbon number distribution in both the liquid and gaseous products. The gas was primarily composed of C_3 compounds, but the distribution was shifted in favor of C_4 and C_5 species as compared to the products from thermal degradation. The liquid products showed a peak at C_8 and were shifted significantly to lower carbon numbers compared to thermal degradation.

Ishihara *et al.* also carried out decomposition of polyethylene at 703 K over a silica–alumina catalyst with 13 wt% alumina but used a fixed bed reactor and a low density reactant.[76] In their experiments, 72 wt% of the products were gaseous, while only 15 wt% of the products were liquids. Only 7 wt% gaseous products were formed without the catalyst present. However, the product yields were a strong function of residence time in the system. As the residence time was varied from 2.2 to 13.2 s, the gas yield passed through a maximum at 4.7 s and reached a low of 47 wt% at the highest residence time examined. The decrease in the gas yield was attributed to carbon deposits on the catalyst surface, as significant deactivation of the catalyst was observed in this reactor geometry.

To address the deactivation behavior in more detail, Uemichi *et al.* recently examined the change in activity of a silica–alumina catalyst with 13 wt% alumina as a function of time on stream.[85] At a reaction temperature of 723 K, the SA catalyst accumulated over 12 wt% coke on the catalyst after 250 min time on stream. The liquid yield increased slightly from 60 wt% to approximately 70 wt% as the coke built up on the catalyst. The limited effect of the coke on the reaction was attributed to the inability of coke deposits to block completely the large pores ($d_{p,ave} = 4.4$ nm) of the amorphous catalyst. Although SA showed no activity toward cracking of *n*-octane, the reactivity of polyethylene was substantially enhanced in the presence of the catalyst. This was attributed to the facile reaction on the catalyst of olefins which could be formed from thermal degradation of polyethylene at the temperatures used in this study.

4.1.1.2 Polypropylene – The studies of degradation of polypropylene in the presence of silica–alumina are more limited than those examining polyethylene.[15,47,52,80] Because of the tertiary centers along the polymer backbone, polypropylene degrades more readily under milder conditions. However, the same overall benefit in rate enhancement with the addition of silica–alumina as compared to thermal degradation has been observed. Key studies in a variety of reactor configurations are summarized below.

Ishihara *et al.* used a semibatch reactor to examine the degradation of polypropylene in the presence of silica–alumina with 13 wt% alumina.[50,51] As the catalyst to polymer ratio increased from 0 to 6 by weight, the conversion of the polymer increased from 22 wt% to more than 90 wt% at a reaction temperature of 493 K and a reaction time of 60 min. The oligomer fraction passed through a maximum, while the gas and liquid yields increased monotonically. This suggested that as the catalyst amount was increased substantially, the oligomers formed from the polymer also contacted the catalyst and underwent further degradation, contributing to the liquid and gas yields.

More detailed product information was provided by Uemichi *et al.* who used a fixed bed flow reactor to examine the decomposition of polypropylene in the presence of silica–alumina.[80] Detailed product analysis was carried out for compounds with 16 or fewer carbon atoms, resolving the products into normal alkanes, isoalkanes, alkenes and aromatics. For a reaction temperature of 750 K, a catalyst weight of 3.0 g and a polypropylene feed rate of 0.150 g min^{-1}, the polymer was converted over 95% to low molecular weight products, forming C_4 and C_5 isoalkanes and alkenes in the highest yields. The overall carbon number distribution for reaction over SA is summarized in Figure 4, in which thermal degradation results for even more severe conditions of 799 K and a feed rate of 0.116 g min^{-1} are included for comparison. The increased rate of conversion and the shift of the product distribution to lower carbon numbers are apparent from this plot.

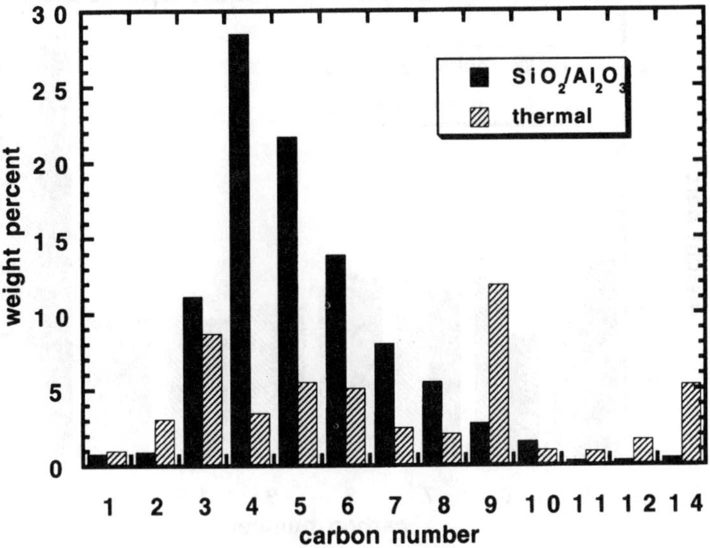

Figure 4 *Carbon number distribution of degradation products of polypropylene reacted thermally and in the presence of a silica–alumina catalyst*[80]

Direct comparison of semibatch reaction of polyethylene and polypropylene can be made using the results of Beltrame *et al.*[40] summarized above and their study with the same catalyst and reactor configuration for polypropylene degradation.[15] At 673 K, polypropylene was completely converted to gas and distillate, affording approximately 5.5 wt% gas, a slightly lower value than that observed for polyethylene. No further improvement or significant change was observed when the temperature was increased to 873 K, which was similar to the result obtained for polyethylene. A direct comparison of the carbon number distribution of the distillate products for polyethylene and polypropylene reacted at 673 K is presented in Figure 5.

4.1.1.3 Polystyrene – The degradation of polystyrene over silica–alumina catalysts has also been extensively investigated using a variety of catalyst–polymer contacting patterns.[22,29,31,37,53,58,59,72,73,84,87,89] In general, if the polymer itself contacts the acid catalyst, then styrene, the principal product from thermal degradation, is observed in low yields. Benzene, formed *via* removal of the phenyl substituents from the polymer backbone, is observed in higher yields, and a significant amount of indan derivatives is also detected. If only the volatile thermal degradation products contact the catalyst, then styrene is observed in the product spectrum, but in lower yields than obtained for thermal degradation.

Semibatch reaction of mixed polystyrene and a silica–alumina catalyst (11.8 wt% alumina) was carried out by Lin and White at 673 K.[59] In contrast to

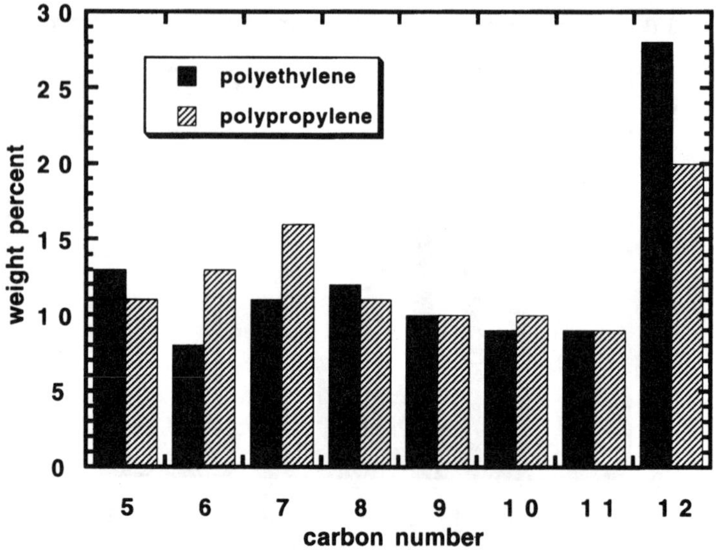

Figure 5 *Comparison of the distribution of products from decomposition of polyethylene and polypropylene at* 673 K *co-milled with silica–alumina catalyst* (13.2 wt% *alumina*)[15,40]

thermal degradation, in which 68 wt% styrene was formed, reaction in the presence of SA only afforded 4 wt% styrene. Benzene was the most abundant product at 30 wt%, while indan and methylindan accounted for 15 and 17 wt%, respectively. Ethylbenzene was also abundant, present at a yield of 14 wt%.

A fixed bed flow reactor study carried out by Ogawa *et al.*[87] was consistent with the results of Lin and White. Using a 13 wt% alumina SA catalyst, the space time was varied, and its effect on the product distribution determined. As the contact time increased, the yield of styrene decreased and the yields of ethylbenzene and methylphenylindans increased. The weight of total liquid products decreased as contact time increased, but the yield of gases remained essentially constant at a value close to zero. This suggests that carbon residue remained on the catalyst, and its value increased with increasing space time. This is similar to the behavior observed for polyethylene conversion over SA catalysts.

The impact of the reactor configuration on the polystyrene product distribution can be clearly seen by comparing these studies with the results of Zhang *et al.*[37] Zhang and co-workers allowed only the thermal degradation products of polystyrene, universally observed to be rich in styrene, to contact the SA catalyst. The amount of oils formed when SA was used to degrade the volatile products further was unchanged compared to thermal degradation alone, but the yield of styrene was reduced from 70 wt% to 60 wt%. The yield of ethylbenzene increased to 9 wt% from 2.4 wt% when the volatile products were exposed to SA.

4.1.2 Zeolites – In contrast to silica–alumina catalysts, zeolites are crystalline materials with regular pore structures. The pores are typically smaller than those comprising the distribution found in amorphous silica–alumina catalysts and are on the order of 5–10 Å. With the narrow, confined architecture, access of a polymeric material to the internal catalytic sites is limited. However, when zeolites and polymers are contacted directly, the product distributions are altered substantially. This suggests that the polymer does indeed interact with catalytic sites, although perhaps only with those on the exterior of the catalyst, or that the products from thermal degradation taking place at the reaction temperatures used penetrate the polymer melt on the surface to gain access to the internal surface area. Although the true nature of the interactions between the polymer and zeolites is not known, this lack of understanding has not prevented a wide range of catalyst/polymer systems from being investigated.

4.1.2.1 Polyethylene – A large number of studies has been carried out reacting polyethylene in the presence of HZSM-5.[35,36,43–45,53,55,62,64,65,77,85,88] HZSM-5 is a crystalline aluminosilicate with the MFI structure with pores on the order of 5.5 Å. HZSM-5 is widely used in hydrocarbon conversion processes, particularly in those which can exploit the narrow pore architecture such as toluene disproportionation.[90]

Ding *et al.* degraded high density polyethylene in the presence of HZSM-5 (Si/Al = 35) at a polymer to catalyst weight ratio of 50 at 708 K in a 150 cm^3 autoclave.[62] The presence of the catalyst decreased the oil yield slightly as compared to thermal degradation and increased the gas yield from 7.3 to 16.5 wt%. In examining the oil yield in more detail, the addition of HZSM-5 clearly affected the type of products, increasing the weight percentage of aromatics from 4.38 to 23.8. The quantity of *n*-paraffins decreased significantly, and the yield of naphthenes increased. These general trends were also observed by Mordi *et al.* who reacted polyethylene in the presence of an HZSM-5 sample (Si/Al = 21) at 623 K for 2 hours in a batch reactor.[65] The yield of gaseous products increased from 34 to 54 wt% with the addition of the catalyst, and the yield of aromatics also increased.

Fernandes, Jr. *et al.* evaluated HZSM-5 as a catalyst for reaction of polyethylene by monitoring the degradation kinetics in semibatch mode by thermal gravimetry.[44] The activation energy, evaluated using an integral kinetics method, decreased 60.6% in the presence of the zeolite. Furthermore, the product distribution shifted substantially to lower carbon numbers. No volatile products with a carbon number greater than 16 were observed, suggesting that the products in the range C_{17}–C_{26}, which were observed in thermal degradation and are therefore volatile at the conditions employed, were completely converted in the presence of the catalyst.

The decomposition of polyethylene in the presence of zeolite Y has also been extensively studied using a variety of catalyst–polymer contacting patterns.[13,40,45,48,54,85] Zeolite Y is a component of commercial fluidized catalytic cracking catalysts that is very stable under hydrothermal conditions and has relatively large pores (7.8 Å pores and 11.8 Å supercages). Since the pores are larger than those in ZSM-5, it is suggested that bulky polymer chains will have greater access to the internal structure of zeolite Y.

Using a tubular fixed bed flow system, Uemichi *et al.* studied the reaction of low density polyethylene over an HY catalyst derived from Linde SK-40.[85] The authors observed that HY zeolite showed high activity for degradation of polyethylene, but a large amount of undesirable coke was deposited on the catalyst surface, achieving a value greater than 0.20 g of coke per g of catalyst after 50 min time on stream at 723 K. The products at a space time of 13 g of catalyst min per g of polyethylene over HY are compared to HZSM-5 at the same conditions in Table 6. While HZSM-5 favored the formation of aromatics and gaseous products, HY resulted in a high yield of liquid products rich in isoparaffins.

Table 6 *Comparison of the decomposition products of polyethylene at* 375 °C *and* W/F = 13 *over HZSM-5 and HY catalysts in fixed bed tubular flow reactor*[84]

	HZSM-5	*HY*
	Yield (wt%)	
Gas	61.9	26.3
Liquid	37.2	54.8
Wax	0	9.2
Coke	0.9	9.7
	Liquid composition (wt%)	
Isoparaffins	20.5	56.9
Olefins	0.7	11.3
Aromatics	73.6	13.2
GC-RON	112	97

In a study by Garforth *et al.*, a high coke content was also observed for an HY catalyst co-milled with polyethylene and reacted in semibatch mode.[45] After degradation at 873 K, the catalyst contained over 10 wt% coke. The activation energies for decomposition were very similar over HZSM-5 and HY, which were both lower than thermal degradation by 100 kJ mol^{-1}, while USY decreased the activation energy by an additional 10 kJ mol^{-1}.

Zeolite X was also examined for the decomposition of polyethylene in a series of articles by Ayame and co-workers.[70,75,78] Using both a batch reactor and a fixed bed tubular flow reactor, the rate of conversion of polyethylene was enhanced in the presence of CaX and NaX. Deactivation was observed in the flow studies, as the rate of formation of gaseous products decreased by a factor of three after 2.5 hours time on stream when the reaction was carried out at 750 K with 4.0 g of catalyst and a polyethylene flow rate of 7.23×10^{-2} g min^{-1}. When a CaX catalyst was used, C_4 species were observed in the highest yield. In the batch reaction, the yield of iso-C_4 species was increased dramatically compared with thermal degradation, as thermal degradation afforded no iso-C_4.

4.1.2.2 Polypropylene – The number of studies examining polypropylene degradation in the presence of zeolite materials is more limited[15,57,63,80,91] than that examining polyethylene decomposition. Using a batch apparatus, Mordi *et al.* investigated the reaction of polypropylene with HZSM-5, H-mordenite and H-theta-1.[63] H-mordenite is a one-dimensional zeolite with 6.1 × 7.0 Å channels, and H-theta-1 has a TON structure with smaller one-dimensional channels. The experiments were carried out by placing 0.2 g of catalyst with 1 g of polypropylene pellets in a 300 mL glass tube and reacting for 2 h at 623 K. No solid polymer remained at the end of the reaction when any of the catalysts were used. HZSM-5 formed the highest quantity of gaseous products in the range C_1–C_7, while the highest quantity of liquid products in the range C_5–C_{11} was formed over H-mordenite. Detailed resolution of the product fractions was not provided.

Co-milling polypropylene and zeolite Y was carried out by Audisio and co-workers.[15] The product distributions as a function of carbon number at 673 K are reproduced in Figure 6 for NaY, HY and rare earth Y (REY, 10.7% of rare earth oxides) and compared with those observed for thermal degradation and in the presence of silica–alumina (also reproduced above in Figure 4). The product distributions for HY and REY were very similar, while a significantly lower fraction of C_5–C_{11} products was formed when NaY was used. A detailed analysis of the products over HY and REY revealed that a significant amount of isomerization had occurred and olefins were formed in large yields.

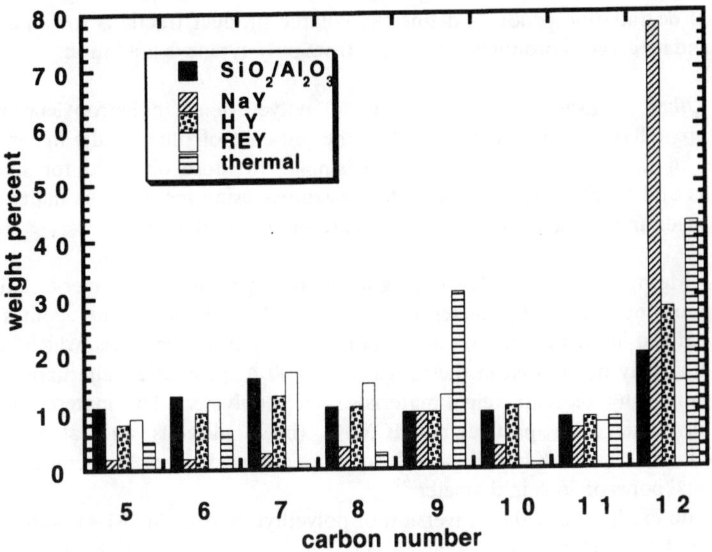

Figure 6 *Comparison of the distribution of products from decomposition of polypropylene at* 673 K *with the addition of zeolite Y catalysts*[15]

4.1.2.3 Polystyrene – The acid-catalysed reaction of polystyrene over zeolites has also been examined over a range of materials.[29,31,37,58–60,72,84] In general, the same shifts in the product distribution as compared to thermal degradation that were observed when silica–alumina catalysts were used were seen when zeolites were employed. The yield of styrene monomer decreased, and the amounts of benzene and indan derivatives increased. The extent to which the catalyst affected the degradation was dependent on the catalyst–polymer contacting pattern. The product distributions observed over particular zeolites are summarized as examples below.

Carniti *et al.* examined the product distributions evolved when polystyrene was reacted in batch mode at temperatures of 623 and 673 K up to 120 min in the presence of HY and REY zeolites. The catalyst concentration was 10 wt% for all runs.[72] The product selectivities for these two catalysts were similar and markedly different from thermal degradation. The yields of benzene and ethylbenzene increased, and the yield of styrene decreased to zero with the addition of the catalyst. Selected product yields from the more comprehensive table provided by Carniti *et al.* are summarized in Table 5. Values for a silica–alumina catalyst with a SiO_2/Al_2O_3 ratio of 6.5 are also included for comparison.

In addition to investigating the reaction of polystyrene thermal degradation products over a silica–alumina (SA) catalyst, Zhang *et al.* monitored the products when an HZSM-5 catalyst was used.[37] The results were similar to those for the SA catalyst: the styrene content in the oil fraction decreased as compared to thermal degradation, while the yields of benzene, ethylbenzene and indan increased. The total amount of oils did not change significantly; this is not surprising since the thermal degradation products define the volatile product fractions for all catalysts, and no gaseous products are formed from polystyrene degradation.

4.1.3 Other Materials – Degradation of polyethylene, polypropylene and polystyrene has also been carried out in the presence of other acidic materials. Trends in the product distributions are similar to those observed for silica–alumina and zeolite catalysts. Degradation studies using mesoporous materials, Lewis acids and sulfated zirconia catalysts are summarized below.

4.1.3.1 Mesoporous materials – Owing to the bulky nature of polymeric chains, conversion over catalytic materials such as silica–alumina and zeolites is governed and limited by the size of the pores. Mesoporous materials, which have pores that may be tailored in the range of 15–100 Å, provide an alternative for processing high molecular weight materials such as polymers. Two materials have been investigated in particular: MCM-41 (both siliceous and aluminum-containing)[45,92] and KFS-16, a folded silica derived from Kanemite with hexagonal pores of 36 Å in diameter.[55]

Aguado *et al.* studied the conversion of polyethylene over MCM-41 with a Si/Al ratio of 29.4 in a semibatch reactor at 673 K.[92] At a polymer to catalyst mass ratio of 4, the polymer was almost completely converted to low molecular weight products with carbon numbers less than 40. The products were centered in the C_5–C_{12} range (59 wt%), but 20 wt% C_1–C_4 products and 8 wt% aromatic products were formed as well. Even with its small pore size, HZSM-5 at the same mass loading was observed to be more active for polyethylene conversion. Garforth *et al.* observed that MCM-41 with a Si/Al ratio of 17.5 resulted in a low activation barrier for polyethylene degradation; in fact, it was lower than for all the other catalysts studied which included HY, USY, HZSM-5 and silica–alumina.[45]

In contrast to the work of Garforth *et al.*,[45] who studied a siliceous form of MCM-41 and observed no enhancement in the rate of conversion of polyethylene, Sakata *et al.* observed degradation rates for polyethylene over KFS-16 that were as fast as those observed over a silica–alumina catalyst. However, the product distribution was similar to that observed for thermal degradation.[55] Since KFS-16 possesses no acid sites, the authors rationalized these observations in terms of a 'radical flask' concept, in which the radicals formed *via* thermal degradation are stabilized by the hexagonal pores of the KFS-16 catalyst, thereby increasing the total radical population and enhancing the overall propagation rate.

4.1.3.2 Lewis acids – Early work on the degradation of polymers in the presence of catalytic materials focused on reactions of polyolefins with strong Lewis acids.[11,71,93–99] It was well established that aluminum chloride and other strong acids attack saturated compounds, causing isomerization, cracking, addition to olefins and other reactions, and therefore numerous investigators sought to extend this chemistry to polymeric materials.

Decomposition of polyethylene with $AlCl_3$ was studied by Ivanova *et al.* at 643 K, affording a higher yield of gaseous products than uncatalysed degradation at 673 K.[97] The selectivity to ethylene monomer was low, and significant isomerization reactions were observed. The product distribution was primarily composed of C_4 species that accounted for over 57 wt% of the gaseous products. By monitoring the intrinsic viscosity, Carrick was able to quantify the degradation of polyethylene as a function of time with the addition of $AlCl_3$, observing a rapid initial decline in viscosity followed by a slower decrease.[96]

The addition of $AlCl_3$ to polystyrene melts was also studied in detail. Pukánszky *et al.* monitored the torque as a function of time for polystyrene at 453 K and observed an immediate decrease in the torque upon addition of $AlCl_3$.[94] Nanbu *et al.* also examined the degradation of polystyrene in the presence of $AlCl_3$ at 323 K for 1–4 h.[71] The only volatile degradation product observed was benzene over the entire course of the reaction, and a linear decrease in the molecular weight with time was reported. A linear relationship was also observed between the decrease in molecular weight and the decrease in the content of the phenyl groups.

4.1.3.3 Sulfated zirconia – The degradation of both polyethylene and polystyrene has been studied in the presence of sulfated zirconia catalysts, commonly referred to as solid superacids, and platinum-stabilized ZrO_2/SO_4.[53,59,60,62,69] Lin and White investigated the cracking reactions of polyethylene in semibatch mode with a sulfated zirconia catalyst that contained 9 wt% sulfate and a surface area of 157 $m^2\ g^{-1}$.[53] The onset temperature for weight loss and the activation energy for degradation with the addition of sulfated zirconia were lower than when either silica–alumina or HZSM-5 catalysts were used. Xiao *et al.* examined polyethylene decomposition in the presence of a ZrO_2/SO_4 catalyst in a hydrogen atmosphere in a batch reactor.[69] The conversion was close to 100% for reaction temperatures above 703 K, and the C_5–C_{12} product fraction was over 60 wt% when the temperature was held at 723 K. The concentrations of gasoline range and C_{13+} fractions were insensitive to hydrogen pressures above 1500 psig.

The results obtained for polystyrene reacted over ZrO_2/SO_4 catalysts were similar to those observed over other acidic materials. Lin *et al.* observed a product distribution for ZrO_2/SO_4 at 673 K that was comparable to that for reaction over a silica–alumina catalyst: the yield of styrene decreased to zero, the yield of benzene increased to 33 wt%, and the amounts of indan and indan derivatives increased.[59,60] At temperatures above 673 K, SO_2 was observed, suggesting that the catalyst decomposed.

4.2 Mechanistic Interpretation – Although the nature of the interactions between the polymers and the catalysts is not exactly known, the products that are formed and the changes that occur when an acid catalyst is used as compared to thermal degradation are consistent with a carbenium ion mechanism commonly used to account for products observed for conversion of small hydrocarbons over acid catalysts. Recent work suggests that the actual species on the surface of the catalyst is an alkoxide,[100] but carbenium ion chemistry can still be used as a model of the transformations that occur.[101,102]

4.2.1 Polyethylene and Polypropylene – For reactions on catalysts with Brønsted acid sites, protolytic cracking of carbon–hydrogen and carbon–carbon bonds initiates the reaction of polymer chains or volatile products of thermal degradation to form a secondary or primary carbenium ion and an alkane. Primary carbenium ions are likely to undergo inter- or intramolecular rearrangements to more stable secondary carbenium ions (in the case of high density polyethylene) or to tertiary carbenium ions (in the case of polypropylene). Carbenium ions decompose *via* β-scission reactions, in which the bond β to the ionic center is cleaved, to form an olefin and a primary carbenium ion. This is an important reaction for decomposition of polymeric materials since the reaction results in a decrease in molecular weight. Olefins formed *via* β-scission may also be protonated and subsequent decomposition of these species may occur. Carbenium ions may also react *via* hydride transfer to form a paraffin and another carbenium ion. This set of likely reactions for the degradation of polyethylene and polypropylene over acid catalysts is summarized in Figure 7.

The reactions of polyethylene and polypropylene in the presence of Lewis acids are proposed to be similar to those for Brønsted acids, *i.e.* carbenium ions are the active intermediates. However, the formation of the protons which initiate the degradation is a result of the interaction of the Lewis acid with polar impurities, *e.g.* water, as illustrated in eqn. (1). It is proposed that the charged complex formed interacts with the polymer chains to form a carbenium ion and its corresponding counter-anion.

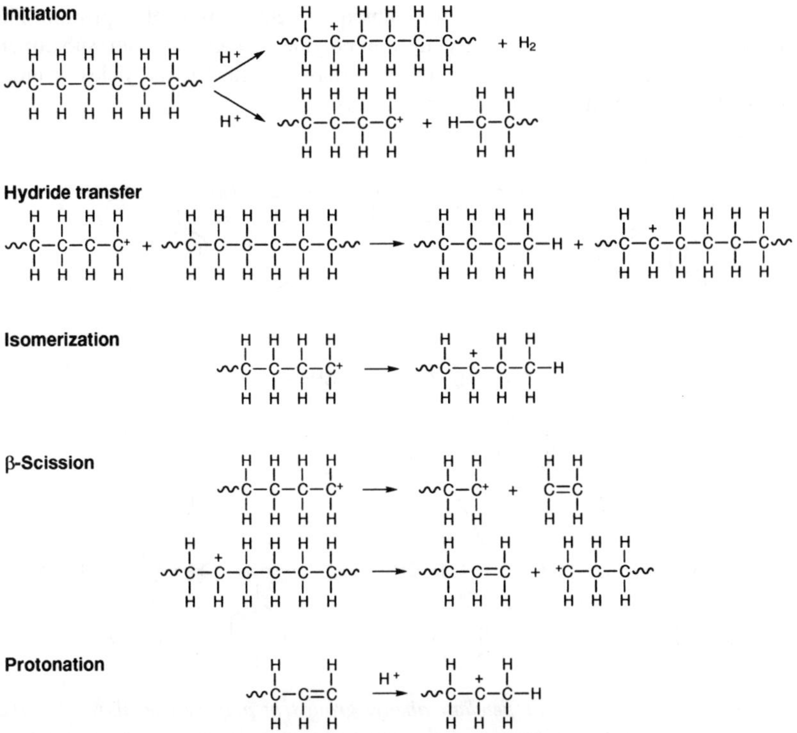

Figure 7 *Proposed mechanism for decomposition of polyethylene in the presence of acid catalysts* via *carbenium ion chemistry*

$$MX_n + HOH \rightarrow H^+[MX_nOH]^- \quad (1)$$

The carbenium ions formed may then undergo the reactions outlined above and in Figure 7.

4.2.2 Polystyrene – The mechanism for polystyrene degradation in the presence of acid catalysts is envisioned to proceed *via* the same types of steps put forth for polyethylene and polypropylene degradation. However, the high yields of benzene formed and the formation of indans observed in acid-catalysed degradation of polystyrene require additional steps to be postulated. As illustrated in Figure 8, a proton may add to the electron-rich pendant phenyl groups to give a carbenium ion intermediate. This intermediate may then undergo β-scission, cleaving the backbone carbon–phenyl carbon bond, affording a different carbenium ion intermediate and a molecule of benzene. The carbenium ion formed may then undergo β-scission, cleaving the carbon backbone of the polymeric chain and resulting in molecular weight decrease. A second option available to this intermediate is to add to an adjacent phenyl ring through a 1,5-addition reaction and subsequent elimination of a proton. This reaction forms an indan moiety in the polymer backbone, and subsequent reactions of the indan-containing structure may account for the yields of indan derivatives observed in the product spectrum.

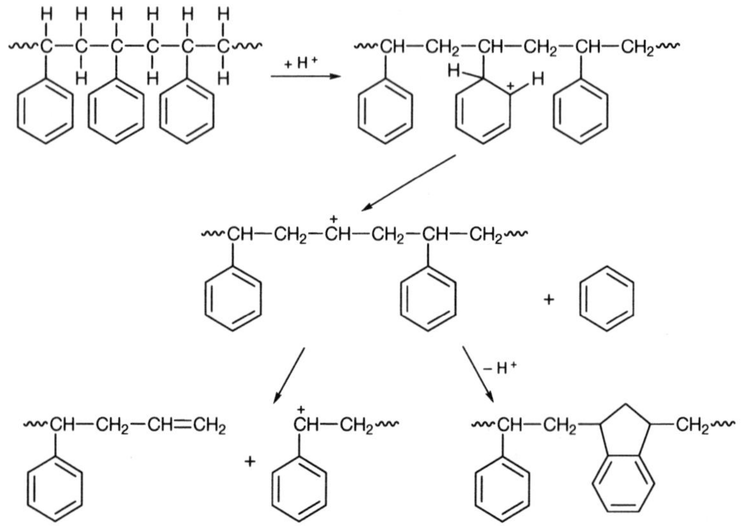

Figure 8 *Protonation of pendant phenyl groups in polystyrene allows for the formation of benzene and indan derivatives observed experimentally*

5 Base-catalysed Degradation

The number of studies carried out to examine the decomposition of polymers catalysed by basic materials is much smaller than the investigations carried out using acid catalysts. However, the motivation to use basic materials is clear. Inorganic compounds such as metal oxides are often dispersed in polyethylene as flame retardants for applications such as electrical cable.[12] It would therefore be attractive to use these materials as *in situ* directors of the degradation chemistry when the polymer is exposed to high enough temperatures to induce reaction. Furthermore, conversion of polystyrene over acidic catalysts has been shown to afford high yields of benzene and low yields of styrene monomer, a higher valued feedstock chemical. A resource recovery strategy capable of handling mixed plastic waste should be equally attractive for polyethylene and polypropylene as it is for polystyrene. Base catalysis provides one possible alternative.

5.1 Product Distributions

5.1.1 Polyethylene and Polypropylene – Ohkita *et al.* used a series of inorganic compounds to degrade polyethylene *via* base catalysis, specifically concentrating on magnesium oxide and zinc oxide.[12,35] Thermal degradation of 15 g of polyethylene at 673 K was carried out, and the volatile degradation products were allowed to contact 1.5 g of catalyst. The results for ZnO and MgO were compared to degradation *via* acid catalysis and thermal degradation alone. In comparison to thermal degradation, the amount of oils was increased, and the weight fraction of gaseous products was decreased. However, by comparing the product distributions as a function of carbon number in more detail, it was clear that the overall effect of the base catalyst was minor, particularly in contrast to acid catalysis, with the exception of the decreased amount of residue in the reactor. The authors attribute this effect to the ability of inorganic oxides to scavenge hydrocarbon radicals, preventing their recombination, but the catalyst does not promote the cracking of these radicals into light hydrocarbons. They concluded that the metal oxides examined are not effective catalysts for polyethylene decomposition.

Using the same reactor configuration, Simionescu *et al.* examined the modification of thermal decomposition products of a series of polymers, including polyethylene and polypropylene, using a series of metal oxide catalysts.[77] The effects of MnO_2/K_2O and Cr_2O_3/K_2O catalysts were similar: the amount of gaseous products was increased and the amount of condensable products was decreased with the addition of the catalyst. However, the effect was not nearly as dramatic as when acid catalysts, such as mordenite and ZSM-5, were used.

An early patent by the Ford Motor Company reported the use of basic salts to degrade automobile shredder residue, a mixture containing 12.7 wt% polypropylene.[103] With the addition of 2 g of sodium carbonate powder to 40 g of auto shredder residue, the total liquid yield increased by approximately 3%, and the overall olefin yield increased by approximately 20%. The benefit observed increased with the amount of the sodium salt added. The same effect was observed when calcium and potassium carbonates were used, but magnesium and zinc carbonate actually increased the amount of char formed.

5.1.2 Polystyrene – A limited number of studies has been carried out reacting polystyrene with basic materials.[31,37] The reactor configuration used in both studies was thermal degradation followed by catalytic degradation of the volatile products over BaO, K_2O, MgO and ZnO catalysts. The reactions were carried out at 623 K in flowing nitrogen. BaO was observed to alter the product distribution most significantly. The amount of oils was increased from 80.1 to 93.4 wt% as compared to thermal degradation, and the overall yield of styrene increased from 56.1 to 71.4 wt%. Since thermal degradation alone affords styrene monomer in high yield, the increase in styrene yield *via* further conversion of the volatile products is noteworthy. It is not possible to discern, given the reactor configuration used, whether the increased yield is due to a selectivity enhancement or simply a rate enhancement of converting higher molecular weight products (*i.e.* tetramer and lower) to afford additional styrene.

5.2 Mechanistic Interpretation – Very little work has been done to probe the mechanism of degradation of polymers on solid bases. However, the catalytic degradation is purported to occur *via* the formation of carbanions through elimination of a hydrogen atom from the polymer backbone.[31,37] The carbanion may then undergo further reaction, including decomposition *via* β-scission to afford monomer. These mechanistic ideas are illustrated in Figure 9.

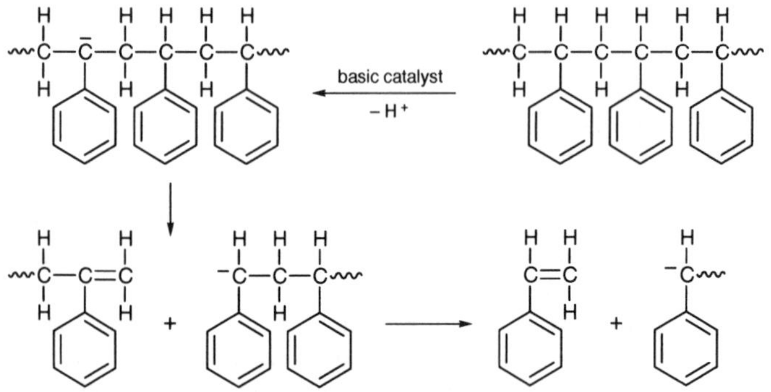

Figure 9 *Proposed mechanism for decomposition of polystyrene in the presence of base catalysts* via *carbanions*[31,37]

6 Co-processing with Other Feedstocks

6.1 Hydrocarbon Solvents – A limited number of studies have been conducted to study the conversion of polymeric waste dissolved in refinery feedstocks and reacted in traditional catalytic hydrocarbon processing units.[18,39,104] The approach has primarily involved direct catalytic cracking of the plastic wastes once they are dissolved in the feed of a fluidized catalytic cracking (FCC) unit. The proposed advantages of this approach are that less capital outlay is required since existing processing units are exploited and handling of the polymeric waste is limited to dissolution in the FCC unit feed.[39] Contaminants in the plastics can be processed directly, and transforming the plastics into a pumpable liquid, as might be required by alternative processes, is not necessary.[18] One potential drawback is the limited solubility of polymers such as polyethylene in the refinery feeds, reducing the amount of polymeric waste that can be processed to as little as 10 wt% of the total feed. However, it is proposed that even at these levels, the capacity would exist to process the amount of plastics waste available through current collection methods. For example, if plastic wastes were fed at 10 wt% in a standard FCC commercial unit (40 000 bbl per day), the maximum quantity of plastics recovered in Spain (220×10^3 Tm year^{-1}) would be treated, and processing of 25% of present plastics waste collection in Germany would be possible.[39]

The conversion of polyethylene blended with vacuum gas oil (VGO) to transporation fuels *via* catalytic cracking was examined by Ng.[18] Blends of VGO and 5 and 10 wt% high density polyethylene were thermally and catalytically cracked in a fixed bed reactor at 783 K and 20 h^{-1} WHSV. Thermal cracking led to low conversion of VGO, but the conversion of polyethylene was calculated to be over 50% by assuming that the VGO gave the same overall product distribution as that when polyethylene was not present. Although this ignores the contribution of kinetic coupling to the conversions and product distribution, it provides a basis to which the catalytic cracking could be compared. In contrast to thermal cracking, high conversions were obtained from catalytic cracking, and they increased with increasing catalyst to oil ratio. The gasoline yields were in the range of 42–54% for all conditions studied, increasing with the catalyst to oil ratio. However, the amount of polyethylene in the blend appeared to be a critical variable. If the amount was too low, the polyethylene was overcracked to dry gas and coke, but this conclusion depended on the method by which the contribution to the product spectrum for each feedstock component was calculated. If the amount of polyethylene in the feed were too high, difficulties in transporting the fluid would be encountered, but technical feasibility at 10 wt% was demonstrated.

A similar study using a variety of polymers was carried out by Arandes *et al.*[39] The thermal and catalytic cracking of polypropylene and polystyrene dissolved in refinery light cycle oil (LCO) was studied in a riser simulator under operating conditions similar to commercial FCC units. A commercial cracking catalyst (Octidyne 1160 BR from Engelhard) based on a REY zeolite was used. The runs were carried out at 1 atm at temperatures of 723–823 K for thermal cracking and 773–823 K for catalytic cracking, with a catalyst to feed ratio of 6 by weight and a contact time of 3 s. Conversion was defined based on the weight percentages of products; the contribution of the LCO to the products was assumed to be constant with or without polymeric components added. The feeds were made up of polypropylene at 5 wt% (higher concentrations resulted in highly viscous mixtures and difficulties in feed injection) and polystyrene at 10 and 25 wt%. At high temperatures, the dissolved plastics were reported to be completely converted *via* thermal cracking. Thermal cracking of polypropylene produced mainly olefins and isoparaffins. From polystyrene, styrene monomer was the major product formed. When catalytic cracking was carried out, the formation of gasoline increased with temperature, resulting in a gasoline with higher molecular weight and a higher content of aromatics. Although the contact time and the catalyst to oil ratio were not studied, the authors noted that these are important variables for incorporating this process into the operation of a refinery.

6.2 Coal – Studies examining co-processing of polymeric waste and coal have been more widespread.[105–120] Coal is a hydrogen-poor material, and hydrogen must be added in some manner to upgrade the coal to usable liquid fuels. One potential advantage of co-processing coal with polymeric materials is the higher hydrogen to carbon ratio in most plastics as compared to coal, suggesting that the polymers are a potential hydrogen source for coal upgrading. Furthermore, this strategy provides a way to consume waste plastics and turn them into high-valued products, while exploiting the existing technological achievements and developments in catalyst formulations over the past few decades for coal liquefaction.

The majority of studies examining the co-liquefaction of waste plastics and coal has focused on polyethylene.[105–108,111,117–120] Joo and Curtis examined the effect of reaction time on the conversion and product distribution of co-processing reactions in systems at 703 K containing low density polyethylene and coal and a pre-sulfided $NiMo/Al_2O_3$ catalyst.[108] The products were classified according to gas, hexane-soluble material, toluene-soluble and hexane-insoluble material, material soluble in tetrahydrofuran and insoluble in toluene, and tetrahydrofuran-insoluble material. For polyethylene alone, the conversion to THF soluble material at 60 min was quite low (39.5%); the majority of the products was gas and hexane soluble material. The conversion steadily increased with reaction time, achieving 52.4% after 240 min. With the addition of coal, similar conversion and product distribution behaviors were observed; however, the overall gas yield was lower than in the case of polyethylene alone and a shift from tetrahydrofuran-soluble to toluene-soluble material was observed. In examining the solid material, Joo and Curtis found that after 60 min of reaction, polyethylene composed 55.8% of the reaction solids, while after 180 min, the percentage of polyethylene decreased to 42.2. In their study, petroleum residue was also added to the reaction mixture, and the addition of residue was observed to be beneficial, promoting conversion and production of hexane-soluble products. As expected, the addition of polyethylene caused the products to have a higher hydrogen to carbon ratio than when coal was reacted alone.

Taghiei *et al.*[120] also examined polyethylene co-liquefaction with coal but used two different types of catalyst: highly dispersed iron-based catalysts and an HZSM-5 catalyst. Typical liquefaction conditions were used, carrying out the reaction at 693–723 K, 60 min and 800 psig (cold) of hydrogen. Using 50:50 mixtures of coal and plastics, the oil yields were as high as 60–80%, while the total conversions reached levels of over 90%. Oil yields for coal/plastic mixtures were higher, typically by about 10%, than the average of oil yields for coal and plastic alone, implying synergistic effects. The HZSM-5 catalyst was much more active for liquefaction of plastics at moderate temperatures, while the iron-based catalyst was more effective at higher temperature. When no solvent was added to the coal and plastic reactants, acceptable oil yields and total conversions were obtained at 723 K, indicating that the polymer plays the role of a hydrogen donor solvent for coal at these conditions. Rothenberger *et al.*[117] also saw synergistic effects, as all of the conversions to heptane-soluble material were increased compared to the assumption of a non-interacting system, but attributed them to artifacts of the analytical procedure, rather than to any beneficial effect of the co-conversion of the coal with plastics.

To probe the presence of synergistic effects in more detail, experiments with polyethylene and coal model compounds have been carried out in our own laboratory.[121] Tetradecane ($C_{14}H_{30}$) was used as a polyethylene mimic, and 4-(naphthylmethyl)bibenzyl (NBBM) was used as a coal model compound. Reaction temperatures were 693 and 773 K, and batch reaction times ranged from 5–150 min. Reactions of binary mixtures of tetradecane and NBBM revealed interactions between the reactants and synergistic effects. Tetradecane conversion was increased in the presence of NBBM, which was rationalized in terms of kinetic coupling. The internal carbon–carbon bonds of tetradecane have a higher bond dissociation energy (90 kcal mol^{-1}) than that of the bibenzyl bond in NBBM (60 kcal mol^{-1}). This has the potential to increase the quantity of radicals in the system with respect to the neat tetradecane experiments at a particular reaction time. The NBBM-derived radicals can abstract hydrogen from the secondary carbons of tetradecane, forming a tetradecane-derived radical and thus allowing more facile conversion of the tetradecane molecule. Once formed, these tetradecane-derived radicals undergo the same decomposition reactions as observed for neat pyrolysis, and similar product yields and selectivities were observed. The interactions between NBBM and tetradecane were further interpreted by examining the products derived from NBBM. Since abstraction of hydrogen from tetradecane is facile and has a high reaction path degeneracy, NBBM radicals are capped and stabilized through this abstraction step before undergoing secondary reactions. This hypothesis was consistent with the observed increase in primary products from NBBM and the decrease in the formation of other products that would result from the addition of primary radicals to NBBM.

The reaction of polystyrene with coal was investigated in detail by Luo and Curtis.[111] Co-processing was investigated at reaction conditions typical of direct liquefaction with polystyrene alone and with coal in the presence and absence of hydrogen donor solvent and catalysts (using slurry phase Mo and Fe hydrogenation catalysts) at 673 K and an initial hydrogen pressure of 5.6 MPa. Polystyrene was readily liquefied at these conditions and produced a reasonable solvent medium for coal during co-processing. The results in the absence of catalyst and donor solvent for a 50:50 weight mixture of coal and polystyrene are summarized in Figure 10. When catalysts were added, both materials were effective in affording THF-soluble material, but Mo-containing catalysts were more effective. The presence of a hydrogen donor solvent promoted coal conversion but did not influence the polymer conversion substantially.

7 Summary

Catalytic resource recovery is a promising technology for the conversion of polymeric waste materials to valuable products that has received considerable attention in only the last 10 years. It has been proposed that catalysts will be able to direct the complex chemistry of high molecular weight materials, and the distribution of products can be tailored by manipulation of the catalyst properties, obtaining useful and valuable products from waste polymers with maximum economic benefit and minimal environmental liability. However, the complexity of the feedstocks can obscure the fundamental underlying chemistry; therefore, there is a need for fundamental studies aimed at unravelling the reaction pathways controlling decomposition of polymers in the presence of catalytic materials.

Three commodity addition polymers, polyethylene, polypropylene and polystyrene, were the focus of this review. Although thermal decomposition of these polymers is one strategy used for tertiary recycling, it is inherently non-selective

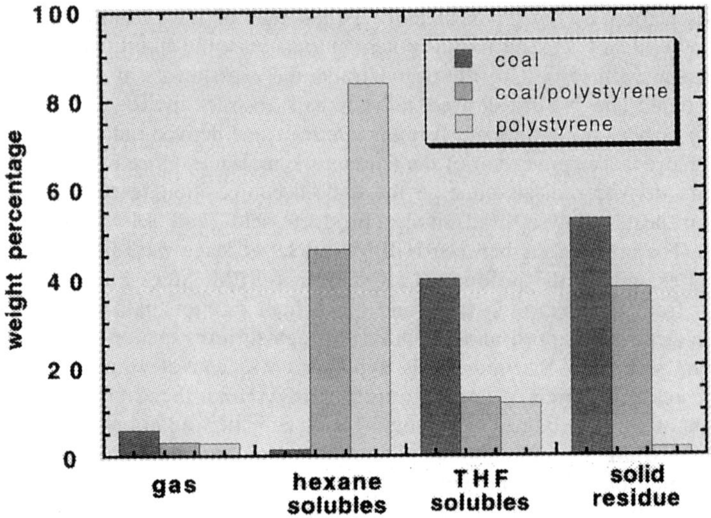

Figure 10 *Product distributions from coal and polystyrene coprocessing reactions in the absence of catalyst and donor solvent*[111]

and affords product streams of lower value. However, it has been extensively studied and provides the baseline to which catalytic degradation studies may be compared. Although the term 'catalytic degradation of polymers' has been widely used in the literature, there is not a consistent definition of the contacting pattern between the catalyst and the polymer that this phrase implies. It is therefore imperative that the catalyst–polymer contacting patterns used be identified and their potential influence on the products be well understood to allow comparison among the diverse results reported in the literature. A wide range of materials has been used to degrade the addition polymers examined in this review, and a comprehensive summary of these studies is provided in Table 2. Acid catalysts have been the most extensively studied because of their known ability to cleave carbon–carbon bonds and their widespread use in the hydrocarbon processing industry. Although the number of studies is more limited, base catalysts have also been examined for decomposing polymeric materials. To address the engineering concerns raised by a number of catalyst–polymer contacting patterns, including viscosity, product recovery and catalyst deactivation, much attention has focused on the catalytic conversion of polymers in the presence of co-reactants, primarily petroleum feedstocks and coal. This approach potentially offers the best opportunity for catalytic degradation to become a viable resource recovery strategy.

Only a small number of the investigations summarized in this review addresses the obvious practical concern of mixed plastics waste. Higher performance products with multiple polymeric layers, the use of additives, fillers and colorants, and the expense of sorting individual polymers for subsequent processing all demand resource recovery strategies capable of handling a complex mixture of components. The prognosis for catalytic resource recovery is excellent, however, if an analogy with the petroleum refining industry is drawn. This mature industry has long been successful at processing complex, multicomponent feeds into high-valued products. Therefore, future research efforts examining catalytic resource recovery of waste polymers must begin to assemble the growing body of fundamental information provided by single component studies and examine the reactions of real feedstocks composed of mixed plastics waste.

References

1 B. Hileman, *C & E News*, 1997, **75**, *46*, 8.
2 B. Hileman, *C & E News*, 1997, **75**, *50*, 9.
3 G.M. Masters, *Introduction to Environmental Engineering and Science*, Prentice Hall, New Jersey, 1998.
4 B. Piasecki, D. Rainey and K. Fletcher, *American Scientist*, 1998, **86**, July/August, 364.
5 A.L. Bisio and M. Xanthos, *How to Manage Plastics Waste. Technology and Market Opportunities*, Hanser, Munich, 1994.
6 M.M. Bitritto, *Polymeric Materials: Science and Engineering News*, 1994, Summer, 22.
7 A. Wood and E.S. Keische, *Chemical Week*, 1993, August 25/September 1, 32.
8 G. Menges, H. Emminger and G. Lackner, *Int. J. Materials and Product Technology*, 1991, **6**, 307.
9 K. Fouhy, I. Kim, S. Moore and E. Culp, *Chemical Engineering*, 1993, December, 30.
10 S. Shelley, K. Fouhy and S. Moore, *Chemical Engineering*, 1992, July, 30.
11 S.R. Ivanova, E.F. Gumerova, A.A. Berlin, K.S. Minsker and G.E. Zaikov, *Russian Chemical Reviews*, 1991, **60**, 225.
12 H. Ohkita, R. Nishiyama, Y. Tochihara, T. Mizushima, N. Kakuta, Y. Morioka, A. Ueno, Y. Namiki, S. Tanifuji, H. Katoh, H. Sunazuka, R. Nakayama and T. Kuroyanagi, *Ind. Eng. Chem. Res.*, 1993, **32**, 3112.
13 S.H. Ng, H. Seoud, M. Stanciulescu and Y. Sugimoto, *Energy & Fuels*, 1995, **9**, 735.
14 M.J. De Witt, D.J. Prekup and L.J. Broadbelt, 'Coal/Polymer Coprocessing with Efficient Use of Hydrogen', Semi-annual report, Department of Energy, DE-FG22-96PC96204, August 1997.
15 G. Audisio, A. Silvani, P.L. Beltrame and P. Carniti, *J. Anal. Appl. Pyrol.*, 1984, **7**, 83.
16 Y. Tsuchiya and K. Sumi, *J. Polym. Sci., Part A-1*: 1969, **7**, 1599.
17 P. Carniti, P.L. Beltrame, M. Armada, A. Gervasini and G. Audisio, *Ind. Eng. Chem. Res.*, 1991, **30**, 1624.
18 S.H. Ng, *Energy & Fuels*, 1995, **9**, 216.
19 W.J. Bailey, *SPE Transactions*, 1965, **5**, 59.
20 R. McGuchan and I.C. McNeill, *Eur. Polym. J.*, 1968, **4**, 115.
21 H. Sinn, W. Kaminsky and J. Janning, *Angew. Chem., Int. Ed.*, 1976, **15**, 660.
22 S. Ide, T. Ogawa, T. Kuroki and T. Ikemura, *J. Appl. Polym. Sci.*, 1984, **29**, 2561.
23 W. Kaminsky and H. Rossler, *CHEMTECH*, 1992, **22**, 108.
24 R.H. Still and O.A. Peters, *J. Appl. Polym. Sci.*, 1993, **50**, 989.

25 L.L. Sharp, J. Ness, R.O., T.R. Aulich and J.M. Sosa, *Annual Meeting of the American Institute of Chemical Engineers*, 1994, **1**,
26 S. Mehta, S. Biederman and S. Shivkumar, *J. Mater. Sci.*, 1995, **30**, 2944.
27 Y.D.M. Simard, M.R. Kamal and D.G. Cooper, *J. Appl. Polym. Sci.*, 1995, **58**, 843.
28 T. Ogawa, T. Kuroki, S. Ide and T. Ikemura, *J. Appl. Polym. Sci.*, 1981, **27**, 857.
29 G. Audisio, F. Bertini, P.L. Beltrame and P. Carniti, *Polym. Degrad. Stab.*, 1990, **29**, 191.
30 J.-K. Koo, S.-W. Kim and Y.-H. Seo, *Resources, Conservation and Recycling*, 1991, **5**, 365.
31 Z. Zhibo, S. Nishio, Y. Morioka, A. Ueno, H. Ohkita, Y. Tochihara, T. Mizushima and N. Kakuta, *Catal. Today*, 1996, **29**, 303.
32 S. Sato, T. Murakata, S. Baba, Y. Saito and S. Watanabe, *J. Appl. Polym. Sci.*, 1990, **40**, 2065.
33 A.R. Songip, T. Masuda, H. Kuwahara and K. Hashimoto, *Energy & Fuels*, 1994, **8**, 131.
34 A.R. Songip, T. Masuda, H. Kuwahara and K. Hashimoto, *Energy & Fuels*, 1994, **8**, 136.
35 Y. Tochihara, H. Ohkita, R. Nishiyama, T. Mizushima, N. Kakuta, Y. Morioka, A. Ueno, R. Itoh, Y. Namiki, H. Katoh, S. Tanifuji, H. Sunazuka and T. Kuroyanagi, *Trans. Mat. Res. Soc. Jpn.*, 1994, **18A**, 713.
36 C. Vasile, M. Sabliovschi and V. Barboiu, *Revue Roumaine de Chimie*, 1995, **40**, 679.
37 A. Zhang, T. Hirose, S. Nishio, Y. Morioka, N. Azuma and A. Ueno, *Ind. Eng. Chem. Res.*, 1995, **34**, 4514.
38 A.R. Songip, T. Masuda, H. Kuwahara and K. Hashimoto, *Appl. Catal., B: Environmental*, 1993, **2**, 153.
39 J.M. Arandes, I. Abajo, D. Lopez-Valerio, I. Fernandez, M.J. Azkoiti, M. Olazar and J. Bilbao, *Ind. Eng. Chem. Res.*, 1997, **36**, 4523.
40 P.L. Beltrame, P. Carniti, G. Audisio and F. Bertini, *Polym. Degrad. Stab.*, 1989, **26**, 209.
41 M. Day, J.D. Cooney and M. MacKinnon, *Polym. Degrad. Stab.*, 1995, **48**, 341.
42 G.S. Deng, W.H. McClennen and H.L.C. Meuzelaar, *Preprints of the Division of Fuel Chemistry of the American Chemical Society*, 1997, **42**, 987.
43 V.J. Fernandes, Jr., A.S. Araujo and G.J.T. Fernandes, *J. Therm. Anal.*, 1997, **49**, 255.
44 V.J. Fernandes, Jr., A.S. Araujo and G.J.T. Fernandes, in *Progress in Zeolite and Microporous Materials*, ed. H. Chon, S.-K. Ihm and Y.S. Uh, Elsevier, New York, 1997, vol. 105, p. 941.
45 A. Garforth, S. Fiddy, Y.-H. Lin, A. Ghanbari-Siakhali, P.N. Sharratt and J. Dwyer, *Thermochim. Acta*, 1997, **294**, 65.
46 M.M. Ibrahim, E. Hopkins and MS Seehra, *Fuel Processing Technology*, 1996, **49**, 65.
47 Y. Ishihara, H. Nanbu, C. Iwata, T. Ikemura and T. Takesue, *Bull. Chem. Soc. Jpn.*, 1989, **62**, 2981.
48 Y. Ishihara, H. Nanbu, K. Saido, T. Ikemura and T. Takesue, *Bull. Chem. Soc. Jpn.*, 1991, **64**, 3585.
49 Y. Ishihara, H. Nambu, T. Ikemura and T. Takesue, *J. Appl. Polym. Sci.*, 1989, **38**, 1491.
50 Y. Ishihara, H. Nanbu, K. Saido, T. Ikemura and T. Takesue, *Polymer*, 1992, **33**, 3482.

51 Y. Ishihara, H. Nanbu, C. Iwata, T. Ikemura and T. Takesue, *Bull. Chem. Soc. Jpn.*, 1989, **62**, 2981.

52 Y. Ishihara, H. Nanbu, K. Saido, T. Ikemura, T. Takesue and T. Kuroki, *Fuel*, 1993, **72**, 1115.

53 R. Lin and R.L. White, *J. Appl. Polym. Sci.*, 1995, **58**, 1151.

54 Y.-H. Lin, P.N. Sharratt, G. Manos, A. Garforth and J. Dwyer, *ICHEME Research Event/Second Eur. Conference for Young Researchers*, 1996, **2**, 576.

55 Y. Sakata, M. Azhar Uddin, A. Muto, Y. Kanada, K. Koizumi and K. Murata, *J. Anal. Appl. Pyrol.*, 1997, **43**, 15.

56 Y.A. Sangalov, K.S. Minsker, A.A. Berlin and Y.A. Prochukhan, *Polymer Science U.S.S.R.*, 1983, **25**, 1676.

57 W. Zhao, S. Hasegawa, J. Fujita, F. Yoshii, T. Sasaki, K. Makuuchi, J. Sun and S. Nishimoto, *Poly. Degrad. Stab.*, 1996, **53**, 199.

58 P. Carniti, A. Gervasini and S. Bernardelli, *Polym. Degrad. Stab.*, 1997, **57**, 301.

59 R. Lin and R.L. White, *J. Appl. Polym. Sci.*, 1997, **63**, 1287.

60 R. Lin, D.L. Negelein and R.L. White, *Preprints of the Division of Fuel Chemistry of the American Chemical Society*, 1997, **42**, 982.

61 M.A. Uddin, K. Koizumi, K. Murata and Y. Sakata, *Poly. Degrad. Stab.*, 1997, **56**, 37.

62 W. Ding, J. Liang and L.L. Anderson, *Fuel Processing Technology*, 1997, **51**, 47.

63 R.C. Mordi, T. Fields and J. Dwyer, *J. Chem. Soc., Chem. Commun.*, 1992, 374.

64 R.C. Mordi, R. Fields and J. Dwyer, *J. Anal. Appl. Pyrol.*, 1994, **29**, 45.

65 R.C. Mordi, J. Dwyer and R. Fields, *Poly. Deg. Stab.*, 1994, **46**, 57.

66 I. Nakamura and K. Fujimoto, *Catal. Today*, 1996, **27**, 175.

67 H. Nanbu, Y. Ishihara, T. Takasue and T. Ikemura, *Polym. J.*, 1986, **18**, 871.

68 R. Ochoa, H. Van Woert, W.H. Lee, R. Subramanian, E. Kugler and P.C. Eklund, *Fuel Processing Technology*, 1996, **49**, 119.

69 X. Xiao, W. Zmierczak and J. Shabtai, *Preprints of the Division of Fuel Chemistry of the American Chemical Society*, 1995, **40**, 4.

70 T. Yoshida, A. Ayame and H. Kano, *Bull. Jpn. Petrol. Inst.*, 1975, **17**, 218.

71 H. Nanbu, Y. Sakuma, Y. Ishihara, T. Takesue and T. Ikemura, *Polym. Degrad. Stab.*, 1987, **19**, 61.

72 P. Carniti, A. Gervasini, P.L. Beltrame, G. Audisio and F. Bertini, *Appl. Catal., A: General*, 1995, **127**, 139.

73 M. Yamamoto, I. Suzuki and S. Yamanaka, *Nippon Kagaku Kaishi*, 1976, **5**, 802.

74 O.S. Woo and L.J. Broadbelt, *Catal. Today*, 1998, **40**, 121.

75 A. Ayame, Y. Uemichi, T. Yoshida and H. Kanoh, *J. Japan Petrol. Inst.*, 1979, **22**, 280.

76 Y. Ishihara, H. Nanbu, T. Ikemura and T. Takesue, *Fuel*, 1990, **69**, 978.

77 C.I. Simionescu, C. Vasile, P. Onu, M. Sabliovschi, G. Moroi, V. Barboiu, D. Ganju and M. Florea, *Thermochim. Acta*, 1988, **134**, 301.

78 Y. Uemichi, A. Ayame, T. Yoshida and H. Kanoh, *J. Japan Petrol. Inst.*, 1980, **23**, 35.

79 Y. Uemichi, A. Ayame, Y. Kashiwaya and H. Kanoh, *J. Chromatogr.*, 1983, **259**, 69.

80 Y. Uemichi, Y. Kashiwaya, M. Tsukidate, A. Ayame and H. Kanoh, *Bull. Chem. Soc. Jpn.*, 1983, **56**, 2768.

81 Y. Uemichi, Y. Kashiwaya, A. Ayame and H. Kanoh, *Chem. Lett.*, 1984, **1**, 41.

82 Y. Uemichi, Y. Makino and T. Kanazuka, *J. Anal. Appl. Pyrol.*, 1989, **14**, 331.

83 Y. Uemichi, Y. Makino and T. Kanazuka, *J. Anal. Appl. Pyrol.*, 1989, **16**, 229.

84 Y. Uemichi, H. Kamibayashi, M. Sugioka and T. Kanazuka, *Kobunshi Ronbunshu*, 1993, **50**, 887.
85 Y. Uemichi, M. Hattori, T. Itoh, J. Nakamura and M. Sugioka, *Ind. Eng. Chem. Res.*, 1998, **37**, 867.
86 S. Ide, H. Nanbu, T. Kuroki and T. Ikemura, *J. Anal. Appl. Pyrol.*, 1984, **6**, 69.
87 T. Ogawa, T. Kuroki, S. Ide and T. Ikemura, *J. Appl. Polym. Sci.*, 1982, **27**, 857.
88 C. Vasile, P. Onu, V. Barboiu, M. Sabliovschi, G. Moroi, D. Ganju and M. Florea, *Acta Polymerica*, 1988, **39**, 306.
89 S. Ide, H. Nanbu, T. Kuroki and T. Ikemura, *J. Chem. Soc. Japan*, 1983, 1657.
90 M.E. Davis, *Ind. Eng. Chem. Res.*, 1991, **30**, 1675.
91 I. Kodaira, Z. Osawa and H. Ando, *J. Chem. Soc. Japan*, 1977, **12**, 1892.
92 J. Aguado, D.P. Serrano, M.D. Romero and J.M. Escola, *Chem. Commun.*, 1996, 725.
93 H.P. Frank, *Monatsh. Chem.*, 1963, **94**, 393.
94 B. Pukánszky, J.P. Kennedy, T. Kelen and F. Tudos, *Polym. Bull.*, 1981, **5**, 469.
95 N.K. Baramboim, A.F. Pinyayev and L.S. Kaurova, *Polymer Science U.S.S.R.*, 1979, **21**, 487.
96 W.L. Carrick, *J. Polym. Sci., Part A-1*: 1970, **8**, 215.
97 S.R. Ivanova, E.F. Gumerova, K.S. Minsker, G.E. Zaikov and A.A. Berlin, *Prog. Polym. Sci.*, 1990, **15**, 193.
98 S.R. Ivanova, E.F. Gumerova, A.A. Berlin and K.S. Minsker, *Polymer Science*, 1991, **33**, 268.
99 T. Yamaguchi, T. Kamiguchi, T. Ito, S. Ohkita and T. Goro, *J. Chem. Soc. Japan*, 1976, 1171.
100 J.F. Haw, B.R. Richardson, I.S. Oshiro, N.D. Lazo and J.A. Speed, *J. Am. Chem. Soc.*, 1989, **111**, 2052.
101 G. Yaluris, J.E. Rekoske, L.M. Aparicio, R.J. Madon and J.A. Dumesic, *J. Catal.*, 1995, **153**, 54.
102 G. Yaluris, J.E. Rekoske, L.M. Aparicio, R.J. Madon and J.A. Dumesic, *J. Catal.*, 1995, **153**, 65.
103 J. Wingfield, R.C., J. Braslaw and R.L. Gealer, Ford Motor Company, United States Patent 4,515,659, 1985.
104 T.-Y. Yan, United States Patent 4,118,281, 1978.
105 L.L. Anderson and W. Tuntawiroon, *Preprints of the Division of Fuel Chemistry of the American Chemical Society*, 1993, **38**, 816.
106 G.P. Huffman, Z. Feng, V. Mahajan, P. Sivakumar, H. Jung, J.W. Tierney and I. Wender, *Preprints of the Division of Fuel Chemistry of the American Chemical Society*, 1995, **40**, 34.
107 H.K. Joo and C.W. Curtis, *Energy & Fuels*, 1996, **10**, 603.
108 H.K. Joo and C.W. Curtis, *Energy & Fuels*, 1997, **11**, 801.
109 K. Liu, W.H. McClennen and H.L.C. Meuzelaar, *Preprints of the Division of Fuel Chemistry of the American Chemical Society*, 1995, **40**, 9.
110 M. Luo and C.W. Curtis, *Fuel Processing Technology*, 1996, **49**, 177.
111 M. Luo and C.W. Curtis, *Fuel Processing Technology*, 1996, **49**, 91.
112 M. Luo and C.W. Curtis, *Preprints of the Division of Fuel Chemistry of the American Chemical Society*, 1997, **42**, 1062.
113 S.R. Palmer, E.J. Hippo, D. Tandon and M. Blankenship, *Preprints of the Division of Fuel Chemistry of the American Chemical Society*, 1995, **40**, 29.
114 V.R. Pradhan, A.G. Comolli, T.L.K. Lee and R.H. Stalzer, *Preprints of the Division of Fuel Chemistry of the American Chemical Society*, 1995, **40**, 82.

115 G.A. Robbins, R.A. Winschel and F.P. Burke, *Preprints of the Division of Fuel Chemistry of the American Chemical Society*, 1995, **40**, 92.
116 G.A. Robbins, R.A. Winschel and F.P. Burke, *Preprints of the Division of Fuel Chemistry of the American Chemical Society*, 1996, **41**, 1069.
117 K.S. Rothenberger, A.V. Cugini, R.L. Thompson and M.V. Ciocco, *Energy & Fuels*, 1997, **11**, 849.
118 K.S. Rothenberger, A.V. Cugini, M.V. Ciocco, R.R. Anderson and G.A. Veloski, *Preprints of the Division of Fuel Chemistry of the American Chemical Society*, 1995, **40**, 38.
119 M.M. Taghiei, F.E. Huggins and G.P. Huffman, *Preprints of the Division of Fuel Chemistry of the American Chemical Society*, 1993, **38**, 810.
120 M.M. Taghiei, Z. Feng, F.E. Huggins and G.P. Huffman, *Energy & Fuels*, 1994, **8**, 1228.
121 M.J. De Witt and L.J. Broadbelt, *Preprints of the Division of Fuel Chemistry of the American Chemical Society*, 1997, **42**, 38.

5
Can Heterogeneous Catalysts Contribute to Clean Syntheses of Fine Chemicals – on the Example of Fragrances

BY WOLFGANG F. HÖLDERICH AND D. HEINZ

1 Introduction

In the development of new chemical processes the market requires two objectives to be fulfilled. One is the highest possible efficiency with respect to the raw materials including feedstock and energy. This leads to processes with high conversion and selectivities which do not require expensive separation procedures downstream. The key to high yields is an active and selective catalyst. The second objective for a new process is concerned with the vast number of environmental regulations. Any by-product that cannot be used as a feedstock in another process immediately becomes an expensive waste which has to be disposed of. This has formed the public opinion that ecology and economy are contradictory objectives. But especially heterogeneous catalysis has proven for bulk chemicals that it can combine both claims.[1]

The use of homogeneous Brønsted and Lewis acids and bases was one of the main reasons for the production of high amounts of inorganic salts as by-products in industrial syntheses. The ratio of mass unit by-product per mass unit of product is explained as an E-factor by Sheldon for several industry segments.[1,2] For example, in the fine chemical industry and for the production of pharmaceuticals, sometimes 50–150 kg of by-products per kg desired product are formed. Therefore, new processes in the production of fine chemicals in order to reduce the high numbers of by-product formation are needed.

A large fraction of the examples cited involves the use of zeolites. This group of catalysts does already play a substantial role in the chemical production. The unique features of zeolites – the possibility to tune acidic and basic sites as well as the creation of multifunctional catalysts – open a wide field of applications in the production of fine chemicals.

In this article we present new heterogeneously-catalysed processes for the synthesis of industrially relevant fine chemicals on the example of fragrances. The emphasis of this review article will remain on solid acids. The less frequent application of solid bases is described in detail in refs. 1 and 2a.

Flavours and fragrances have been playing a dominating role stimulating the human senses. Starting as substances used for cultic and religious purposes, they

Catalysis, Volume 14

found their way to the individual contribution to the improvement of the personal comfort. Among the oldest fragrances, incense and myrrh, as well as sandalwood and cedarwood, are counted. Of the more than 5000 available fragrances nowadays, only 5% are gained from natural sources.[3]

In most cases the demand cannot be provided anymore by natural sources. Therefore, a lot of fragrances, flavours and aromas are based on chemical compounds in the fragrance industry. An important class/category of compounds in this field are aldehydes and ketones.

A very valuable method for the introduction of an aldehyde or ketone group in organic fine chemicals is the epoxidation of olefins and the following rearrangement of the oxiranes in the presence of acidic catalysts.

The activation of epoxides for ring opening reactions can be achieved either by Brønsted acidic catalysts *via* the addition of a proton to the epoxide oxygen or by Lewis acidic catalysts *via* the coordination of the epoxide oxygen to a multivalent cation.

Homogeneous catalysts such as phosphoric acid, BF_3, $FeCl_3$, $ZnBr_2$ and $SnCl_y$ as well as heterogeneous catalysts such as SiO_2, Al_2O_3, $ZnOWO_3$, supported metals and various precipitated phosphates have been applied.

The heterogeneous catalysts used in the past had some drawbacks such as incomplete conversion in most cases, low selectivity because of consecutive aldol condensation to form preferably trimers and low service time [*i.e.* catalyst time on stream (TOS)].

Therefore, the fragrance industry still applies homogeneous catalysts in the manufacturing plants. Furthermore, in respect to the regioselectivity of the ring opening reaction of oxiranes, electronic as well as steric factors can play a role. These general considerations stimulate the use of zeolites and 'non-zeolitic' molecular sieves as heterogeneous catalysts for such rearrangement reactions in the liquid or in the gas phase in a slurry reactor and in a continuous fixed bed reactor, respectively.

2 Isomerization of Epoxides

2.1 Isomerization of Styrene Oxides – The conversion of styrene oxides to phenylacetaldehydes yields a substance which can be used as fragrances directly or as valuable intermediates for the production of fragrances. In addition, such compounds can also be used for manufacturing pharmaceuticals and insecticides, fungicides and herbicides in particular when halogenated derivatives are needed. The reaction is carried out in a fixed bed reactor, *e.g.* under gas phase conditions.

The catalytic performance of various catalysts on the activity, selectivity and service time was screened at 300 °C and WHSV = 2–3 h^{-1}. As shown in Figure 1, zeolites with an MFI-structure are superior to other zeolite types, non-zeolitic molecular sieves and much superior to amorphous metal oxides.

The high regioselectivity to phenylacetaldehydes may be related to the stabilization of a developing α-cation.[4] A possible side product is the trimeric

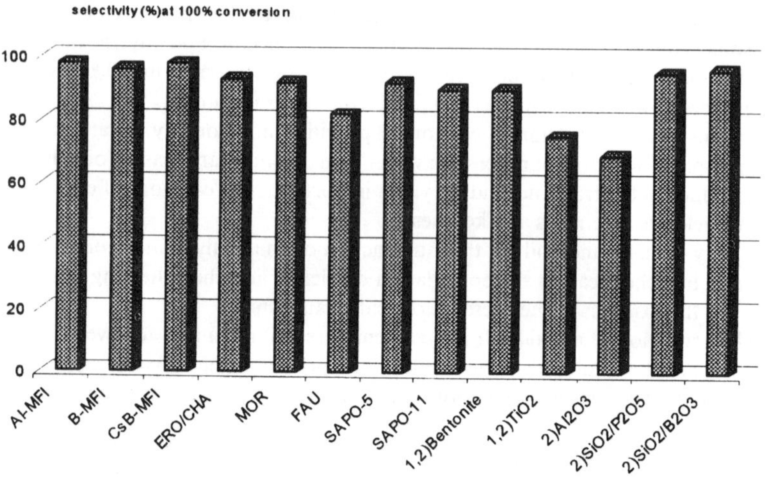

Figure 1 *Catalyst screening at* 300 °C,[4] (1) 91% *or* 98% *conversion,* (2) *short service time*

phenylacetaldehyde, which is formed particularly at lower temperatures (10% at 200 °C *vs.* 2% at 300 °C).

B_2O_3- and P_2O_5-modified silica as well as bentonite show similar selectivities as pentasil zeolites, but they deactivate much faster. The reason for that is the formation of trimers by aldol condensation and aromatisation in the case of the non-shape-selective materials. In contrast, the use of MFI-zeolites avoid the trimer formation due to the steric constraints and consecutive reactions resulting in coke.

The aim to produce a fine chemical with 100% yield is achieved by the use of a Cs-doped boron-pentasil zeolite having an extremely weak acidity.

If the epoxide rearrangement of styrene oxide is carried out in the presence of hydrogen and by use of a bi-functional boron-pentasil zeolite catalyst having a hydrogenation component such as Cu, then phenylethanol is obtained in one step. This hydro-isomerization renders high yields (> 85%) at 250 °C under the gas phase conditions. It is an example for multifunctional catalysis in a one-pot reaction that means simultaneous rearrangement and hydrogenation.

This route might be an environmentally benign alternative to the conventional method for the production of the very important fragrance phenylethanol which has the smell of roses. This compound is still manufactured out of ethylene oxide and benzene in the presence of a high excess of $AlCl_3$, yielding unwanted toxic by-products.

This knowledge was applied for even more complicated, bi-functional compounds such as phenyl glycidic acid methyl ester to yield α-ketone carboxylic acid esters such as 3-phenylpyroracemic acid methyl esters. These esters can be used as intermediates for herbicides, *e.g.* triazinones, and for the syntheses of L-amino acids. Also in this reaction, the weakly acidic boron-pentasil zeolite and even better the Cs-doped boron-pentasil zeolite are the most favored catalysts and are superior to other heterogeneous catalyst systems.

The advantages of the new process for α-ketocarboxylic acid ester under gas phase conditions are high yields, high service time of the catalyst, and a one-step reaction using easily available starting materials. In contrast, the known processes are very complicated, costly and environmentally hazardous *via* Grignard reactions of oxalic acids.

By increasing the temperature over $\geq 300\,^{\circ}C$, the phenyl glycidic acid esters are converted to phenyl acid esters. That occurs in a consecutive decarbonylation of α-ketone carboxylic acid esters as intermediates. In the case of *p-tert*-butyl glycidic acid ester in the presence of a boron-pentasil zeolite at 350 °C, the ratio between α-ketone carboxylic acid ester and the phenyl acetic acid ester is 65:35.

The phenyl acetic acid esters are highly valuable intermediates for a lot of applications. Among them they are applied to the synthesis of fragrances. The present industrial process for manufacture of phenyl acetic esters starts from benzyl chloride and uses the reactions with KCN to form benzyl cyanide and followed by hydrolytic cleavage. The drawbacks of the conventional route are multistep reaction and use of toxic cyanides. The new route is an environmentally benign alternative.

2.2 Isomerization of α-Pinene Oxide – α-Pinene oxide **(1)** is known as a reactive molecule which rearranges easily under the influence of an acid catalyst.[5,6] Thereby, many products can be formed. For example, compounds such as the isomeric campholenic aldehyde **(3)**, *trans*-carveol **(4)**, *trans*-sobrerol **(5)**, *p*-cymene **(6)** or isopinocamphone **(7)** are observed as main by-products (Figure 2). The most industrially-desired compound among these is campholenic aldehyde **(2)**. It is the key molecule for the synthesis of various highly intense sandalwood-like fragrance chemicals.[6,7]

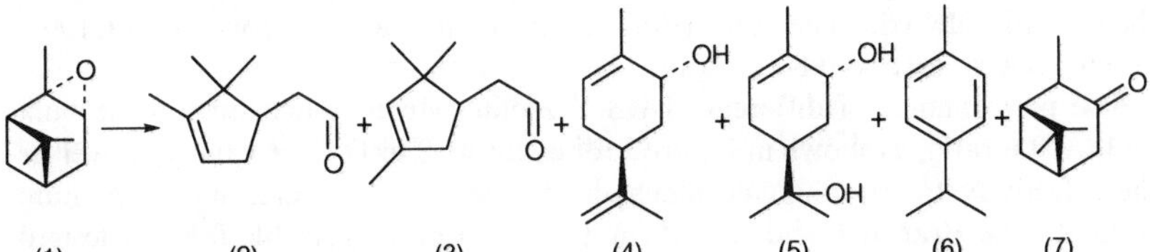

Figure 2 *Product range of isomerization of α-pinene oxide*

Although $ZnBr_2$ in benzene is well known as an effective homogeneous catalyst with selectivities of about 85% to **(2)**,[8] a lot of efforts have been made to find a truly heterogeneous catalyst system.[9] All of the selectivities and yields obtained catalytically never exceeded 60% and have not been competitive with the homogeneous system. The mechanism of the formation of **(2)** is still being discussed in the literature.[10] The main disadvantage is water pollution by zinc halides which cause severe problems in the sludge treatment by destroying the bacteria. To overcome these well-known disadvantages of the homogeneous process, a multiple recyclable, heterogeneous catalyst must be found which leads to high yields of **(2)** at environmentally more benign conditions.

Surprisingly, it was found[6] that some H-US-Y zeolites having a high amount of mesopores are very suitable for this kind of reaction. Under the given conditions at a temperature of 0 °C, best selectivities of about 75% at a conversion of 30% were obtained with a commercially available, highly dealuminated H-US-Y zeolite (A), kindly provided by The Zeolyst Corporation (Figure 3).

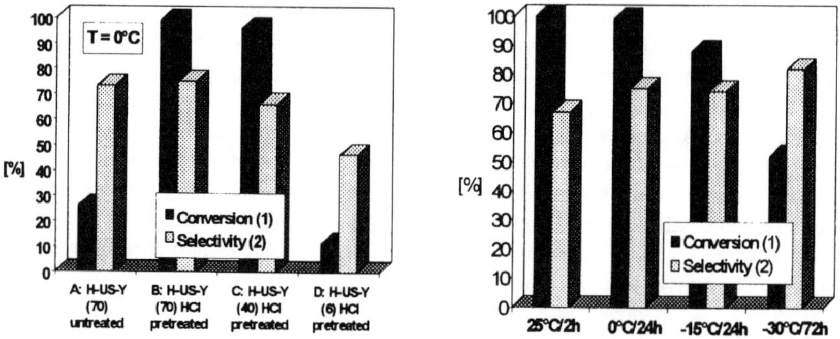

Figure 3 Left: *influence of acid treatment and SiO_2/Al_2O_3 ratio*, right: *influence of temperature with catalyst B, batch reactor*. T = 0 °C, t = 24 h, *catalyst loading* = 7.5 g epoxide/g catalyst

As also shown in Figure 3, a pretreatment of the catalyst with diluted acid (0.01 molar HCl at 25 °C for 24 h) and subsequent washing and calcination at 550 °C yields a major enhancement in activity without any loss in selectivity for the desired aldehyde. The conversion achieved with catalyst B was about 100% and the selectivity about 80% within 24 h.

The performance of different H-US-Y zeolites strongly depends on the bulk SiO_2/Al_2O_3 ratio, as shown in Figure 3 for catalysts B to D. The activity, as well as the selectivity to campholenic aldehyde, increases with decreasing aluminium content. The Brønsted acid sites do not seem to be responsible for the desired reaction since the number of these sites is equal to the number of aluminium atoms in the framework. From experiments we cannot rule out that some Brønsted active sites are required, but the performance of these catalysts, especially the good selectivities observed, appear to be caused by Lewis acid sites. They can arise from the dealumination process by steaming and are generally described as extra-framework-aluminium (EFA) species.[11] The first intention of washing with HCl was to remove the EFA species to make the inner part of the zeolite framework more accessible and to diminish the influence of diffusion constraints.

With respect to the temperature influence, Figure 3 shows increasing selectivity for campholenic aldehyde at lower reaction temperatures with catalyst B. By running the isomerization at lower temperatures down to −30 °C, selectivities can be improved to 85%. Surprisingly, our catalyst B shows an unexpectedly high activity, even at such unusually low reaction temperatures. The conversion at −30 °C reaches values up to 55% after 72 h (Figure 3). The enhancement in selectivity at lower temperatures can be explained with an improved kinetic reaction control and/or by the avoidance of side reactions of campholenic aldehyde.

It was found that the selectivity and conversion are optimal if the pretreatment of the highly dealuminated Y zeolite is pursued at pH = 2. Lower pH values in the pretreatment of the catalyst A cause a decrease of both conversion and selectivity in the catalytic run.[6] It is well known that the application of strong acid treatment (pH < 1) leads to removal of framework as well as extra-framework-alumina from dealuminated Y zeolite.[12]

To elucidate the reason for the better performance of the acid-treated catalyst (B) we examined samples A and B with various analytical methods such as AAS, FTIR, BET, pyridine adsorption, ^{29}Si and ^{27}Al solid NMR investigations.

AAS of aluminium and FTIR indicate that there must be a large amount of EFA species in the zeolite. By measuring the outer T–O–T stretching frequency, it was calculated using the equations disclosed in ref. 13 that only *ca.* one Al atom per unit cell is incorporated into the zeolite framework of catalyst B. On the other hand, experimental results derived from Al-AAS show that hypothetically a higher number, *i.e.* all together four Al atoms, could be present in the unit cell on the assumption that all Al is incorporated in the framework. This difference tells us that most of the aluminium must exist in the form of EFA. Nitrogen adsorption shows that both zeolites have a large mesopore volume. This allows sufficient diffusion transport of organic molecules even at low temperatures. FTIR measurements of zeolite samples loaded with pyridine at 150 °C were performed.[14] The measurements show a strong presence of Lewis sites and a few Brønsted sites: these findings are in agreement with the results from Al AAS.

The ^{27}Al and ^{29}Si NMR measurements[6] showed that after treatment with 0.01 molar HCl most of the amorphous silica-containing material is removed from the parent catalyst A. This can easily be understood since the maximum solubility of silica[15] is reached at pH = 2. Although the improved performance of the treated catalyst cannot be entirely explained by the removal of less active material, *i.e.* the increase of the number of Lewis acid sites per mass unit, it is believed that these silica species block most of the catalytically active centres, *i.e.* the highly dispersed Lewis acidic alumina sites in the micro- and mesopores of the parent US-Y zeolite.

After complete reaction, the catalyst can be re-used without loss of performance. The catalyst can be successfully reactivated by calcination under an air atmosphere at 550 °C.

A new process for the heterogeneously-catalysed production of campholenic aldehyde from α-pinene oxide has been found which is competitive with the homogeneous $ZnBr_2$ system and gives yields up to 85%.

2.3 Isomerization of Isophorone Oxide – Terpene epoxides are very reactive compounds. Some products formed by isomerization of such epoxides are valuable raw materials for perfumes, synthetic flavourings and pharmaceuticals, and also provide useful intermediates in organic syntheses. The isomerization of isophorone oxide **(1)** (Figure 4) was originally investigated by House and Wasson using boron trifluoride etherate as a homogeneous catalyst.[16]

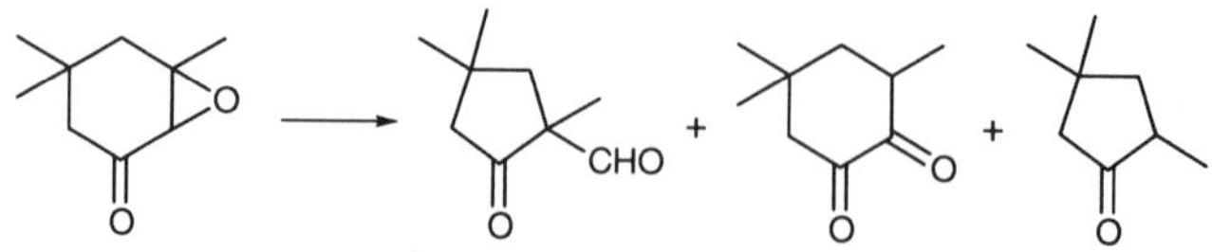

Figure 4 *Products of the isomerization of isophorone oxide*

They found that at room temperature using benzene as the solvent only 3% consisted of the diketone **(3)** and mainly the ring-contracted products were obtained: 33% of keto aldehyde **(2)** and 28% of ketone **(4)**. From an industrial point of view the desired compound is the keto aldehyde **(2)**, which is an interesting intermediate for fragrance chemicals. The acid-catalysed reaction mechanism leading to the synthesis of keto aldehyde **(2)** has been discussed earlier.[15,17] Therefore, it is of interest whether the product distribution changes in the presence of a heterogeneous catalyst system, and also whether the decarbonylation of the compound **(2)** to compound **(4)** can be suppressed.

In the presence of zeolites, the keto aldehyde **(2)**, accompanied by the keto-enolic form of the α-diketone **(3)**,[18] is mainly formed as reported by Sheldon *et al.* and Hölderich *et al.*[19–22]

The reports on catalytic isomerization using various zeolitic catalysts in comparison with the conventional catalysts previously used gives results of reactions carried out discontinuously in a batch reactor in the liquid phase, as well as for those carried out continuously in a fixed bed reactor in the vapor phase. The results in the liquid phase over heterogeneous catalysts are summarized in Table 1.

The numbers in parentheses correspond to the SiO_2/Al_2O_3 or SiO_2/B_2O_3 ratios. A blank test in the liquid phase yields less than 1% conversion. Unlike non-zeolitic catalysts, except for H-Nafion, most zeolites yield complete conversion. A high yield of the keto aldehyde **(2)**, up to 81%, was surprisingly attained by using H-FER as a heterogeneous catalyst. For the rearrangement of isophorone oxide, the presence of acidic sites is necessary for the catalytic activity. The reactivity of H-FER can be explained by the acidic outer surface of the catalyst. Molecular modelling showed that the isophorone epoxide is too bulky for the small pore size of ferrierite.

The ratio between the two main products **(2)** and **(3)** depends on the type and acidity of zeolite used. For example, H-BEA with a higher acid outer surface, demonstrated a high activity but a low selectivity of **(2)** in comparison to H-FER or H-US-Y catalysts.

High amounts of **(2)** were also obtained by the use of H-US-Y (96) followed by H-US-Y (96)-HCl. The H-US-Y (96)-HCl used a modified highly dealuminated

Table 1 *Conversion and selectivities of used zeolites in liquid phase isomerization of isophorone oxide* **(1)**

		Selectivity (%)		
Catalyst	*Conversion* (%)	**(2)**	**(3)**	**(4)**
H-ZSM-5 (28)	88	68	14	1
H-ZSM-5 (60)	95	68	15	0
H-BEA (25)	100	68	13	14
H-US-Y (70)	96	73	12	0
H-US-Y (96)	100	74	11	1
H-US-Y (96)-HCl	100	71	17	4
H-FER (18)	100	81	12	0

$T = 110\,°C$; $t = 6\,h$; solvent: toluene; loading: 10 g isophorone oxide/g catalyst; 1 atm. pressure.

ultrastable Y zeolite, and was pretreated with diluted acid according to the method described before.[20,23] This zeolitic catalyst unlike many others, remains active at lower temperatures and also at high loadings, as was previously demonstrated in the isomerization of α-pinene oxide using this heterogeneous catalyst.[20,23]

Further evidence for the excellent activity of H-US-Y (96)-HCl and H-BEA (25) was derived when investigating the influence of temperature. At 80 °C, a maximum selectivity to **(2)** of 75% (71%) was obtained with the H-US-Y (H-BEA) at complete conversion within 2 h of reaction time. The solvent in this reaction was toluene.

The influence of solvents was also investigated. A high selectivity was achieved when using less polar solvents, such as heptane or toluene. More polar solvents caused lower activity and selectivity; except for anisole with which selectivity of **(2)** reached 85% using H-FER. An aromatic character of the solvent seems to favor the formation of keto aldehyde **(2)**, as the results obtained with toluene and anisole suggest (Table 2). As expected, the solvent acetonitrile totally inhibited the reaction.[21]

The data were correlated to that obtained by Sheldon *et al.* in the isophorone oxide rearrangement.[23] There, a dealuminated mordenite was used and a selectivity of 85% to **(2)** in benzene as a solvent was found. The results are comparable to our findings over H-FER or US-Y-zeolites.

To our knowledge, no vapor phase isomerization utilizing heterogeneous catalysts has been previously described in the literature. Continuous vapor phase reactions should be preferred over discontinuous liquid phase reactions, as this yields a higher efficiency (space time yield) as well as an easier separation of the catalyst from the reaction mixture.

A general view of the zeolites used in the vapor phase isomerization and their conversions and selectivities is shown in Table 3. The blank test yields less than 2% conversion.

Table 2 *Conversion and selectivities by using H-FER* (18) *with different solvents*

		Selectivity (%)		
Solvent	*Conversion* (%)	**(2)**	**(3)**	**(4)**
Heptane (98 °)	100	81	13	0
Toluene (110 °C)	100	81	12	0
Toluene (80 °C)	92	74	11	0
Toluene (60 °C)	37	62	5	1
Anisole (110 °C)	100	85	12	0
THF (67 °C)	20	62	5	1
Ethanol (78 °C)	19	9	11	3
Acetonitrile (81 °C)	1	75	0	0
Without solvent (110 °C)	100	75	11	1

t = 6 h; loading: 10 g isophorone oxide/g catalyst; 1 atm pressure.

Table 3 *Conversion and selectivities by using zeolites in gas phase isomerization*

		Selectivity (%)		
Catalyst	*Conversion* (%)	**(2)**	**(3)**	**(4)**
H-ZSM-5 (28)	99	69	18	2
H-ZSM-5 (60)	100	72	18	1
H-[B]-ZSM-5 (38)	90	82	10	1
H-US-Y (70)	100	66	19	6
H-US-Y (96)	99	69	21	5
H-US-Y (96)-HCl	100	68	23	3
H-FER (18)	100	66	24	3

TOS = 4 h; *T* = 250 °C; solvent: toluene; Toluene/Isophorone oxide 2:1 wt/wt; nitrogen flow 8 l h^{-1}; WHSV = 1 h^{-1};1 atm. pressure.

Using zeolites with the same structure but in different ratios of SiO_2/Al_2O_3, we found an increased tendency towards selectivity to **(2)** with higher SiO_2/Al_2O_3 ratios, as was previously observed in liquid phase isomerization.

All tested zeolites yielded high conversions, except for Na-ZSM-5 and silicalite. A shorter contact time seems to have a positive effect on the aldehyde formation, possibly due to reduced decarbonylation. With a residence time of 1.9 s at a temperature of 250 °C using the H-BEA (25), a selectivity to the aldehyde **(2)** of 34% was observed. Reducing the residence time to 1.3 s, the selectivity rises to 49%. For both residence times a complete conversion was obtained. The highest selectivity obtained for the keto aldehyde **(2)** was 82% using H-[B]-ZSM-5 in the vapor phase and 72% selectivity with H-ZSM-5 (60).

The influence of 'solvents' in the gas phase is similar to that of the liquid phase. Best yields were obtained with toluene, anisole and *o*-xylene. In acetonitrile the catalysts were found to be deactivated. The amount of solvent in the reaction mixture did not appear to influence the results, neither in liquid nor in vapor phase.

In conclusion, the use of zeolites as catalysts in the isomerization of isophorone oxide **(1)**, yields up to 86% keto aldehyde. The formation of **(4)** by decarbonylation of **(2)** could be reduced by increasing the catalyst loading in a liquid phase batch reactor or by conducting the reaction under short contact time in the gas phase. The heterogeneously-catalysed isomerization of terpene epoxides over zeolites is a suitable, non-polluting method of preparing relevant and useful aldehydes for the synthesis of perfumes and synthetic flavours.

2.4 Isomerization of 1,5-Dioxaspiro-[2.5]octane – 1,5-Dioxaspiro[2.5]ooctane can be easily converted to 4-formyltetrahydropyran in the presence of zeolites. The product is a valuable intermediate for the production of fragrances. This compound is hardly obtainable by known synthesis routes (Rosemund reduction). Using Al- or B-pentasil zeolite at 300 °C the 4-formyltetrahydropyrans can be synthesized in an easy manner and high yields over 97%.[24,25,28] Also, this reaction was carried out in the gas phase in a fixed bed reactor[4] with a WHSV = $2h^{-1}$ and a ratio of substrate/solvent (THF) = 50/50. Using SiO_2 as solid catalyst gave additionally a yield of 96.5%.[25] The time on stream in all testings was more than 6 hours.

Over B-MFI zeolite and SiO_2 as catalysts, more than 95% yield of 4-formylpyran could be achieved in a fixed bed reactor under gas phase conditions.[4,25,28]

2.5 Isomerization of 4,4,5,8-Tetramethyl-1-oxa-spiro[2.5]octane – 4,4,5,8-Tetramethyl-1-oxaspiro[2.5]octane **(1)** is obtained in a multistep synthesis from easily available (−)-(*S*)-β-citronellol or (+)-*R*-pulegone.[26] It reacts to the industrially-desired 2,2,3,6-tetramethyl-cyclohexane-carbaldehyde **(2)** (Figure 5) in the presence of acidic catalysts beside many other by-products.

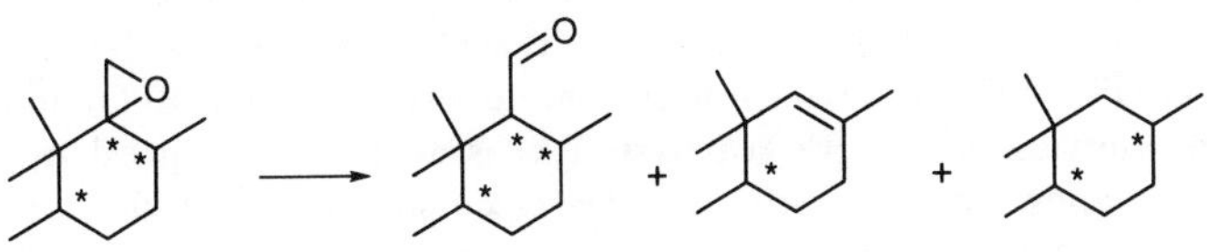

Figure 5 *Main products of the ring opening reaction of* **(1)** *under gas–phase conditions*

The aldehyde **(2)** is described as a key molecule for the synthesis of different highly active fragrances such as the saturated alcohol 1-(2,2,3,6-tetramethyl-1-cyclohexyl)-3-hexanol, appreciated by perfumers because of its woody-/ambra-like character[27] (Figure 6).

The suitable stereochemistry, not focused herein, is described to play an important role.[27] As reported in patents of Firmenich S.A., the ring opening

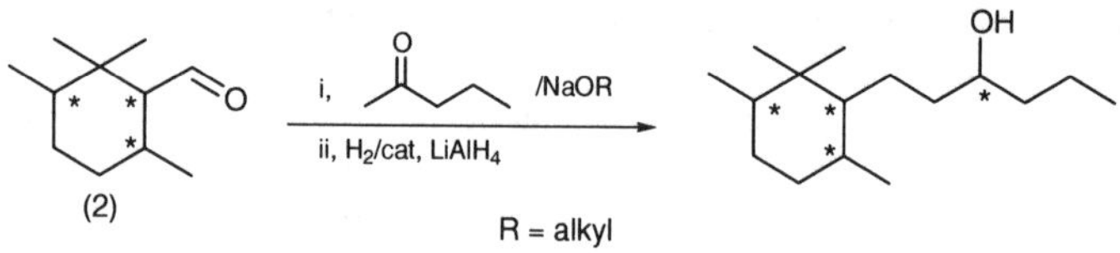

Figure 6 *Synthesis of the ambralwoody fragrance*

reaction of **(1)** was preferably carried out with Lewis acidic homogeneous catalysts such as BF_3-etherate,[4] MgI_2[26] or $SnCl_4$[26] with almost complete yield at room temperature in toluene as solvent. These catalysts, however, imply a process limitation to discontinuous liquid phase conditions. Major drawbacks are given by a relatively complicated separation procedure of products and catalyst in addition to corrosion, loss of catalyst and environmental problems such as high salt formation.

The heterogeneously-catalysed rearrangement of **(1)** has received no attention in literature. First, following the results of the 1,6-dioxa-spiro[2.5]octane rearrangement,[4,25] continuous gas phase conditions were applied in a fixed bed reactor and secondly under liquid phase conditions in a slurry reactor. The catalytic experiments carried out showed that two main reactions took place: rearrangement of **(1)** to the aldehyde **(2)** and an oxidative decarbonylation reaction to the olefin 1,3,3,4-tetramethyl-cyclohex-1-ene **(3)**, which is assumed to be caused by a formaldehyde elimination reaction. Also observed was a deoxygenation reaction to the alkane 1,1,2,5-tetramethylcyclohexane **(4)**, explained by elimination of CO. There are several other side products such as 2,2,3,6-tetramethylcyclohex-1-enyl-methanol, ring-contracting compounds and double bond isomers of dimethyl-isopropylene-cyclopentene.

The first tests were carried out to evaluate the behavior of different catalysts under gas phase conditions (Table 4). It was observed that conversion of **(1)** decreased rapidly in the presence of various acidic zeolites H-B-ZSM-5, H-ZSM-5 and H-US-Y. This behavior was even more distinctive with BPO_4 and Nb_2O_5. The low service time of the catalysts are assumed to be caused by strongly adsorbed compounds as well as coke precursors blocking the acidic sites. Surprisingly, a silica catalyst having gentle acidity showed the best performance. With selectivity to **(2)** of about 40% there was no drop in conversion after 8 h TOS, even at 230 °C.

On account of high activity and selectivity of this silica catalyst D11-10, the influence of temperature was studied in the range 150–300 °C (Figure 7). It is advantageous to run the reaction at about 250–300 °C. In this range, complete epoxide conversion is achieved within 6 h TOS in combination with the highest aldehyde selectivities of 45%. The formation of **(2)** and **(4)** are increased by increasing the temperature. Compound **(3)** passes a maximum at around 200 °C.

Adjusting the residence time with respect to the WHSV, a maximum value can be found for WHSV = 2 h^{-1}, which results in an increase of the selectivity to 47% at 100% conversion, compared to a WHSV of 1 h^{-1} (3 h^{-1}) with selectivities of

Table 4 *Conversion of* **(1)** *and selectivity to* **(2)–(4)** *with different catalysts in a fixed bed reactor*

Catalyst (module)	*Conv.* **(1)**(%)	*Sel.* **(2)** (%)	*Sel.* **(3)**(%)	*Sel.* **(4)**(%)	*Drop of conv.* (TOS 2/6 h or 2/8 h*)
H-B-ZSM-5 (34)	88.9[a]	42.1	15.0	0.1	94/87
H-ZSM-5 (60)	95.6[b]	39.8	22.5	0.2	99/83*
H-US-Y (70)	98.5[b]	29.2	25.5	0.4	100/96*
BPO_4	3.4[b]	29.4	<0.1	<0.1	52/2*
Nb_2O_5	71.7[a]	25.2	17.4	1.0	81/65
SiO_2D11–10**	100[a]	39.9	32.5	1.6	100/100

Catalyst 2.0 g (1.0–1.6 mm); 20 wt% **(1)** in THF, WHSV = 1 h^{-1}, T = 180 °C; p = 1 bar, [a] TOS = 4 h, [b] TOS = 6 h, N_2 carrier gas = 5 l/h, ** T = 180 °C, TOS (time on stream).

45.1% (44.6%) respectively. This is a very low sensitivity of the system to this parameter.

A reduced pressure also had an influence on the selectivity (45% at 1 bar and 50% at 100 mbar). A further optimization step includes a change of the solvent THF to the more apolar solvent toluene. Thereby, selectivity to **(2)** could be enhanced to 53% at complete conversion and no drop of deactivation is observed within 8 h TOS. The positive effect of toluene in comparison to THF might be explained by a different adsorption behavior towards the active sites of the silica catalyst.

It is surprising that silicagel D 11-10 shows better results than zeolites in the gas phase rearrangement reaction. In the case of 1,6-dioxaspiro[2.5]octane, excellent results were obtained in the presence of this silicagel as well as of pentasil zeolites.[28] In the present case, the molecular dimensions of the starting material seem to be too bulky owing to the additional methyl groups and cannot enter the pore volume of the middle sized weakly acid boron-pentasil zeolite.

In contrast to the gas phase reactions, surprisingly, zeolitic catalysts show a superior catalytic performance in the liquid phase in a slurry reactor. Good conversions and selectivities are obtained over H-US-Y (70) and H-US-Y (70) treated with HCl whose catalytic features are described elsewhere (Table 5).[6]

Highest selectivities are obtained at very low temperatures, around 0 °C. In order to achieve high conversion, long residence times are necessary. This means that the space time yield (STY) in the liquid phase at low temperatures is less than in the gas phase. The HCl-treated H-US-Y zeolite shows a better catalytic performance than the untreated one. The reason for that might be the high Lewis acidity and high mesoporosity in this particular material.[6]

H-B-ZSM-5, H-Al-ZSM-5, Nb_2O_5, BPO_4 and SiO_2 are not suitable catalysts for the conversion in the liquid phase and H-Al-ZSM-5 (60) shows a good catalytic performance with a conversion of 91.1% and a product selectivity of 31% at a high temperature of 65 °C.

In summary, we have established a facile heterogeneous catalytic method for

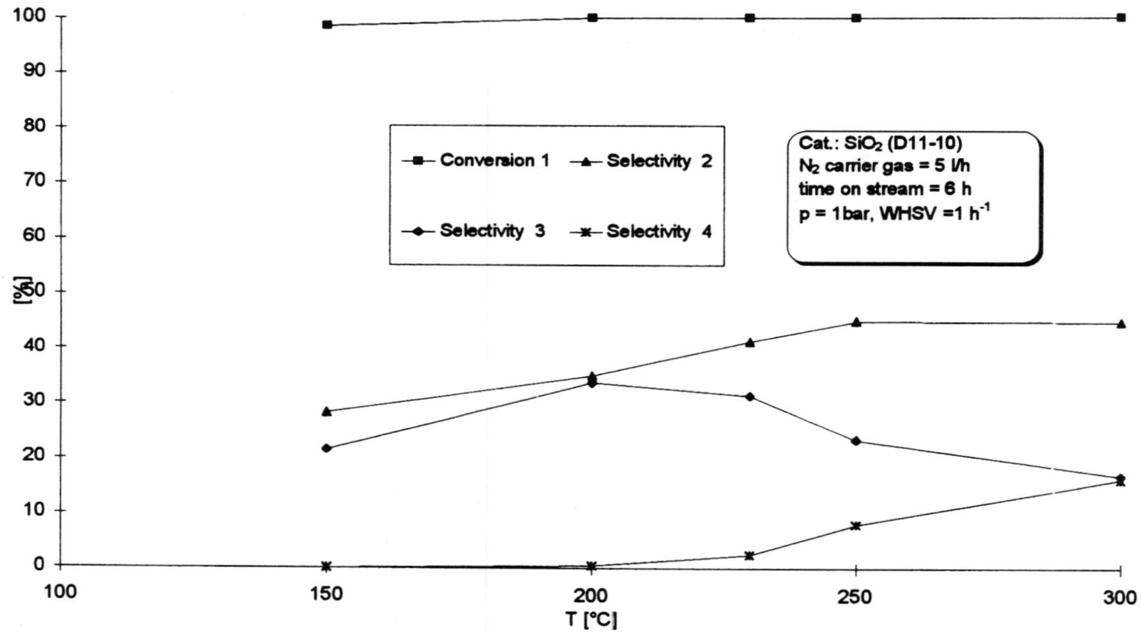

Figure 7 *Influence of temperature on conversion of* **(1)** *(*20 wt% *in THF) and selectivity to* **(2)–(4)** *over silica catalyst in a fixed bed reactor*

Table 5 *Liquid phase reactions for the rearrangement of* **(1)** *to* **(2)** *using Catalyst A and acid treated Catalyst B in toluene*

Catalyst	T(°C)	t(h)	*Conv.* **(1)**(%)	*Sel.* **(2)**(%)
H-US-Y (70)	0	6	12.5	63.4
		24	22.0	59.8
		72	29.9	55.2
H-US-Y (70)/HCl	0	2	47.1	59.8
		24	79.2	58.6
		72	95.1	47.2
H-US-Y (70)/HCl	25*	6	99.7	41.0
		24	100.0	44.0

Catalyst 2.0g; load = 7.5g; solvent: 30.0 g toluene; $t = 24$ h; ratio: 10 ml acid/g catalyst.

the synthesis of the aldehyde intermediate 2,2,3,6-tetramethyl-cyclohexane-carbaldehyde **(2)** with easy handling, lack of corrosion and ease of separation of the products from catalyst. A silica catalyst is most suitable for the conversion in the gas phase, whereas a H-US-Y (70) treated with HCl shows the best performance in the liquid phase. Furthermore, it was realized that the introduction of an alkyl residue in position 5 of the ring system leads to a clear drop in the 5-alkyl-4-formylpyran selectivity in comparison to the conversion of the compound without this alkyl group. The yield obtained in the presence of heterogeneous catalysts is not as high as in the presence of conventionally used catalysts. Improvements of the heterogeneous catalysts are still necessary to become competitive with the homogeneous system.

3 Alkoxylation and Mercaptanization of Limonene and α-Pinene

1-Methyl-4-[α-alkoxyisopropyl]-1-cyclohexenes are used as flavours and fragrances for perfume and cosmetic products, as additives for pharmaceuticals and agricultural chemicals, as well as in the food industry.[29–32] They are also very important for the synthesis of specialty fine chemicals and natural-like products.[33,34]

Among 1-methyl-4-[α-alkoxyisopropyl]-1-cyclohexenes (Figure 8) the methyl ether smells mangofruit-like and might be used as a perfumery material (Figure 9). The ethyl derivative has a green citrus fragrance, the propyl one has a fresh grass flavour and the butyl one smells woody.

Several ways to produce such α-terpinyl alkyl ethers are known, but limonene and α-pinene are the most used feedstock molecules (Figure 8). Alkoxylation of limonene and α-pinene over homogeneous or heterogeneous catalysts as, for example, strong acids HCl, H_2SO_4 or *p*-toluenesulfonic acid,[35–40] aluminium trichloride and boron trifluoride etherate[41] and acidic cation exchange resin[42] has

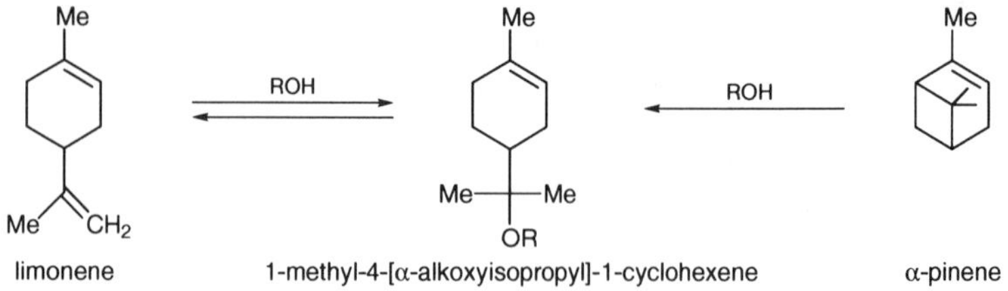

Figure 8 *Formation of 1-methyl-4-[α-alkoxyisopropyl]-1-cyclohexene*

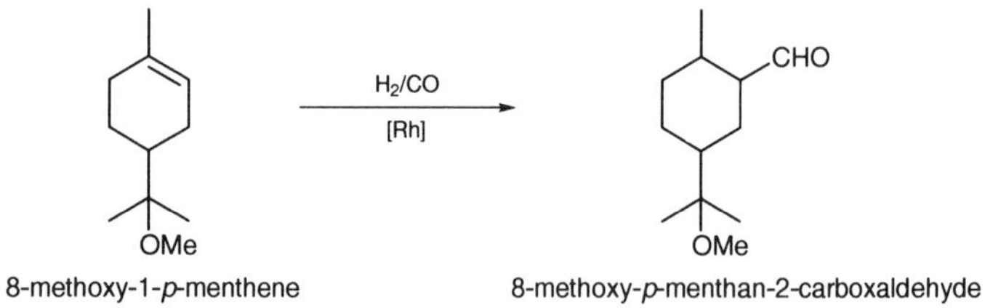

Figure 9 *Synthesis of* 8-*methoxy*-p-*menthan-2-carboxaldehyde having mangofruit flavour*

already been reported. The reaction of α-terpinyl chloride (which can be obtained by the addition of dry hydrogen chloride to limonene) with protic solvents (C_1–C_5 alcohol) over zinc or silver compounds as catalysts leads to the corresponding ethers with good yields (up to 95%).[43–45] Owing to corrosion and environmental problems, almost all of the catalysts for the production of 1-methyl-4-[α-alkoxyisopropyl]-1-cyclohexenes are unacceptable for the industry.

Another process to produce alkyl terpinyl ethers is the oxymercuration–demercuration of limonene.[46,47] Owing to the above mentioned problems and the high toxicity of the mercury salts, this process is not feasible for an industrial application.

The zeolite-catalysed alkoxylation of limonene[48,49] or α-pinene[50,51] over acid-treated mordenite, clinoptilolite and ferrierite as catalysts has already been reported in the literature.[50,51] The best results were obtained for methoxylation of limonene in the presence of a clinoptilolite-type zeolite (60% yield). The alkoxylation of α-pinene with methanol in the presence of mordenite achieved also the highest yield of 66% for 1-methyl-4-[α-methoxyisopropyl]-1-cyclohexene. Syntheses of 1-methyl-4-[α-alkoxyisopropyl]-1-cyclohexenes *via* zeolite-catalysed alkoxylation of other terpenes were reported in a review paper.[52]

The aim is still (a) improvement and optimization of the already existing zeolite-catalysed processes with regard to conversion and selectivity, (b) regeneration of the zeolite, and (c) running of the reaction in a continuous process to make an industrial application feasible.

3.1 Alkoxylation of Limonene

3.1.1 Reactions in the Batch Reactor – Methanol reacts with limonene over acidic catalysts in a batch reactor to 1-methyl-4-[α-methoxyisopropyl]-1-cyclohexene (α-terpinyl methyl ether) as the main reaction product (see Figure 8; R = Me). Besides the desired methoxylation, isomerization reactions leading to terpinolene and traces of α- and γ-terpinene can be observed. Furthermore, the addition of methanol to the terpinyl methyl ether leads to the undesired *cis*- or *trans*-1,8-dimethoxy-*p*-menthane. The amount of unidentified products does not exceed 1%. At high temperatures and long reaction times the reverse reaction of the α-terpinyl methyl ether and the other addition products to limonene and its isomers can be observed. The reaction scheme of the alkoxylation of limonene is illustrated in Figure 10

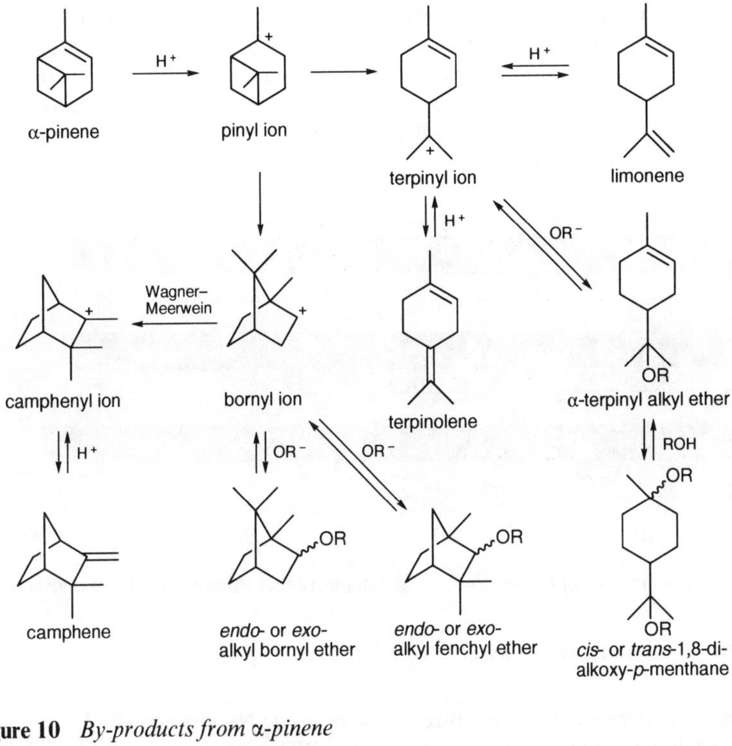

Figure 10 *By-products from α-pinene*

The reaction pathway is identical to that of α-pinene alkoxylation except that no bicyclic compounds are formed. As mentioned in the literature,[48,49] alkoxylation of limonene to 1-methyl-4-[α-alkoxyisopropyl]-1-cyclohexene can be carried out only in the presence of acidic catalysts. After a catalyst screening using various zeolitic and non-zeolitic acid heterogeneous catalysts we found that β-zeolite is the best candidate.

Addition of methanol to limonene in the presence of a β-zeolite produces the highest selectivity to 1-methyl-4-[α-methoxyisopropyl]-1-cyclohexene of about 93% at 91% conversion. Surprisingly, the highest yield of about 85% has been obtained at room temperature. The other zeolites and solid acids applied for the alkoxylation of limonene reveal considerably lower conversion and selectivity.

It is interesting to notice that the destruction of the structure of β-zeolite by treatment with strong acids or high temperature leads to a complete deactivation of the catalyst for limonene alkoxylation. By using a higher reaction temperature only isomerization and polymerization products have been obtained. 1-Methyl-4-[α-methoxyisopropyl]-1-cyclohexene or other addition products cannot be found.

The most likely reason for the high activity of β-zeolite is the relatively high BET surface area of the catalyst (750 $m^2\ g^{-1}$). Furthermore, there are hints by temperature-programmed desorption (TPD) of absorbed ammonia that a large amount of acid sites is present. We assume that the alkoxylation of limonene takes place inside the pore structure of the β-zeolite. The high selectivity of β-zeolite might be originated from suitable acid sites in pores of its defined size and shape.

Several β-zeolite samples have been used as catalysts (A,B,C). All of them show the XRD-diffraction pattern of β-zeolite. However, it is an astonishing result that all of them showed excellent selectivities (> 92%) while the conversion varied from 40 to 90% (Figure 11).

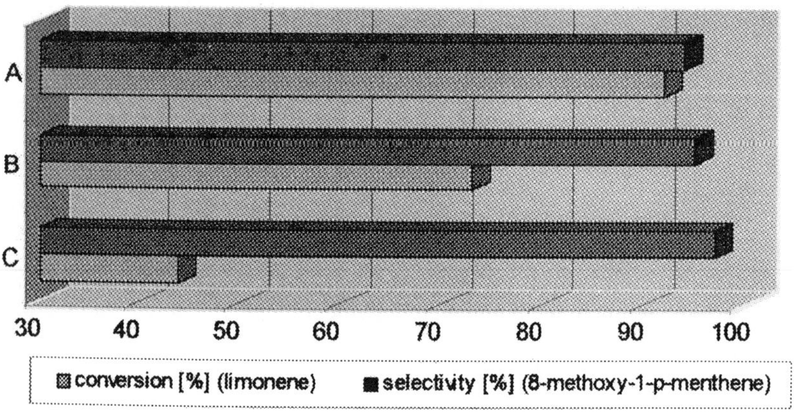

Figure 11 *Conversion and selectivity with different β-zeolites*

The characterization of these β-zeolite samples *via* N_2-sorption (BET) revealed, that catalyst C only had micropores and a BET surface area of 640 $m^2\ g^{-1}$ whereas the two catalysts with a higher conversion had meso- and macroporous volumina resulting in a faster mass transport. The difference in the performance of samples A and B can be attributed to the highest BET surface area of A (750 $m^2\ g^{-1}$) in contrast to B (680 $m^2\ g^{-1}$). Additionally, TPD measurements revealed that the total amount of acidic sites on sample A was still higher than on sample B.[53] A shape-selective effect in the microporous system might be also involved since the dialkoxylation of limonene occurs only to a minor extent. In order to achieve high selectivity it is also necessary that the reaction products can leave the zeolitic material very fast *via* the meso- and macropores. On account of high activity and selectivity of β-zeolite and good product yields for limonene alkoxylation in the presence of this catalyst, the influence of various reaction conditions to optimize the reaction procedure is studied by using this β-zeolite.[54] The dependence of limonene conversion and selectivity to 1-methyl-4-[α-methoxyisopropyl]-1-cyclohexene over β-zeolite in a batch reactor on temperature shows that the optimum has been found at 40 °C for the reaction time of 5 h. At this temperature, the highest selectivity of about 92% at 93% conversion has been obtained. Above 40 °C, the selectivity is considerably reduced due to either consecutive reactions and decomposition of 1-methyl-4-[α-methoxyisopropyl]-1-cyclohexene or to other side reactions occurring at an even shorter reaction time. At a reaction temperature of 60 °C and a reaction time of 2 h, a maximal conversion of 94% is reached. A higher temperature of 80 °C promotes increasingly the reverse reaction of the α-terpinyl methyl ether to limonene and its isomers with elimination of the methanol, so that the conversion remains almost constant while the selectivity to the desired product decreases. Lower temperature results in a drop of the limonene conversion. However, the latter can be compensated by a longer reaction time. The selectivities obtained at lower reaction temperatures are up to 98%.

As in the case of limonene methoxylation, the addition of higher alcohols delivers also the corresponding α-terpinyl alkyl ethers as the main reaction products. However, the conversions of limonene by using ethanol, propan-1-ol, butan-1-ol or pentan-1-ol are suprisingly lower than that with methanol.[54] It would have been expected that in a electrophilic addition the longer-chain alcohols would be more reactive. The higher conversion when methanol is used is due possibly to fewer diffusion constraints. The maximum selectivity obtained for ethoxylation of limonene is 94% at 42% conversion. By increasing the chain length of the linear alcohols, a decrease of the selectivity was observed. This might be caused by steric hindrance in the zeolite pore structure in such a way that longer alcohols do not easily form the transition state. Good product yields are achieved only in the case of the linear alcohols, whereas for iso-alcohols, such as propan-2-ol, butan-2-ol or pentan-2-ol, a remarkable drop in limonene conversion and selectivity to corresponding α-terpinyl alkyl ethers is observed; with 2-methyl-2-propanol, no addition products have been found. The reason for those effects might again be steric hindrance and diffusion restrictions in the elliptical pores of β-zeolite resulting in a lower reactivity.

3.1.2 Reaction in an Integral Continuous Flow Fixed Bed Reactor – For methoxylation of limonene in the presence of a β-zeolite in a continuous flow-type apparatus with an integral fixed bed reactor, the same tendencies as for the batch reactor are observed in general. The great advantage of the reaction in an integral continuous fixed bed reactor is the easy separation of the products from zeolite and a higher productivity than in a batch process. Surprisingly, at the highest (80 °C) and the lowest (25 °C) reaction temperatures the conversion is almost constant. The highest limonene conversion obtained is about 95% at

80 °C. The selectivity to the α-terpinyl methyl ether increases within 9 h TOS. However, the selectivity is lowered to 75% at 80 °C while at a lower temperature it increases to 95% during 9 h TOS.

At the limonene conversion of about 60% and a reaction temperature decreased to 40 °C, selectivities higher than 95% are achieved. A temperature lower than 40 °C causes a noticeable decrease of the limonene conversion. In spite of this, selectivity to α-terpinyl methyl ether remains on the same level of about 95%. The product distribution is nearly the same as in the batch reactor.

The chosen mass ratio between methanol and limonene of 2:1 is advantageous, because the feedstock and the products are well dissolved at this concentration. However, the excess of methanol can react with 1-methyl-4-[α-methoxyisopropyl]-1-cyclohexene to *cis*- or *trans*-1,8-dimethoxy-*p*-menthane in a consecutive reaction. This reaction is strongly accelerated when the limonene conversion exceeds 90%. That consecutive reaction together with the decomposition of the desired product and other side reactions decrease the selectivity at a high limonene conversion.

The selectivity to the α-terpinyl methyl ether increases with the reduction of the residence time. The experiments have revealed that not only higher reaction temperature and the limonene conversion above 90%, but also a longer contact time caused by a lower WHSV favor the consecutive reaction of the excess methanol with 1-methyl-4-[α-methoxyisopropyl]-1-cyclohexene to *cis*- or *trans*-1,8-dimethoxy-*p*-menthane.

The application of various linear, iso- and branched alcohols for the addition to limonene in the integral fixed bed reactor gives the same results as those in the batch reactor.[54]

The catalyst service time can be increased by using a higher temperature as deactivating compounds desorb more easily at elevated temperatures. However, after an initially small increase, the limonene conversion decreases in one to two days. It was found that the catalyst could be regenerated.[54]

3.2 Alkoxylation of α-Pinene

3.2.1 Reactions in the Batch Reactor – The addition of methanol to α-pinene in the presence of the above mentioned β-zeolite as catalyst in the batch reactor results in the cleavage of the cyclobutane ring and yields 1-methyl-4-α-methoxyisopropyl]-1-cyclohexene (α-terpinyl methyl ether) as the main reaction product. The most common by-products to be found are isomerization compounds like camphene, limonene and terpinolene, and several bicyclic and double addition products, *e.g.* endo- or exo-methyl bornyl ether, endo- or exo-methyl fenchyl ether and *cis*- or *trans*-1,8-dimethoxy-*p*-menthane.

Depending on the reaction conditions, over 50% selectivity to the desired product can be achieved. The other methoxy compounds formed by the secondary addition of methanol to the product or to various isomerization products are obtained with 28% yield. By summarizing all addition products, the selectivity of the methoxy compounds increases to about 79%. The reaction mixture contains only 20% of terpenes including the unreacted α-pinene. The amount of unidentified products does not exceed 1% and can therefore be neglected.

The influence of temperature has been investigated within the range from 25–80 °C. The optimum temperature for α-pinene methoxylation has been found at 40 °C at a reaction time of 5 h. At this temperature, the highest selectivity obtained over β-zeolite is about 54% at 92% conversion. A temperature above 40 °C gives significant reduction in selectivity due to consecutive reactions or decomposition of 1-methyl-4-[α-methoxyisopropyl]-1-cyclohexene and other side reactions at even shorter reaction times. A lower temperature causes a drop of the α-pinene conversion. However, this can be compensated by a longer reaction time.[54] The dependence of temperature and reaction time is consistent with the above observations for limonene methoxylation. However selectivities up to 98% are obtained for limonene methoxylation, whereas the highest selectivity achieved for methoxylation of α-pinene amounts to 54%. On account of the above mentioned additional formation of various bicyclic addition products, the selectivities for all methoxy compounds increase to about 80%. This indicates that for the production of 1-methyl-4-[α-methoxyisopropyl]-1-cyclohexene β-zeolite is more selective for the addition of methanol to limonene than to α-pinene. The activity of the catalyst, however, is nearly the same for both reactions.

Methoxylation of α-pinene in the continuous flow-type apparatus with an integral fixed bed reactor over β-zeolite as catalyst shows very similar results as in the batch reactor. The dependence of α-pinene conversion and selectivity to the α-terpinyl methyl ether on temperature and time on stream are in agreement with observations for limonene methoxylation.

The heterogeneously-catalysed alkoxylation of α-pinene and limonene over β-zeolite provides excellent results in both a discontinuous batch reactor and a continuous flow-type apparatus with a fixed bed reactor. In both reactors, the use of methanol as addition compound and limonene as feedstock gives 1-methyl-4-[α-methoxyisopropyl]-1-cyclohexene with the yield of 85% (conversion: 93%, selectivity: 92%). By means of variation of the reaction parameters, the limonene conversion can be adjusted within the range 40–90%. The selectivity to 1-methyl-4-[α-methoxyisopropyl]-1-cyclohexene remains always about 95%.

For mixtures of methanol and α-pinene, the highest yield of corresponding α-terpinyl methyl ether is only of about 50% (conversion: 99%, selectivity: 51%) in both reactor types, as several bicyclic and double addition products are formed in parallel. The selectivity of all addition products can reach values of about 85%.

The most likely reasons for the high activity of β-zeolite are the relatively high BET surface area of the catalyst (750 $m^2\ g^{-1}$) and the large amount of acid sites measured by temperature-programmed desorption (TPD) of ammonia.[53] We assume that the alkoxylation of limonene takes place inside the pore structure of the β-zeolite. We also assume that the high selectivity of β-zeolite originates from suitable acid sites in pores of its defined size and shape.

3.3 Sulfur-containing Fragrances and Flavours – 1-*p*-Menthene-8-thiol, which is a part of the grapefruit aroma, and 1-*p*-menthene-8-thioethers, which are used as ingredients for perfumes and cosmetics, can be synthesized from terpenes (*e.g.* α-pinene and limonene) with hydrogen sulfide using $AlCl_3$ and $AlBr_3$ as catalysts.[55,56] The disadvantages of this type of catalyst are well known and frequently named herein.

Acid zeolite catalysts offer a very good alternative for the clean synthesis of these sulfur-containing substances. A suitable feedstock is the 4-isopropenyl-1-methyl-1-cyclohexene. In the presence of a commercial β-zeolite (25) hydrogen sulfide is added to the autoclave at a reaction temperature 50 °C at a pressure of 17 bar. The conversion of limonene is 65.1% and the selectivity to 1-*p*-menthene-8-thiol is 43.9%. These are very promising results and they can be improved by using a commercial H-US-Y zeolite which rendered a conversion of 76.8% and a selectivity of 64.3%.[57]

4 Esterification of Dicyclopentadiene with Carboxylic Acids

The esterification of cyclic olefins with carboxylic acids, such as dicyclopentadiene or naturally occurring terpenes, yields compounds of industrial value.[58] For instance, the esterification of dicyclopentadiene with saturated carboxylic acids, such as acetic acid, leads to a starting material for the flavour and fragrance industry, as well as the resulting alcohols in consecutive hydrolysis.[59] By using unsaturated carboxylic acids as nucleophiles, the resulting esters are also useful compounds in the manufacture of binders for polymer concrete (Scheme 1).[60]

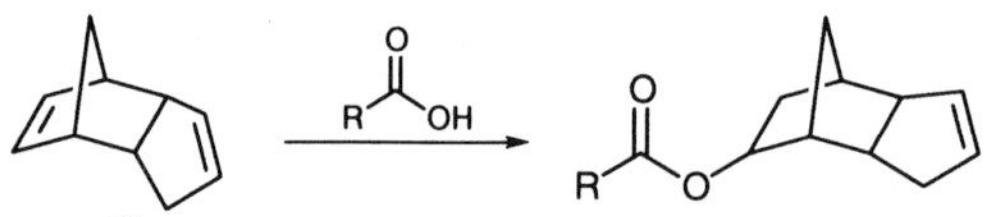

(1)

Scheme 1

The esterification is known to be catalysed quite well by homogeneous catalysts.[61] Several traditional catalysts such as BF_3 render the desired compounds in high yield.[62] However, toxicity, corrosion and difficult separation procedures of the catalysts are well-known disadvantages of the homogeneously-catalysed reaction in industrial applications. Up to now, several heterogeneous catalysts have been found, which show a good performance in this reaction.[63] However, only very reactive olefins are effectively catalysed heterogeneously.[64] In the case of less active olefins in order to obtain high yields very high carboxylic acid olefin ratios or very high catalyst loadings are needed.[65] Unfortunately, raising the reaction temperature to enhance the reaction rate causes a decrease in selectivity.

The use of Nafion/silica composite and zeolite catalysts in this kind of reaction has been studied. The comparison of zeolites with these new materials seems interesting because excellent catalytic performance of such ion exchange resins in several reactions were found.[66–70] This new material can be considered as nano-sized Nafion particles entrapped within a porous silica matrix. Thus the surface area of the dispersed Nafion in the composite is much larger than that of the original Nafion material, resulting in a much improved accessibility of the acid sites on the catalyst.

These composites containing different amounts of Nafion catalyst are compared with the original material and with Amberlyst A 15, a well-known macroreticular ion-exchange resin, which achieves remarkable results in several acid-catalysed reactions.[71]

4.1 Esterification of Dicyclopentadiene with Saturated Acids – During the acid-catalysed esterification of dicyclopentadiene **(1)**, a rapid Wagner–Meerwein-type rearrangement occurs simultaneously, according to the reaction Scheme 2. As a result, the ester compound contains exclusively the dicyclopentenyl component in the exo-form.

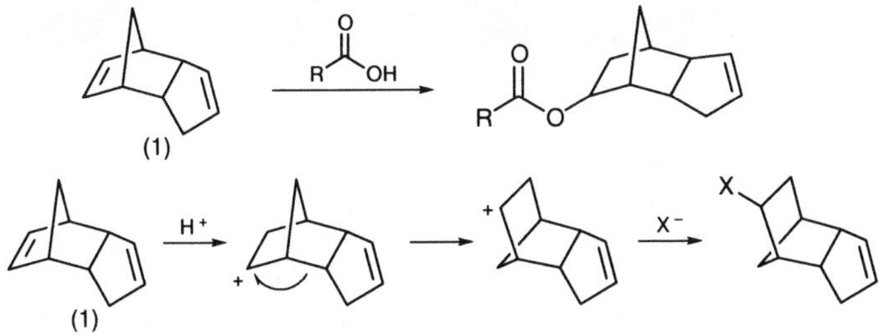

Scheme 2

Figure 12 shows the esterification of different saturated carboxylic acids with dicyclopentadiene **(1)**. In all depicted experiments the carboxylic acid/dicyclopentadiene molar ratio is about 4 and the amount of catalyst is 10% by weight of dicyclopentadiene at a reaction temperature of 80 °C. In this reaction the Nafion/silica composite catalyst is more active compared to the Amberlyst resin, in particular with respect to the amount of the acid groups on the resin. The ion-exchange capacity of pure Nafion is about 0.89 mmol g^{-1}; thus, the composite capacity is only about 0.12 mmol g^{-1}, whereas the Amberlyst resin provides 4.8 mmol g^{-1}. As a result, in the case of acetic acid as nucleophile the turnover number (TON) of the sulfonic acid groups in the Nafion-based material are two magnitudes higher than the ones of the Amberlyst material. By using carboxylic acid with longer chain length the difference in activity between the Nafion/silica composite and the Amberlyst seems to be even higher. With propionic acid the Amberlyst catalyst renders only just about 5% yield, and with both isobutyric and caproic acid only traces of the products are obtained. The Nafion-based material leads to more than 45% yield by using propionic acid and almost 20% for caproic acid.

Carrying out experiments with equal ratios of starting material **(1)** to sulfonic acid group of the specific catalysts reveal the same pattern as obtained in the experiments described above. Figure 13 shows the results obtained over some Nafion/silica composite materials with different amounts of Nafion compared to the Amberlyst ion-exchange resin. All Nafion/silica composite catalysts show

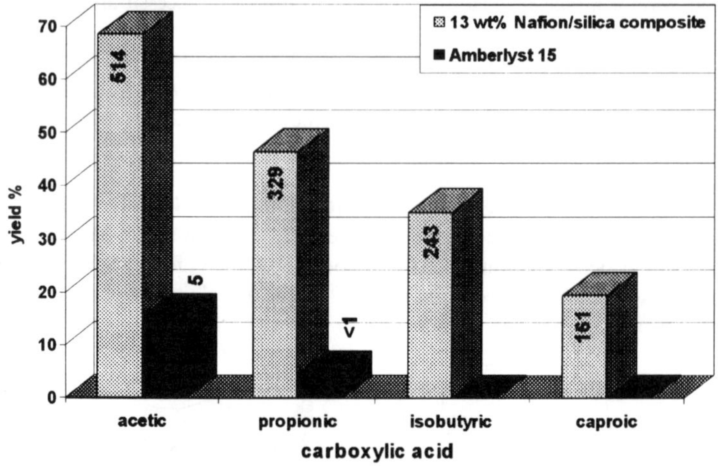

Figure 12 *Esterification of* **(1)** *with different carboxylic acids over* 13 wt% *Nafion/ silica composite (SAC 13) and Amberlyst 15; the figures on the columns depict the TON mol/mol(acid group)*

complete conversion independently of the amount of Nafion resin. In contrast, a conversion of only about 80% is obtained over the pure polymer. Again, the Amberlyst material achieves only poor conversion of about 12%.

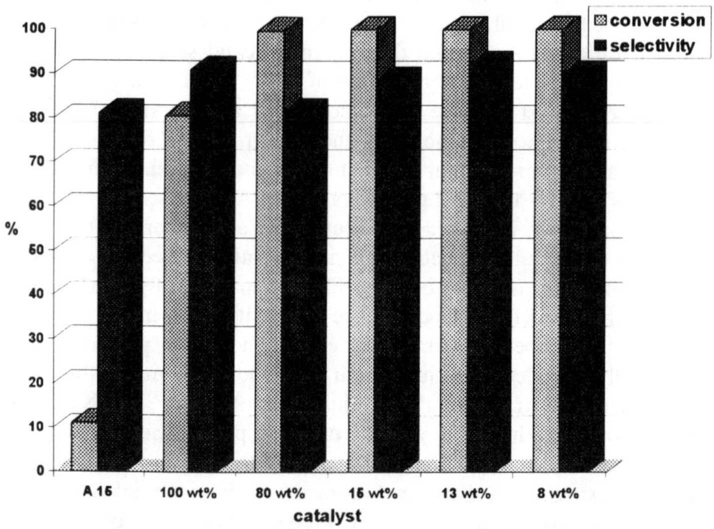

Figure 13 *Esterification of* **(1)** *with acetic acid over Nafion/silica composites and A 15;* T = 120 °C, t = 2 h, *acetic acid in* 4 *molar excess, molar ratio of* **(1)** *to* SO_3H *group* = 411

It seems, owing to the very high activity of the Nafion/silica-based catalysts, that these reaction conditions are not suitable to observe any differences in activity between the various Nafion/silica materials. All composites with a ratio of starting material to sulfonic acid group of about 1000 reveal complete conversion.

4.2 Esterification of Terpenes – Under standard conditions and temperatures from 40 to 100 °C limonene and α-pinene were esterificated with acetic and acrylic acid, respectively, over the Amberlyst resin as well as the Nafion composite. The conversion is high, however, the reaction is very unselective and no distinct main product is obtained. It seems that both terpenes rearrange to several different products before the final esterification takes place. Using camphene **(2)** as a starting material, much better results are obtained (Scheme 3).

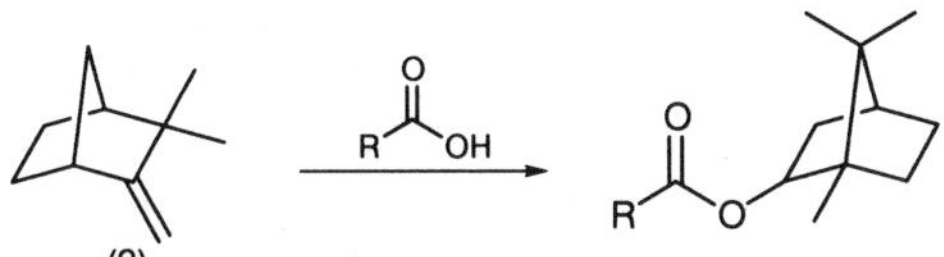

Scheme 3

The esterification of camphene **(2)** with acetic acid at a reaction temperature of 80 °C showed a conversion of 88% with a selectivity of 87% with the catalyst SAC 13 and a selectivity of 88% with the Amberlyst A 15. The carboxylic acid was in four-fold molar excess with 10 wt% catalyst. Both, the SAC 13 and the A 15 show high activity (each 88% conversion) and high selectivity in this reaction. This olefinic compound seems to be so reactive as to render high conversion independently of the acid strength of the catalyst. This could be due to the terminal double bond and to the very stable carbocation formed by protonation.

In the esterification of all studied cyclic olefins the Nafion/silica composite catalysts are much more active than the original pure resin and than the A 15 resin. That is also valid for cyclization with unsaturated carboxylic acids.[72] Even at comparatively low temperatures the composite materials achieve high conversion. This is due to the much-improved accessibility of the acid sites and to their higher acid strength. Although the ion-exchange capacity of the 13 wt% Nafion-containing material (SAC 13) is more than 40 times lower than Amberlyst the resulting activity per mass of the catalyst is much superior to the Amberlyst resin. Only, when very reactive olefins are used do all catalysts reveal high reactivity.

5 Heterogeneous Catalysed Synthesis of *p*-Cymene

p-Cymene is an important starting material for the production of intermediates such as *p*-cresol and can also be used, for example, in the manufacturing of fragrances, herbicides, pharmaceuticals and heat transfer media.[73–75] Conventially, cymenes are synthesized by Friedel–Crafts alkylation of toluene with propene using hazardous catalysts such as $AlCl_3$.[76] Moreover, the process only leads to mixtures of cymene isomers in thermodynamic equilibrium composition.

Recently, selective synthesis of *p*-cymene from toluene and propene or isopropanol over zeolite catalysts has been thoroughly investigated.[77–79] Zeolite-based processes avoid the disposal of spent catalyst, product contamination by the catalyst, separation of the catalyst from the product and corrosion of the reactor and tubes. The results using various zeolite types were promising. However, the formation of undesired *n*-propyl toluene is observed particularly in the presence of MFI type zeolites, whereas large pore zeolites yield significant amounts of *m*- and *o*-cymene besides the desired *p*-cymene. However, low conversion and relatively low selectivity are the drawbacks of these investigations.

Other approaches to synthesize selectively *p*-cymene start from α-limonene, a material with the same carbon structure as *p*-cymene. α-Limonene is widely available as a cheap by-product in the orange juice production and paper industry. In early studies[81,82] it was shown that α-limonene can be disproportionated to *p*-menthane and *p*-cymene over Pt and Pd catalysts on carrier materials. According to Scheme 4, 33% of the α-limonene is lost as a saturated compound. Also, Se promoted PdO on charcoal as carrier material has been suggested.[82] However, such a catalyst is not favourable from an environmental point of view.

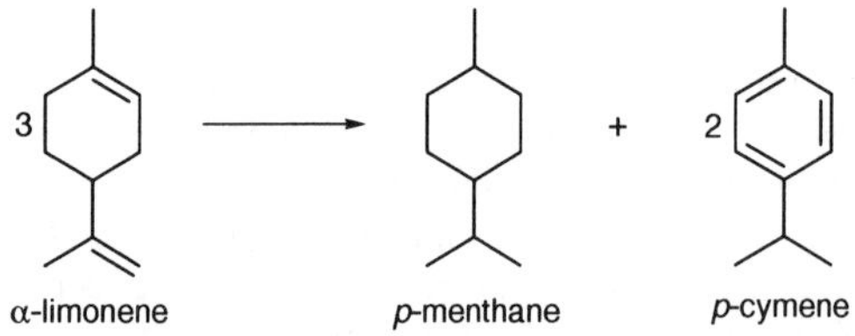

Scheme 4

Using zeolite-supported, Ce-promoted Pd catalysts brought enhanced dehydrogenation activity during α-limonene conversion.[83] Thereby, they used preferably weakly acidic boron-pentasil zeolites and obtained *p*-cymene yields around 85%.

5.1 Dehydrogenation of α-Limonene

5.1.1 Conversion over Unmodified H-ZSM-5 and H-US-Y – The influence of the pore structure of the zeolites on the product distribution of *m*- and *p*-cymene isomers has been studied in the presence of an unmodified shape-selective medium pore-sized H-ZSM-5-(55) and the unmodified H-US-Y zeolite at temperatures between 200 and 300 °C.[86] In contrast to the H-US-Y, the *p*-cymene formation in the presence of H-ZSM-5 is favoured due to the shape selectivity.[86]

5.1.2 Conversion over Pd-modified H-ZSM-5 – Dehydrogenation of α-limonene to cymenes is only obtained in the presence of strong acidic sites of H-ZSM-5 with a very low selectivity for *p*-cymene (18%). In such a case very strong deactivation occurs. Therefore, the dehydrogenation activity has to be enhanced, for example, by insertion of noble metals. The results of α-limonene conversion for 6 h TOS over Pd-exchanged H-ZSM-5-(55) are shown in Table 6.

Table 6 *α-Limonene conversion over Pd-containing H-ZSM-5-(55), 1 wt% Pd*

TOS (h)	0.2	1	2	4	6
Conversion (%)	99.8	98.8	95.3	93.6	92.4
Selectivity (%)					
p-Cymene	16.9	8.2	5.6	4.7	3.8
Σ*m*-, *o*-Cymene	1.2	0.7	0.3	0	0
Σα-, γ-Terpinene	22.5	36.6	44.3	49.7	54.3
Terpinolene	2.3	5.0	7.2	12.2	14.5
Other $C_{10}H_{16}$-terpenes	12.0	16.5	16.9	18.1	18.1
Σ*p*-Menthane, *p*-menthene	13.1	6.3	4.1	2.9	2.2
Coke precursor ($\geqslant C_{20}$)	3.7	2.4	1.9	1.1	0.8
Cracked products (toluene + lower alkanes, propanes)	28.3	24.3	19.7	11.3	6.3

$T = 250\,°C$; $p = 1$ bar; carrier gas N_2 4 Nl/h; WHSV = $1.3\,h^{-1}$.

α-Limonene conversion decreases from 99.8 to 92.4% after 6 h time on stream (TOS). Dehydrogenation activity monitored by the *p*-cymene selectivity declines strongly within 6 h. Thus, the loss of dehydrogenation activity results in an increase of isomerized $C_{10}H_{16}$ terpenes. The share of terpinolene, a product of the isomerization of the isopropylene group of α-limonene, on the $C_{10}H_{16}$ terpenes in the product mixture rises with increasing TOS. Accordingly, the ratio of terpinolene to α- and γ-terpinene, products of the double bond shift into the cyclohexene ring increases, indicating an ongoing decline of the acid sites responsible for the isomerization. The Pd sites of the catalyst promote the dehydroisomerization to *p*-cymene, whereas the acid sites of the catalyst promote a ring opening and cracking to cleavage products. Therefore, for short time on stream (< 2 h) cracking of *p*-cymene to toluene and propane/propene on the acid sites is the dominant reaction. With increasing TOS, the acid sites of the catalyst deactivate and thus the formation of cleavage products decreases.

5.1.3 Conversion over Pd/Ce-modified H-ZSM-5 – A way to promote the noble metal function is the addition of rare earth metals to the catalyst system. It has been reported that rare earth metals can act as anchors for noble metals and thus enhance dispersion and accessibility of noble metal sites in a zeolite.[84–87] The product distribution for α-limonene conversion over first Ce-exchanged, afterwards Pd-exchanged, H-ZSM-5-(55) is given in Table 7.

p-Cymene selectivity and service time are drastically enhanced when adding a Ce promotor. Initial *p*-cymene selectivity is up to 50.1%, *versus* 16.9% for the unpromoted catalyst. However, *p*-cymene selectivity decreases after 6 h time on

Table 7 *α-Limonene conversion over Ce- and Pd-containing H-ZSM-5-(55), 2 wt% Ce, 1 wt% Pd*

TOS (h)	0.2	1	2	4	6
Conversion (%)	99.8	99.6	99.2	98.7	98.0
Selectivity (%)					
p-Cymene	50.1	45.1	25.3	25.5	25.9
Σ*m*, *o*-Cymene	0	0	0	0	0
Σα-, γ-Terpinene	15.3	30.7	37.3	38.4	36.6
Terpinolene	2.5	5	11.4	12	13.3
Other $C_{10}H_{16}$-terpenes	6.3	7.2	12.5	13.4	14.5
Σ*p*-Menthane, *p*-menthene	16.6	14.2	8.7	8.5	8.5
Coke precursor ($\geq C_{20}$)	2.6	1.8	1.3	0.4	0
Cracked products (toluene + lower alkanes, propane)	19.2	7.1	3.5	2.1	1.2

$T = 250\,°C$; $p = 1$ bar; carrier gas N_2 4 Nl/h; WHSV = 1.3 h^{-1}.

stream. The catalyst is much too strongly acidic and catalyses not only formation of unwanted split by-products from α-limonene but also leads to deactivation of the Pd sites due to coke deposition.

5.1.4 Conversion over Pd/Ce-modified Na-ZSM-5 – Surprisingly, employing a non-acidic Na-ZSM-5 as support for Pd catalyst and Ce promotor does not reduce the formation of dealkylated by-products (Table 8).

Table 8 *α-Limonene conversion over Ce- and Pd-containing Na-ZSM-5-55-1, 2 wt% Ce, 1 wt% Pd*

TOS (h)	0.2	2	4	6	8	24	48
Conversion (wt%)	100	100	100	100	98.8	90.0	67.2
Selectivity (wt%)							
p-Cymene	30.4	69.1	76.7	79.5	80.8	77.2	67.5
Σ*m*, *o*-Cymene	0.1	0	0	0	0	0	0
Σα-, γ-Terpinene	0	0.3	0.4	1.2	1.5	3.1	11.3
Terpinolene	0	0	0	0	0	0.7	1.7
Other $C_{10}H_{16}$-terpenes	0	0	0.1	0.4	0.7	2.0	6.2
Σ*p*-Menthane, *p*-menthene	1.6	6.3	7.5	8.9	10.8	15.6	15.0
Coke precursor ($\geq C_{20}$)	0.4	0	0	0	0	0	0
Cracked products (toluene + lower alkanes, propane)	67.5	24.3	15.3	9.0	6.1	1.4	0.3

$T = 250\,°C$; $p = 1$ bar, carrier gas N_2 4 Nl/h, WHSV = 1.3 h^{-1}.

It is well known, *e.g.* by the work of Sachtler *et al.*,[84,85] that ion-exchange of noble metals into zeolites generates Brønsted acidity *via* two routes:

(1) during calcination of the NH_3 form of exchanged zeolites by decomposition of NH_4^+ ions;[84,85]

$$2\ NH_4^+ \rightarrow N_2 + 3\ H_2 + 2\ H^+$$

(2) during metal reduction:

$$Pd^{2+} + H_2 \rightarrow Pd^0 + 2H^+$$

Thus, with each exchanged Pd ion, two protons will be generated in the zeolite. However, number and strengths of acid sites should be lower for the ion exchanged Na-ZSM-5-(55) than for its counterpart H-ZSM-5-(55).

As shown in Table 8, longer service time and higher selectivities for *p*-cymene are observed. The initial amount of cracked products formed is 67.5%, compared with only 19.2% for the acidic support H-ZSM-5-(55).[86] The amount of cracked products decreases significantly in the first 8 h TOS. An increase of *p*-cymene selectivity is observed with declining cracking activity. Hydrogen transfer is initially complete; no $C_{10}H_{16}$ isomers are found in the product. For short TOS (< 8 h), deactivation of acid sites is dominant, whereas for a longer TOS (> 8 h), deactivation of Pd sites (incomplete hydrogen transfer) becomes more important. The first effect leads to rising *p*-cymene selectivities, whereas the second effect yields lower *p*-cymene selectivities. The maximum for hydrogenated substances is reached at 36 h. This indicates a kinetic control of the ratio of *p*-cymene/*p*-menthenes and *p*-menthanes. In addition to these experiments, the catalysts have been characterized in refs. 86 and 87.

5.1.5 Influence of Hydrogen as Carrier Gas – The dehydrogenation activity of all zeolite supported, Pd-containing catalysts start to decrease after a few hours or days TOS, depending upon support and metal loading. A means to avoid the deactivation is the addition of hydrogen to the carrier gas flow. Figure 14 depicts the selectivity to *p*-cymene for α-limonene conversion over catalyst Pd/Ce-Na-ZSM-5-(55) at 300 °C for a period of 600 h. Carrier gas composition was 50 vol% N_2/50 vol% H_2 and 70 vol% N_2/30 vol% H_2. A temperature of 300 °C was chosen since catalyst deactivation is slower at higher temperatures. Temperatures above 300 °C lead to complete destruction of the *p*-menthane carbon structure. The H_2 to N_2 ratio was random.

The conversion of α-limonene remains complete. The selectivity to *p*-cymene increased from initially 20% to 77% over the whole observed period of 600 h. This can again be attributed to the slow decrease of the formation of cracked products, indicating a much slower deactivation of the acid sites of the catalysts. Formation of $C_{10}H_{16}$ isomers (terpenes) was not observed. No coke precursors were formed.

5.2 Dehydrogenation of Terpene Mixtures

5.2.1 Conversion over Zeolitic Catalysts – The examined starting material was dipentene provided by Arizona Chemical, Florida, USA. This vendor offers a large product spectrum of several dipentenes. For the present work dipentene Sylvapine DP-738 (formerly named Acintene DP-738) was used. Its components

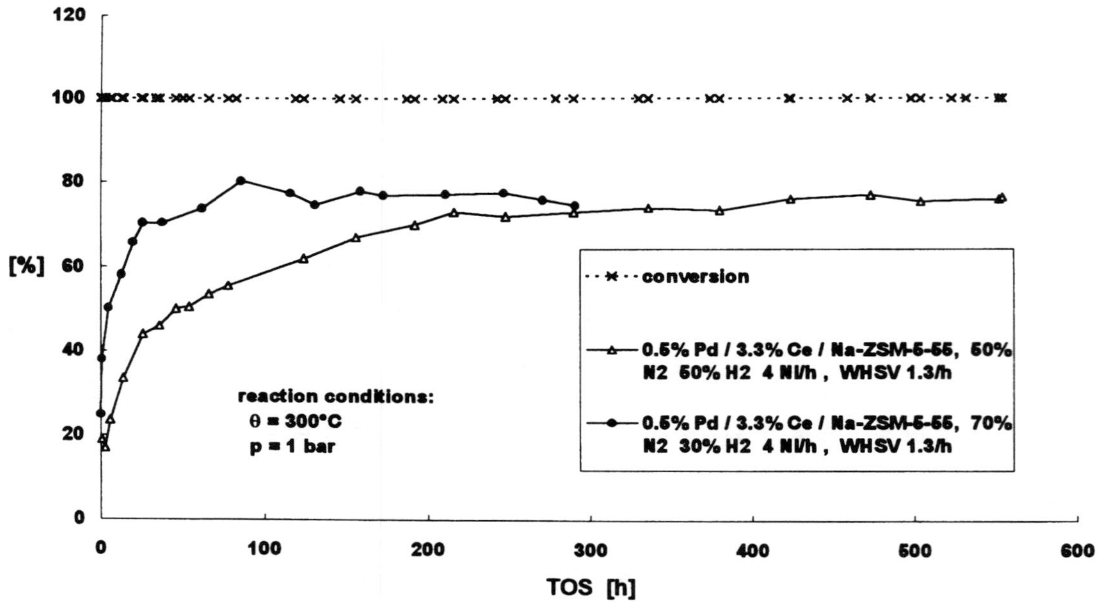

Figure 14 *Impact of hydrogen addition upon α-limonene conversion*

are α-limonene (24%), α-pinene (17%), 1,4-cineole (16%), terpinolene (12%), 1,8-cineole (11%), α-terpinene (7%), camphene (7%) and small amounts of other terpenes.

The experiments were conducted with the following catalysts (Table 9):

Table 9 *Composition and characterization of catalysts A, B and C*

Catalyst	*SiO_2/Al_2O_3*	*Acid strength* *Ho*(p*K*a)	*Acid strength*(*) TPD($10^{-4}n_{NH3}$/g)	*Ce content* (wt%)	*Pd content* (wt%)	*BET area* (m^2/g)
A	27	1.5	7.11	1.7	0.5	311
B	55	1.5	4.91	3.3	0.5	234
C	51	1.5	4.37	1.1	0.5	436

The catalytic results of the conversion of dipentene Sylvapine DP-738 over catalysts A, B and C are depicted in Figure 15. As the starting material is a mixture, a conversion is not determined. Therefore the curves presented indicate the fraction of *p*-cymene in the product.

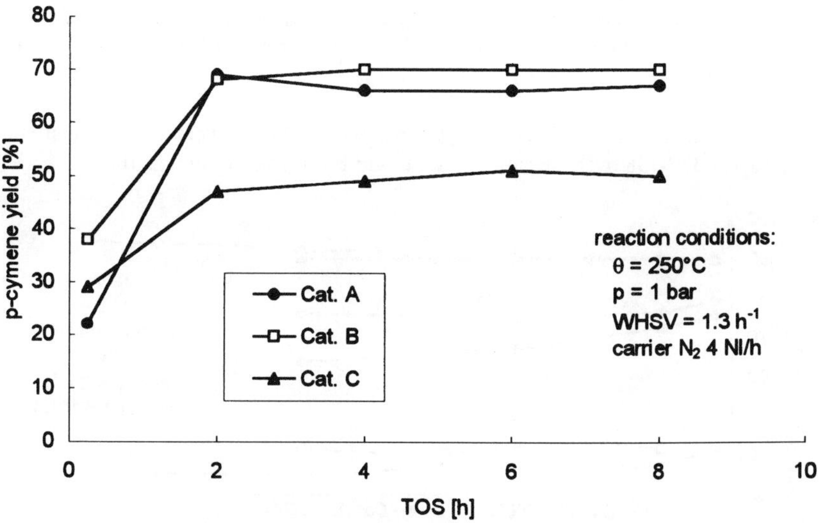

Figure 15 p-*Cymene yield for dipentene conversion over various Ce-promoted, Na-ZSM-5 supported Pd-catalysts*

The monocyclic terpenes are easily disproportionated to the dehydrogenated *p*-cymene and the hydrogenated *p*-menthane. For a short time on stream TOS (< 2 h), cracking of *p*-cymene to toluene and propane/propene on the strongly acidic sites is the dominant reaction. With increasing TOS, these strong acid sites of the catalysts deactivate and, thus, the formation of products obtained by cracking decreases.

The conversion of α-pinene over the Ce-promoted, zeolite-supported Pd catalysts proceeds *via* an acid-catalysed ring opening of the bicyclic terpene to a monocyclic terpene, *e.g.* α-limonene. This is followed by dehydrogenation, possibly *via* isomerization to α-terpinene or γ-terpinene. 1,8-Cineole is dehydrated on the acid sites to *p*-menthadiene prior to dehydrogenation to *p*-cymene on the Pd sites of the catalyst. The conversion of all reactants is complete during the test run of 8 h. The results are quite similar to α-limonene conversion,[86] as expected from the reaction pathway *via* *p*-menthenes and *p*-menthadienes.

The yield of *p*-cymene (Figure 15) is lower for the conversion over catalyst C than over catalysts A or B, although C has the highest BET surface area of 436 $m^2 g^{-1}$. This indicates that not only the Pd content, but also other factors play a key role in this reaction. For example, catalyst B yields slightly higher values than A, even though the BET surface of B (234 $m^2 g^{-1}$) is smaller than that of catalyst A (311 $m^2 g^{-1}$). Catalyst B has also the highest Ce loading (3.3 wt%) among the three catalysts (Table 9). As can be seen, the conversion is not so much influenced by the SiO_2/Al_2O_3 ratio or the Ce content. The further characterization of the catalysts described is published in ref. 87.

As in the case of limonene the carrier gas used in the conversion of the terpene mixtures has a strong influence on the *p*-cymene selectivity. After TOS = 2 h the yield has a constant value of about 70% for a period of 8 h and shows no deactivation.[87]

5.2.2 Silica Support – The catalytic results of the conversion of dipentene Sylvapine DP-738 over the catalysts D, E and F are given in Figure 16.

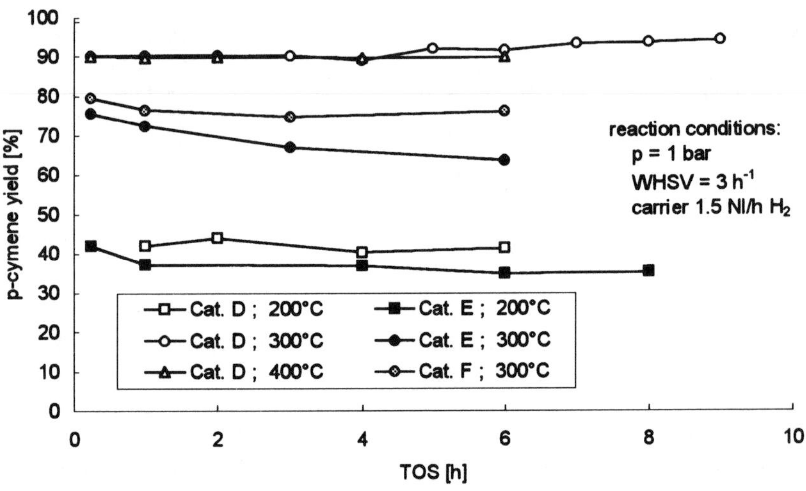

Figure 16 p-*Cymene yield in the conversion of dipentene Sylvapine DP-738 over catalysts D, E and F at various reaction temperatures*

The investigation of the catalytic behaviour over various Pd-modified silica catalysts (D, E and F, Table 10) at different reaction temperatures deliver interesting results. One can see that the worst results are at $\theta = 200\,°C$ (Figure 16). Catalyst D produces the best results both at 300 and 400 °C; thus 300 °C is favoured.

Table 10 *Composition and characterization of catalysts D, E and F*

Catalyst	*Acid strength* Ho(p*K*a)	*Acid strength*(*) TPD($10^{-4} n_{NH_3}$/g)	*Pd content* (wt%)	*Pore size* (Å)	*BET area* (m^2/g)
D	3.3	1.18	0.5	115	120–160
E	4.8	0.12	0.5	80	200
F	6.0	0.65	0.5	170	200

The silica pore size is up to 30 times higher (catalyst F) than the zeolitic pore size. Catalysts D, E and F both differ in pore size and acidity. The effect of these characteristic data is not clear. An optimum seems to exist for catalyst D. However, it cannot be excluded that different manufacturing procedures may result in different catalytic performances.

When comparing the performance of the zeolitic and amorphous supports of the catalytically active Pd, one has to take into account that the WHSV for the latter is almost 2.5 times higher (silica carriers: WHSV = 3 h^{-1}, zeolitic carriers: WHSV = 1.3 h^{-1}). Therefore, the presented data clearly show that the silica-based catalysts offers a higher space-time yield.

The superiority of the silica catalyst in comparison with the catalysts based on a zeolite carrier becomes clear by comparing the results presented above.[88]

The lifetime behavior of the most promising catalyst D shows a TOS of nearly 100 h at 300 °C and atmospheric pressure with 1.5 Nl/h hydrogen as carrier gas.

6 Conclusion

With several examples we could demonstrate that heterogeneous catalysts are becoming very valuable for the clean production of fine chemicals on the example of the fragrance and flavour industry.

The isomerization of styrene oxide to phenylacetaldehyde yields 100% using modified ZSM-5 zeolites, thereby the highest target achieved by catalysis has been fulfilled. A new process as well has been found for the heterogeneously-catalysed production of campholenic aldehyde from α-pinene oxide. By using low reaction temperatures of 0 °C and below in combination with HCl-treated H-US-Y zeolites, up to 85% yield is achieved. This process is competitive with the homogeneous $ZnBr_2$ system.

Also Nafion composites give promising results. In the esterification of dicyclopentadiene the Nafion/silica composites are superior to pure Nafion and ion exchange resins. We find high conversions at low temperatures, which can be attributed to accessible acidic sites with high acidic strength. This results in high turn over numbers (TON).

The use of renewable feedstocks, *e.g.* terpene mixtures, for the production of *p*-cymene was very successful. With Pd-promoted silica, yields of 90% could be achieved over a period of 100 hours. This process might have a good chance for commercialization.

Heterogeneous catalysts, in particular zeolites with their various properties contribute extensively to the environmental protection in the synthesis of fine chemicals. For that a broad and very impressive range of acidic and basic catalysts is already available having all levels of properties between super acidity and super basicity. Furthermore the zeolitic materials may have shape-selective behavior. Also the possibility to prepare bi-functional catalysts will gain in importance.

Catalyst research is better equipped than ever to meet the challenges of the next century.

7 References

1 W.F. Hölderich, in 'New Frontiers in Catalysis' Proceedings of the 10th Int. Congress on Catalysis, eds. K.L. Guczi *et al.*, Budapest 1992, *Stud. Surf. Sci. Catal.*, 1993, **75**, 127.

2 R.A. Sheldon, *Chem. Ind.*, 1997, **1**, 12.

2a Tanaka, H. Yanashima, M. Minobe and G. Suzukamo, *Appl. Surf. Sci.*, 1997, **121/122**, 461.

3 H. Aebi, E. Baumgartner, H.P. Fiedler and G. Ohloff, in *Kosmetika, Riechstoffe und Lebensmittelzusatzstoffe*, G. Thieme Verlag, Stuttgart, Germany, 1978, 32.

4 W.F. Hölderich, N. Goetz, *Proceedings of the 9th International Zeolite Conference*, Montreal 1992, eds. von Ballmoos *et al.*, Butterworth-Heinemann, USA, 1993, 309.

5 J. Kaminska, M.A. Schwegler, A. Hoefnagel and H. van Bekkum, *Rec. Trav. Chim. Pays-Bas*, 1992, **111**, 432.

6 A. Liebens, C. Mahaim and W.F. Hölderich, *Proceedings of the 4th Int. Symposium on Heterogeneous Catalysis and Fine Chemicals*, Sept. 8–12, Basel, Switzerland, *Stud. Surf. Sci. Catal.*,1997, **108**, 587.

7 Firmenich S.A., EP 155.591 1988.

8 B. Arbusow, *Chem. Ber.*,1935, **68**, 1430.

9 K. Arata and K. Tanabe, *Chem. Lett.*, 1979, 1017.

10 G. Carr, G. Dosanjh, A.P. Millar and D. Whittaker, *J. Chem. Soc., Perkin Trans. 2*, 1994, 1419.

11 D. Freude, T. Fröhlich, M. Hunger and G. Scheler, *Chem. Phys. Lett.*, 1983, **98**, 263.

12 R. Szostak, *Stud. Surf. Sci. Catal.*, 1991, **58**, 153.

13 J.R. Sohn, S.J. DeCanio, J.H. Lunsford and D.J.O. Donnell, *Zeol.*, 1986, **6**, 225.

14 L.M. Parker, D.M. Bibby and G.R. Burns, *J. Chem. Soc., Faraday Trans.*, 1991, **87**, 3319.

15 A. Holleman, N. Wiberg, *Lehrb. Anorg. Chem.*, de Gruyter, New York,1985.

16 H. O. House and R. L. Wasson, *J. Am. Chem. Soc.*, 1957, **79**, 1488.

17 R. D. Bach and R. C. Klix, *Tetrahedron Lett.*, 1985, **26**, 985.

18 M. T. Langin-Lanteri and J. Huet, *Synthesis*, 1976, **8**, 541.

19 R. A. Sheldon, J. A. Elings, S. K. Lee, H. E. B. Lempers and R. S. Downing, 4th Int. Symposium on Heterogeneous Catalysis and Fine Chemicals, Abstr. O27, Basel, 1996.

20 A. Liebens, C. Mahaim and W. Hölderich, 11th Int. Congr. Catal., Baltimore 1996.

21 J. Kaminska, M. A. Schwegler, A. J. Hoefnagel and H. van Bekkum, *Recl. Trav. Chim Pays- Bas*, 1992, **111**, 432.

22 J. A. Elings, H. E. B. Lempers and R. A. Sheldon, *Stud. Surf. Sci. Catal.*, 1997, **105**, 1165.

23 W. F. Hölderich, J. Röseler, G. Heitmann and A. T. Liebens, *Catal. Today*, 1997, **37**, 353.

24 W.F. Hölderich, in *Acid–Base Catalysis*, eds. K. Tanabe *et al.*, Kodansha–VCH, Tokyo, 1989, pp. 1–20.

25 W. Spiegler, W.F. Hölderich, N. Goetz, L. Hupfer and J. Wild, BASF AG, EP 259 814 and US 4 824 973, 1986.

26 C. Chapuis, C. Margot, K.H. Schulte-Elte and H. Pamingle, EP 374509, 1990.

27 K.H. Schulte-Elte, C. Chapuis, D. Simmons and D. Reichlin, EP 457022, 1991.

28 N. Goetz, W.F. Hölderich and L. Hupfer, BASF AG, EP 332 981, US 4 968 831, 1988.

29 C.J. Mussinan, B.D. Mookherjee, M.H. Vock, F.L. Schmitt, E.J. Shuster, J.M. Sanders, B.M. Light and E.J. Granda, US Patent 4 163 068, 1979; 4 173 543, 1979; 4 246 287, 1981; 4 252 828, 1981; 4 255 460, 1981.

30 M.R. Kare, *J. Agric. Food Chem.*, 1969, **17**, 677.

31 P.E. Shaw and R. Caleman, *J. Aric. Chem.*, 1971, **19**, 1276.

32 S.S. Chang, *J. Food Sci.*, 1977, **42**, 298.

33 G. Ohloff and E. Klein, *Tetrahedron*, 1962, **18**, 37.

34 F. Delay and G. Ohloff, *Helv. Chim. Acta*, 1979, **62**, 369.

35 E.E. Royals, *J. Am. Chem. Soc.*, 1949, **71**, 2568.

36 K.P. Johny and J. Verghese, *J. Curr. Sci.*, 1970, **39**, 63.

37 B. Burczyk, *Przem. Chem.*, 1970, **49**, 225.

38 B. Burczyk, *Chem. Stosow. Ser. A*, 1970, **14**, 91.

39 B. Burczyk, *Rocz. Chem.*, 1971, **44**, 1381.

40 J.E. Lyons and P. Hosler, DE 2 706 879, 1977.

41 C. Bordenca, SCM Corp., U.S. Patent 3 401 051, 1968.

42 M. Yoshiharu and M. Masahiro, Japan. Kokai, JP 75 131 948, 1976.

43 C.A. Bunton, O. Cori, D. Hachey and J.P. Leresche, *J. Org. Chem.*, 1979, **44**, 3238.

44 S. Anandaraman, K.N. Gurudutt and B. Ravindranath, *Tetrahedron Lett.*, 1980, **21**, 2189.

45 K.N. Gurudutt, B. Ravindranath and P. Srinivas, *Tetrahedron*, 1982, **38**, 1843.

46 G.E. Gream and C.F. Pincombe, *Aust. J. Chem.*, 1974, **27**, 589.

47 N. Bluthe, J. Ecoto, M. Fetizon and S. Lazare, *J. Chem. Soc., Perkin Trans. 1*, 1980, **8**, 1747.

48 S. Xiao, W. Jiang and P. Zhou, *Linchan Huaxue Yu Gongye*, 1989, **9**, 9.

49 M. Nomura, M. Kyouda, Y. Fujihara, K. Tajima and T. Otani, *Yukagaku*, 1994, **43**, 1089.

50 P. Zhou, W. Jiang, Y. Guo, X. Chen and S. Xiao, *Chin. Sci. Bull.*, 1989, **34**, 125.

51 W. Cao, F. Li, X. Li and L. Yiang, *Huaxue Yu Nianhe*, 1995, **1**, 14.

52 Z. Tan and S. Xiao, *Youji Huaxue*, 1993, **13**, 1.

53 K. Hensen, C. Mahaim and W.F. Hölderich, Proceedings of the 11th Int. Zeolite Conf., Aug. 12–17, 1996, Seoul, South Korea, *Stud. Surf. Sci. Catal.*, 1997, **105**, 1133.

54 K. Hensen, C. Mahaim and W.F. Hölderich, *Appl. Catal., A: General*, 1997, **149**, 311.
55 J. Janes, I. Marr, N. Unwin, D. Banthorpe and A. Yusuf, *Flavour Fragrance J.*, 1993, **8**, 289.
56 H. Masuda, H. Kikuiri and S. Mihara, Jpn. Kokai Tokyo Koho, JP 63 201 162, 1988.
57 W.F. Hölderich, K. Hensen, Firmenich S.A., CH 0326-97, 1997.
58 A. Riondel, Elf Atochem, EP 759423, 1997.
59 F. Gude and H. Bellut, Huels AG, DE 3723891, 1987.
60 Sandoz-Patent-GmbH, DE 3822203, 1988.
61 H.A. Bruson and T.W. Riener, *J. Am. Chem. Soc.*, 1945, **67**, 723.
62 L. Schuster, T. Dockner and E. Ambach, BASF AG, DE 4136660, 1993.
63 W.H. Kleine, DE 3619797, Huels AG, 1987.
64 M. Gscheidmeier, Hoechst AG, US 5596127, 1997.
65 J.-M. Ndong Mebah, J.-L. Mielosynski and D. Paquer, *New J. Chem.*, 1993, **17**, 835.
66 Q. Sun, M.A. Harmer and W.E. Farneth, *J. Catal.*, 1996, **164**, 62.
67 A. Heidekum, M.A. Harmer and W.F. Hölderich, *Catal. Lett.*, 1997, **47**, 243.
68 A. Heidekum, M.A. Harmer and W.F. Hölderich, *Prepr.-Am. Chem. Soc., Div. Pet. Chem.*, 1997, **42**, 763.
69 A. Heidekum, M.A. Harmer and W.F. Hölderich, *J. Catal.*, 1998, **176**, 260.
70 M.A. Harmer, W.E. Farneth and Q. Sun, *J. Am. Chem. Soc.*, 1996, **118**, 7708.
71 A. Chakrabarti and M.M. Sharma, *React. Polym.*, 1993, **20**, 1.
72 A. Heidekum, M.A. Harmer and W.F. Hölderich, *J. Catal.*, submitted.
73 H. Fiege, Cresols and Xylenols, Ullmann, VCH, Vol. A8.
74 W.J. Welstaed, Jr., *Kirk-Othmer Encycl. Chem. Technol.*, 1978, **9**, 544.
75 J. Derfer and M. Derfer, *Kirk-Othmer Encycl. of Chem. Technol.*, 1978, **22**, 709.
76 K. Ito, *Hydr. Proc.*, 1973, **89**.
77 P.A. Parikh, N. Subrahmanyam, Y.S. Bhat and A.B. Halgeri, *Chem. Eng. J.*, 1994, **54**, 79.
78 B. Wichterlova, A. Kapustin and J. Cejka, *Appl. Catal., A*, 1994, **108**, 187.
79 P.A. Parikh, N. Subrahmanyam, Y.S. Bhat and A.B. Halgeri, *Appl. Catal., A*, 1992, **90**, 1.
80 J. Zelinski, *Ber.*, 1924, **57**, 2058.
81 R.P. Linstead, K.O. Michaelis and S.L.S. Thomas, *J. Chem. Soc.*, 1940, 1139.
82 R. Martin and W. Gramlich, BASF AG, DE 3607448, 1987, US 4720603, 1987.
83 W.F. Hölderich, R. Fischer, W.D. Mross and F.-F. Pape, BASF AG, DE 3 513 569, 1986; EP 0199 209, 1988; US 4 665 252, 1987.
84 W.M.H. Sachtler, *Acc. Chem. Res.*, 1981, **26**, 411.
85 W.M.H. Sachtler and Zongchao Zhang, *Adv. Catal.*, 1993, **39**, 129.
86 P.A. Weyrich and W.F. Hölderich, *Appl. Catal., A*, 1997, **158**, 145.
87 P.A. Weyrich, H. Treviño, W. Hölderich and W. Sachtler, *Appl. Catal., A*, 1997, **163**, 31.
88 D. Buhl, P.A. Weyrich, W. Sachtler and W.F. Hölderich, *Appl. Catal., A*, in press.

6
Catalytic Combustion for Gas Turbine Applications

BY E. MAGNUS JOHANSSON, DENNIS PAPADIAS, PHILIPPE O. THEVENIN, ANDERS G. ERSSON, ROLF GABRIELSSON, P. GOVIND MENON, PEHR H. BJÖRNBOM AND SVEN G. JÄRÅS

1 Introduction

Catalytic combustion has received an increasing interest as an ultra-low emission technology for gas turbines. The main advantage with catalytic combustion is the possibility to decrease, and for some fuels to increase, the fuel-to-air ratio beyond the limits of flammability. Hence the combustion process and the temperature evolution in a gas turbine combustion chamber can be better controlled and stabilized. This flexibility gives the possibility to reduce emissions of unwanted pollutants compared with conventional flame combustion. Depending on fuel characteristics, emissions such as thermal and fuel-NO_x, CO and hydrocarbons as well as soot can be reduced to ultra-low levels. Other advantages are lower noise levels and less vibrations.

Catalytic fuel combustion for gas turbine and heating applications, originally proposed by Pfefferle in the early seventies[1,2] has been under development during the last 25 years. During the late seventies and early eighties, several concepts and applications of catalytic combustion in gas turbines were presented and explored for several fuels and combustor sizes. However, high temperature stability of the catalytic materials was too low to gain any real commercial interest, and hence several promising projects were abandoned. In parallel to high temperature catalytic combustion development, ceramic[3] and metal[4,5] honeycomb-shaped monolithic substrates for three-way-catalysts (TWC) for vehicle exhaust control and selective catalytic reduction (SCR) have been developed successfully during the last 20–30 years. Similar ceramic and metal substrates have been put to use in catalytic combustors during the last 15 years. At the same time, materials development has been successful to find thermally stable catalyst materials, *e.g.* hexa-aluminates developed by Japanese researchers.[6] Also, the design of combustion chambers to limit material temperatures have been developed, *e.g.* the approach with passive and active channels used by Catalytica Combustion Systems Inc. among others.[7] Design of catalytic combustors is further discussed in Section 5.2.

During this time period, an increasing number of reviews and general articles on catalytic combustion have been published.[8–17] Recently, basic principles as

Catalysis, Volume 14

well as advanced descriptions of specific features were published in a book by Hayes and Kolaczkowski.[18] It can be concluded, that the main obstacle is to find novel catalytic materials that could withstand for at least one year of operation in the harsh combustion environment, *e.g.* temperatures between 300 and 1400 °C, steam, gas velocities around 25 m s^{-1}, temperature gradients, and thermal shocks around 500 °C in fractions of a second.

More recently, new applications of catalytic combustion have been explored. Particularly the use in small-scale gas turbines to be used in mobile applications, such as rail locomotives, road vehicles[19] and ships is anticipated. In addition, an increased use of small gas turbines for industrial applications such as offshore installations, pipeline compressors and power generation is expected.

The aim with the present paper is to survey the literature on catalytic fuel combustion for high temperature gas turbine applications with emphasis on the progress during the last five years. Reference to work before 1993 can be found in an earlier review from our laboratory.[12] Following a brief introduction to catalytic combustion and a discussion on formation and abatement of emission, state-of-the-art in materials development will be reviewed in Section 3. Recent results from mathematical modelling are covered in Section 4. An update of new concepts of catalytic combustors and advanced pilot-scale tests will be presented in Section 5, where also a case study on a recently finished European project is reported. Finally, deactivation of combustion catalysts is discussed in Section 6 and a spin-off effect of catalytic combustion is recapitulated in Section 7.

1.1 Principles of the Catalytic Combustor – It is necessary to consider here how a gas turbine works and where a combustion catalyst can be brought in. Turbines today have usually a coaxial centrifugal air compressor. The preheated and compressed air at 200–500 °C meets the fuel stream in the combustion chamber, *cf.* Figure 1. The fuel burns and the temperature rises up to 1600–2000 °C. That is too high temperature for the turbine, hence part of the air is by-passed to cool the combustion gas before it goes to the turbine. If a catalyst is used, a more fuel-lean combustion can be stabilized and the flame temperature is lowered to 900–1300 °C. Then there is no need for a by-pass of cooling air, this burnt gas can go directly to the turbine, as shown in Figure 1.

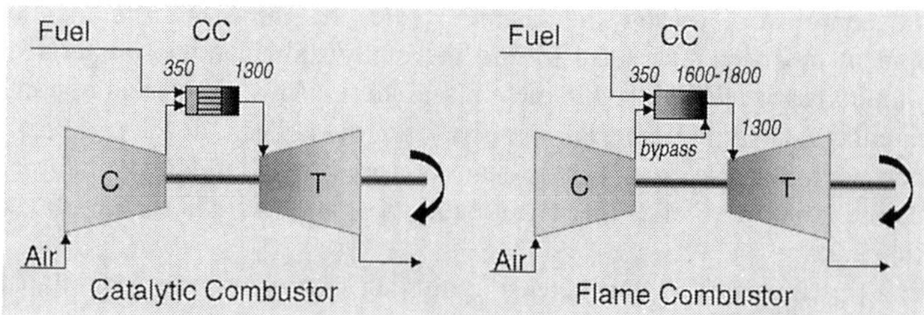

Figure 1 *A catalytic versus a flame combustor in an open-cycle gas turbine; C = compressor; T = turbine; CC = combustion chamber; temperatures in* °C

2 Formation and Abatement of Emissions

Nitrogen oxides (NO_x) can be formed by three mechanisms in methane flames: that is, the thermal or extended Zeldovich mechanism,[20,21] the so-called 'prompt' mechanism and the nitrous oxide mechanism.[22] Thermal NO_x is formed by reactions between atmospheric nitrogen and oxygen. The formation of thermal-NO_x is significant at temperatures above approximately 1500 °C, since the activation energy is high. Prompt NO_x, identified by Fenimore,[23] refers to reactions that occur between hydrocarbons and atmospheric nitrogen in hydrocarbon flames. Typically, formation of prompt NO_x is only significant at close to stoichiometric conditions in flame combustors. The third NO_x formation mechanism is the formation of nitrous oxide as an intermediate,[24] and is significant at low temperatures in lean-premixed laminar flames. The formed nitrous oxide is unstable at higher temperatures and reacts further to molecular nitrogen or NO_x.

Some fuels, such as diesel, heavier oils and low-heating value (LHV) fuels, *e.g.* gasified biomass or gasified coal, contain bound nitrogen. These fuel-N species react further, through several intermediate reactions described by, among others, Miller and Bowman[25] and Glassman.[24] Fuel-bound species form molecular nitrogen and significant amounts of NO_x. The intermediate reactions, starting from fuel-N to form NO_x or molecular nitrogen, most likely starts with thermal decomposition of the bound nitrogen species upstream of the combustion zone. Therefore, precursors to NO_x formation will be N-containing species with low molecular weight, *e.g.* NH_3, NH_2, NH, HCN, CN, *etc.* In the case of LHV-fuels, bound nitrogen species in the raw fuel fed to the gasifier will in general be decomposed to molecular nitrogen as well as significant amounts of NH_3 and HCN, which will be present in the gas fed to the gas turbine. In flame combustors, fuel-N can be reduced by reactions with CH and NH radicals present in high concentrations, *i.e.* in fuel-rich combustion. Therefore, staged combustors, with a fuel-rich zone followed by a fuel-lean zone, can be applied to reduce NO_x from combustion of N-containing fuels. However, although fuel-NO_x is reduced, other emissions such as carbon monoxide and hydrocarbons are typically high. Therefore, staged combustion is not likely to reach ultra-low emission levels.

The reaction route from NH_3 to molecular N_2 or NO_x is not fully understood, but it is clear that it involves several steps. One possible route from NH_3 to N_2 can be the formation of NO_x that is consecutively reduced to N_2 by remaining fuel components. These reactions are usually very temperature-sensitive, both in homogeneous[25] and heterogeneous[26] combustion, which implies that it will be important to control the temperature in the different segments in a catalytic combustor.

2.1 Low-NO_x Technologies for Gas Turbine Applications – The existing methods for NO_x control in gas turbines may be broadly divided into primary measures to control the combustion and secondary measures to clean up the flue gas containing NO_x. Primary measures include, beside catalytic combustion, water or steam injection, lean-premixed combustion and staged combustion. To begin with, water or steam injection measures require a distribution system for clean water or steam generation. Although emission levels are reduced, these methods are not as efficient as lean-premixed combustion. Lean-premixed combustion has been introduced lately and gives low emission levels, typically around 25 ppm thermal-NO_x. Here, homogeneous combustion must operate on the limits of flammability and, consequently, fine-tuned and expensive control systems must be used. Staged combustors operate in two stages, first fuel-rich followed by a quick quench to fuel-lean. This system is more efficient for fuels with large amount of bound nitrogen species, which can be reduced in the fuel-rich stage. Emissions other than fuel-NO_x are usually higher than for other alternatives.

The most important secondary measures include the effective selective catalytic reduction (SCR) and the less effective selective non-catalytic reduction (SNCR). The major drawbacks with SCR are the high capital and operating costs, the large bulky catalytic reactor, NO_x/NH_3 analytical equipment as well as an ammonia or urea storage and distribution system.[17] SNCR on the other hand has a lower capital cost, but is less effective in NO_x reduction.

The catalytic combustor is basically a lean-premixed combustor, in which the combustion is stabilized by a catalyst. Here, the combustion can be ultra-lean, since the catalyst can operate below the limits of homogeneous flammability. Hence the thermal-NO_x emission can be even lower than for lean-premixed combustors. Clearly, catalytic combustion is one of the most promising alternatives for ultra-low emission gas turbines.[27] It has been shown that catalytic combustion should be the lowest-cost and preferred technology for applications that require emission levels below 5 ppm.[28]

2.2 Legislation for, and Expected NO_x Emissions from Gas Turbines – Nitrogen oxide (NO_x) emissions are emitted primarily from transportation sources, power production, utilities, and other industrial sources and contribute to a variety of environmental problems: the formation of acid rain and acidification of aquatic systems, ground level ozone (photochemical smog), *etc.* For these reasons, NO_x emissions are controlled in many ways. In the United States, under the Clean Air Act, the EPA (Environmental Protection Agency) sets national standards for air quality. The EPA is directed to review the standards every 5 years. Until recently, NO_x regulation focused almost exclusively on automobiles and other mobile sources. Now, the focus is shifting toward stationary sources, primarily power plants. The benchmark value today for natural gas is 25 ppm NO_x; however, in California and Japan, 9 ppm NO_x is required for large-scale gas turbines (at 15% O_2). In addition, CO emissions are regulated in some areas to 50 ppm. Moreover, the US Department of Energy's Advanced Turbine System (ATS) program, aims towards single-digit NO_x emissions. Also, several gas turbine manufacturers are expected to introduce single-digit gas turbines during the next decade. Consequently, single-digit emission levels are to be expected for new gas turbine installations in urban areas. The catalytic combustor is one the most promising cost-competitive alternative to reach these ultra-low emission levels.[28]

3 High Temperature Materials

Today, combustion catalysts that can operate up to 900–1000 °C have been developed and studied in both laboratory- and pilot-scales. Still, two catalyst features have not been fully developed. To begin with, a catalyst system that can operate above 1000 °C for one year of operation or more. Secondly, a catalyst system that can ignite natural gas at compressor outlet temperatures of approximately 200–400 °C. However, several combustion chamber designs have been proposed that utilize the features of catalytic combustion, but which operate the catalyst module at approximately 500–1000 °C. Here, a homogeneous zone is used to increase the temperature of the gas to the final maximum temperature. These designs are described in detail in Section 5 of this review.

In this section, recent development of high temperature stable support materials as well as washcoat and active materials is reviewed. Some of the most promising materials to be used as supports in catalytic combustion for gas turbine applications are summarized in Table 1. These monolithic support materials, such as alumina or zirconia, could also be used as washcoat materials with another preparation method. Typical light-off temperatures and specific surface areas for some of the interesting catalyst compositions are summarized in Table 2.

3.1 The Support – To avoid pressure drop and have a large geometric surface, honeycomb-shape monoliths are used in the gas turbine combustion chamber. The desired properties of the support are:

- low thermal expansion;
- high thermal shock resistance;
- high melting point;
- stable in oxidizing atmosphere containing steam;
- chemical inertness towards other components of the catalytic system;
- high area to volume ratio;
- low pressure drop, *i.e.* the cell walls should be as thin as possible.

In addition, it is important to find a washcoat that has a thermal expansion of the same order of magnitude as the support, to prevent rupture or cracks between the washcoat and support layers. If the monolithic support has a sufficient activity without applying a washcoat, this problem may be avoided.

Several different materials have been studied. Metallic monoliths have been used extensively since their first application for automobile converters.[29] They allow very thin walls and have a very high thermal conductivity. However, their thermal expansion gives rise to some problems when looking at the coating and stability of the washcoat on the metallic surface,[12] compared with the ceramic monolith. Furthermore, their maximum operation temperature is limited to 1200–1400 °C, *cf.* Table 1. Probably, the maximum temperature is somewhat lower for long-time exposure. However, several ceramic monoliths that can stand higher thermal conditions have been developed, as reported in Table 1.

Table 1 *Materials for monolith supports*

Material	*Composition*	*Maximum temperature* (°C)	*Thermal expansion* (10^{-6}cm/°C)	*Remarks*	*Ref.*
Dense alumina	α-, γ-Al_2O_3	1500–1600	8	Low cost	165, 166
Mullite aluminum titanate	–	1650		Good thermal shock resistance	166
Aluminium titanate	–	1800	2		165
Silica	SiO_2	1100	0.6		165
Zirconia	ZrO_2	2200	10	Good thermal shock resistance	166
Zirconia-spinel		1700	–		166
Silicon carbide	SiC	1550–1650	5	Good thermal shock resistance, low–high cost	165
Silicon nitride	Si_3N_4	1200–1540	3.7	Good thermal shock resistance, low–high cost	165, 166
Mullite	3 Al_2O_3·2SiO_2	1350	2	Good corrosion resistance	165
Zirconium mullite		1550	4		165
NZP	Na_2O–Zr_2O–P_2O_5–SiO_2	<1500	<2		42
Magnesia	MgO	1800 *	10		37
Spinel	MgO·Al_2O_3	1400	–		166
Cordierite	2 MgO·2 Al_2O_3·5SiO_2	1200–1400	1	Excellent thermal shock resistance, moderate cost	165, 166
FeCr alloy	15% Cr, 5% Al, Y traces, balance Fe	1250–1350	11	Excellent thermal shock resistance, low cost	165, 166

* Value based on $\frac{2}{3}$ of the melting point.

Table 2 *Summary of light-off temperatures, specific surface areas and various substitutions for some combustion catalysts*

Catalyst material	*Formula*	*Typical substitutions*	*Surface area* [62]	*Typical ignition temperature for methane**	*Application*
Hexa-aluminate	$ABAl_{11}O_{19}$	A = alkali, alkaline earth or rare earth metal; B = transition metal (*e.g.* Mn, Fe, Cu, Co)	$8\ m^2\ g^{-1}$ (1400 °C)	500–700 °C [34] (T_{10}-T_{90})	Active phase/support
Perovskite	ABO_3	A = large ion; B = metal ion	$1\ m^2\ g^{-1}$ (1400 °C)	390–690 °C [34] (T_{50})	Active phase
Palladium on alumina	$Pd–PdO/Al_2O_3$	–	–	340–460 °C [167] (T_{50})	Active phase
Platinum on alumina	Pt/Al_2O_3	–	–	590–710 °C [167] (T_{50})	Active phase
Spinel	AB_2O_4	A = large ion (Mg, La, *etc.*); B = metal ion	$8\ m^2\ g^{-1}$ (1400 °C)		Active phase/support
Pyrochlore	$A_2B_2O_7$	A = rare earth or element with a lone electron pair; B = transition or post transition metal	$3\ m^2\ g^{-1}$ (1000 °C) $2\ m^2\ g^{-1}$ (1200 °C)		Active phase/support
Zeolite		Several different ions could be ion-exchanged into the structure	–	756 °C^{+} [69] 608 °C^{++} [69] (T_{50})	Active phase

* T_{50} denotes temperature for 50 % conversion of methane; + Natural gas, Cu-ZSM; ++ Natural gas, Pd-ZSM.

In applications where the temperature range of operation is between 1000 and 1400 °C, there is still a lack of heat-resistant materials. For these applications, a ceramic catalyst system, extruded and completed with support and active phase in one piece, would be the ultimate solution. A surface area-enhancing washcoat is probably not needed at these temperatures, since both mass transfer limitations and reaction rates are high. Probably, only a surface area around 1–10 $m^2\ g^{-1}$ would be sufficient, which could be achieved with fine-tuned extrusion techniques. Hence, complicated washcoat–support interactions can be avoided. Among the several materials that are reported suitable for support extrusion in this review, there is a possibility for some of them to be used as the active component. For example, promising support materials like NZP may be active, depending on the specific ionic substitution. On the other hand, metal structures probably have too low a surface area to be used without washcoat.

In addition, thermally stable oxides, like perovskites or hexa-aluminates, which are used as active material or washcoat in catalytic combustion, can be used to manufacture high temperature supports. Consequently, promising materials that are reported here could be used as support and active phase in one and the same material or as a washcoat on a support made of another material.

3.1.1 Alumina (Al_2O_3) – Alumina is one of the most common ceramics used in catalysis as a support and washcoat, due to its high temperature stability and large specific surface area. Alumina, which has a large specific surface area at temperatures below approximately 850 °C, is not stable when submitted to higher temperature. Owing to dehydration, irreversible changes of the crystallographic phase can be observed, see Figure 2, which lead to a strong decrease of the specific surface area.[30] However, the preparation method of alumina has an influence on the sintering temperature and phase changes. Fibrous alumina shows a higher resistance to sintering, compared with conventional samples.[31] Another cause of deactivation of alumina-supported catalyst is the spinel formation.[32] Some combustion catalysts are based on alumina-supported single metal oxides. When submitted to high temperature treatment, the single metal oxides react with the support *via* a solid-phase reaction to form a spinel phase MAl_2O_4 (M = Co, Ni, Cu, Fe, *etc.*). Consequently, alumina supports could react with some washcoat materials at high temperatures.

$$\text{Bohemite} \xrightarrow{500\text{–}850} \gamma \xrightarrow{850\text{–}1050} \delta \xrightarrow{1050\text{–}1150} \theta \xrightarrow{>1150} \alpha$$

$$\text{Bayerite} \xrightarrow{300\text{–}500} \eta \xrightarrow{850\text{–}1050} \theta \xrightarrow{>1150} \alpha$$

Figure 2 *Crystallographic changes of alumina with temperature*

3.1.2 Cordierite ($2MgO{\cdot}2Al_2O_3{\cdot}5SiO_2$) – Cordierite is, together with alumina, the most common substrate for monolith extrusion. This material has one of the lowest thermal expansions among the proposed materials. This parameter is rather important in order to avoid cracks and loss of washcoat during operation. Furthermore, the cost of cordierite is moderate. However, the thermal stability is the limiting factor,[33] since the melting point is about 1450 °C.

3.1.3 Mullite ($3Al_2O_3{\cdot}2SiO_2$) – Arai and co-workers observed formation of the mullite structure by adding high content of Si to Al_2O_3.[34] The purpose of adding Si was to stabilize Al_2O_3 at temperatures above 1000 °C. Mullite, prepared by supercritical drying, retained a surface area around 50 $m^2 g^{-1}$ after calcination at 1300 °C.

3.1.4 Magnesia (MgO) – Magnesia is also a possible support material having a good thermal stability that can be used as washcoat material as well. Hashimoto *et al.* have described a preparation by vapour oxidation[35] that can lead to a large surface area and ultra-fine single crystal MgO. It is very resistant to thermal treatment and has a surface area above 72 $m^2 g^{-1}$ after calcination at 1500 °C.[36]

Magnesium oxide has been reported to be active in methane complete oxidation by Berg and Järås.[37] The activity was somewhat lower compared with a Ba-substituted hexa-aluminate. The difference between the two catalysts decreased after calcination to 1500 °C.

3.1.5 Zirconia (ZrO_2) and Hafnia (HfO_2) – Zirconia is one of the most promising materials for washcoats. It has a high thermal stability and can be used at temperatures above 2000 °C. This oxide is also inert towards solid-state reactions with other metals and metal oxides at high temperatures.[8] Hence the washcoat may remain stable, and solid-state reactions between the support and the washcoat could be prevented, in contrast to alumina supports. Zirconia could also be used as washcoat, and has been reported both in Catalytica patents and CRIEPI pilot-scale tests, *cf.* Sections 5.2.3 and 5.2.4. Addition of SiO_2 has been tested to enhance the surface area, but the surface area could not be stabilized under severe conditions. The addition of HfO_2 to SiO_2 or to SiO_2 and ZrO_2 has been reported by Catalytica to increase the thermal stability at temperatures above 1000 °C.[38] This hafnium-containing refractory material has a surface area greater than 20 $m^2 g^{-1}$ after 10 h calcination at 1000 °C. It has been used as a support for noble metal based catalysts.

3.1.6 Silicon Carbide (SiC) – Silicon carbide compounds have been described as high temperature materials.[39] They are interesting for high temperature catalytic combustion, since they can withstand operation temperatures above 1400 °C. Therefore, Inoue *et al.*[40] have studied the coating of SiC compound with active hexa-aluminate for use in catalytic combustion for gas turbine applications. The prepared samples have been submitted to thermal treatment up to 1200 °C. A decrease of the thermal stability of the material due to a diffusion of SiO_2 within the coating was observed. However, it was shown that the thermal stability could be improved by depositing an $Al_6Si_2O_{13}$ layer between the support and the washcoat. This intermediate layer avoids silica diffusion and maintains the thermal stability of the active washcoat. The coating on SiC has been further improved by Inoue *et al.*[41]

3.1.7 NZP (Na_2O–Zr_2O–P_2O_5–SiO_2) – NZP is a family of materials that has a very low thermal expansion and high thermal shock resistance. The crystal structure consists of a three-dimensional network of PO_4 tetrahedra and ZrO_6 octahedra, connected at the corners.[42] Their crystallographic structure allows several ionic substitutions in different positions in the NZP lattice, as shown in Table 3. The effect of different preparation methods has been studied in detail with respect to the material properties. NZP materials with very low thermal expansion can be prepared by careful selection of the ionic substitutions.[43] This family of materials has been used for high temperature applications, but not in the field of catalysis so far. However, the thermal properties give this material a certain interest in the case of high temperature catalytic combustion for gas turbine applications.

Table 3 *Possible substitutions within the NZP structure*[42]

	Structural formula $[M'][M'']Zr_2^{VI}P_3^{IV}O_{12}$			
Substitution scheme	*M′*	*M″*	*Zr_2*	*P_3*
Isovalent substitution at M′	H, NH_4, Li, K, Rb, Cs, Cu	□	–	–
Heterovalent substitution at M′	–	□	–	–
Isovalent substitution at Zr	–	□	Ge, Ti, Sn, Hf	–
Balanced substitution at Zr and M″	–	Na	Cr, Fe, Al, Sc, Y, Gd	–
Balanced substitution at P and M″	–	Na	–	SiP_2
Balanced substitution at M′ and Zr	□	□	Nb,Zr	–
Balanced substitution at M′ and P	□	□	–	P_2S

□ Vacancies in the lattice structure.

3.1.8 FeCr Alloy – Metallic monoliths have their most important use in car exhaust converters. They exhibit some properties that can be very interesting in the field of catalytic combustion, with applications in stationary gas turbine and aircraft turbine engines.[44] Nowadays, the metallic monolith is also connected to the use of electrically preheated catalysts to avoid cold-start problems.[4,5]

Compared with ceramic monoliths, the metallic ones are more durable towards mechanical stresses and vibrations. Furthermore, the smaller wall thickness (typically 0.05 mm in the metal *versus* 0.25 mm in the ceramic) enables a lower pressure drop, which is an important parameter in gas turbine applications. Moreover, the higher wall conductivity and lower heat capacity of metallic substrates may reduce the warm-up time during a transient response. Also, the higher wall conductivity shifts the light-off position further upstream of the monolith, increasing the bulk temperature as a result of the enhanced heat release by the catalytic walls.[45] Another advantage with metallic monoliths concerns their higher ability to conduct heat in the radial direction. This is important since hotspot phenomena due to inlet fuel- and flow-maldistributions can be disintegrated thus minimizing local thermal degradation of the catalytic activity.[46]

However, the maximum operating temperature is lower compared with some ceramic materials, *cf.* Table 1. Nevertheless, they can be used in combustor designs where the catalyst temperature is limited, like in the hybrid combustor described in Section 7.2.3. In that case, the catalyst temperature is limited to 900–1000 °C.

The coating of metallic monolith, which was more problematic than for ceramics, has been extensively studied and improved. Zwinkels *et al.*[47] have shown that special treatment of metallic alloy generates an alumina whisker-covered surface. Those alumina whiskers act as an anchor and facilitate the stable washcoating with ceramics.

Among the different metallic alloys, FeCr Alloy is the most commonly used in catalytic combustion. It contains 15% Cr and 5% Al, traces of rare earth metals, and balance Fe.

3.2 Washcoat and Active Material

3.2.1 Hexa-aluminates – One of the most promising groups of materials for combustion catalysts today is the hexa-aluminates. These materials have the general formula $AB_xAl_{12-x}O_{19}$ where the A position could be an alkali, alkaline earth, or a rare earth metal and the B position could be a metal with similar size and charge as the Al ion. The two main features, that have drawn attention to the use of hexa-aluminates in catalytic combustion, are resistance to sintering and a high catalytic activity. However, the thermal expansion is high, hence hexa-aluminates have a poor resistance to thermal shock. But even so, the hexa-aluminates seem to be a promising material for high temperature catalytic combustion, and during the last decade a large number of different substituted hexa-aluminate materials have been described. Groppi *et al.*[48] have reviewed the present state of hexa-aluminates used in combustion catalysis.

One of the keys to the properties of the hexa-aluminates is the crystal structure. The hexa-aluminate structure consists of blocks of spinel-structure divided by mirror planes in which the large A ions are situated. This results in a layered structure. There are two different structure types, the magnetoplumbite and β-alumina structure where, for example, La and Sr in the A-position will yield magnetoplumbite while Ba yields the β-alumina structure.[49]

Secondary ion mass spectroscopy (SIMS) studies have showed that the diffusion resistance of oxide ions is much higher along the *c*-axis than along the *a,b*-axes.[50] This suppresses the crystal growth along the *c*-axis, as the oxide ions are often the slowest diffusing species in metal oxides, and will yield thin plate-like crystals. As the crystal grows, the aspect ratio (ratio between the *a,b*-plane and the *c*-axis) will increase and thereby increases the surface energy of the crystal. The crystal will, owing to the laws of thermodynamics, try to keep its surface energy as low as possible, hence keeping the crystal growth to a minimum even at high temperatures.

Several ions have been substituted into the hexa-aluminate structure both in the A and B positions. The B ions seem to have the most pronounced effect on the catalytic activity of the hexa-aluminate materials. This catalytic effect seems to be closely correlated to the change in oxidation state of B ions between +2 and +3. Several different B ions have been studied.[6,49] For $BaMAl_{11}O_{19}$ (where M is a transition metal) the trend for methane combustion, measured as the temperature for 10% conversion, was Mn, Cu<Fe<Ni, Cr, Co. Similar results have been shown for $LaMAl_{11}O_{19}$ also. The high activity for the Mn could be correlated to its ability to shift oxidation state.[6]

Even the influence of different ions in the A position has been thoroughly investigated, both regarding the catalytic activity as well as crystal-phase and surface area. That the choice of A ions has an impact on the catalytic activity have been shown.[49,51–53] Additions of Pr, Nd, Sm or Gd in $Sr_{1-x}Ln_xMnAl_{11}O_{19}$ improve the catalytic activity for methane combustion in the low temperature region; this is supposed to be related to the increase in structural defects and the oxidation state of Mn.[53] Eguchi *et al.* concluded that the change in activity for different substitutions was more dependent on the valence of the A ions than of the structural changes in the hexa-aluminate structure (*i.e.* between magnetoplumbite and β-alumina).[49] The valence state of the A ion will affect the oxidation state of the catalytically active B ion. The thermal stability of the surface area is also dependent on the A ion. For the above mentioned $Sr_{1-x}Ln_xMnAl_{11}O_{19}$ system, Inoue *et al.* showed that the highest surface was retained, after calcination, for the pure Sr-substituted sample.[51] The surface area decreased with increasing amounts of Sm and Gd, while Nd and Pr showed a maximum and a minimum, respectively. Pr showed the highest surface area of the different substitutions.

Spivey *et al.* have prepared a number of $Sr_{1-x}La_xMnAl_{11}O_{19}$ samples and studied the influence of different La substitutions on the surface area and activity for methane conversion.[54] The largest areas after calcination at 1300 °C were shown for high and low La substitutions, while only high La substitutions showed an increased thermal stability after calcination at 1400 °C. The methane activity was the highest for $x = 0.4$. The results were in contradiction to the ones reported by Machida *et al.*[53] who have reported a peak in activity for $x = 0.2$ which also showed the largest surface area. The cause of the discrepancy between the results is probably the difference in preparation method; Spivey *et al.* have prepared the samples by co-precipitation, whereas Machida *et al.* have used a sol-gel method.

The hexa-aluminates could be prepared by several different techniques, *e.g.* solid-state reactions, hydrolysis of alkoxides, precipitation as insoluble metal salts, *etc.* The preparation method has been shown to have an influence on the thermal stability of the resulting hexa-aluminate material. The morphology of the low temperature hexa-aluminate precursor is particularly important for the ability of preparing high surface area hexa-aluminate materials. Pores with a diameter below 50 Å collapse more readily than larger pores when heated to high temperatures. Thus, it is important to minimize the formation of pores with a diameter below 50 Å, *e.g.* through the use of modifiers in the alkoxide hydrolysis.[55]

The catalytic activity of substituted hexa-aluminates for methane combustion is low compared with noble metals, like Pd. For use in a fully catalytic combustor, these materials will have to be combined with a first stage with higher activity. For other fuels the differences are not so pronounced, *e.g.* for diesel.[56] Hexa-aluminates also show a high activity in CO and H_2 combustion[57] and gasified biomass.[26,58] For these kinds of fuels the ignition temperatures over hexa-aluminates are in the same area as the outlet temperature from the gas turbine compressor.

3.2.2 Perovskites – The perovskites are a group of materials with the general formula ABO_3 where A is a large cation. The materials have a structure similar to $CaTiO_3$, which is the mineral that gave the name to the group, and are ideally cubic. A wide range of different perovskite-type materials have been prepared, since almost 90% of the metallic elements of the periodic table could be substituted into the perovskite lattice. They have been found to be very promising catalytic materials.[59,60]

It is also possible to prepare more complex perovskites, such as $(A_xA'_{1-x})(B_y\_B'_{1-y})O_3$. The partially-substituted A and B sites could adsorb and desorb large amounts of oxygen. This oxygen is more weakly bound than the lattice oxygen.[61] The catalytic activity and high thermal stability of the perovskites have drawn attention to their potential use as combustion catalysts. Several types of perovskites have been tested in combustion applications. In particular, the cobalt, manganese and iron-substituted perovskites showed high activity, *i.e.* the same ionic substitutions that gave the highest catalytic activity with hexa-aluminates, reported in Section 3.2.1.

The main drawback with the use of perovskites is the poor thermal stability of the materials compared with hexa-aluminate based catalysts. Zwinkels *et al.*[62] have compared the thermal stability of two different hexa-aluminates with a $SrZrO_3$-perovskite, a pyrochlore, see Section 3.2.4, and two spinels, see Section 3.2.3. The perovskite, as well as the pyrochlore and one of the spinels decreased in their surface area significantly. One of the explanations of the much lower stability of the perovskites compared with the hexa-aluminates is that the crystal growth will take place in three dimensions and thereby yield a material with a low surface area. Lowe *et al.* have studied several different perovskites and their thermal stability and conclude that the surface area of the perovskites is not sufficient for use in high temperature catalytic combustion.[63] Similar results have been shown by Cristiani *et al.*[64]

3.2.3 Spinels – The general formula of spinels is AB_2O_4. The A position represents a divalent metal ion such as Mg, Fe, Ni, Mn and/or Zn. The B position represents trivalent metal ions such as Al, Fe, Cr and/or Mn. Ti may also occupy this site with a +4 charge. The spinel has been described as a heat-resistant material. Different groups have studied the thermal properties and application in high temperature catalytic combustion.[62,63,65,66] The $MgAl_2O_4$ spinel, prepared by the sol-gel method, showed a large surface area after calcination under severe conditions (16 h, 1400 °C, 10% H_2O).[62]

3.2.4 Pyrochlores – The pyrochlores are a group of materials with the general formula $A_2B_2O_7$. They have been mentioned as a material for catalytic combustion.[61] The structure allows vacancy at the A site and the O sites to some extend. The A position can be a rare earth metal or an element with lone pair of electrons and the B position can be a transition metal or a post-transition metal. This make the structure rather flexible as the oxidation state of the transition metal B can be varied as well as the nature of the A and B metal ions. Subramanian[67] and Castro *et al.*[68] have prepared several pyrochlores. When studying the thermal stability of different complex oxides, Zwinkels *et al.*[62] have shown that $La_2Zr_2O_7$ pyrochlores have a surface area lower than 5 $m^2\ g^{-1}$, already after calcination at 1000 °C. Hence, such materials are probably not suitable for high temperature applications unless the preparation method is improved. However, pyrochlore compounds have been patented for catalytic combustion applications, see Section 5.5.

3.2.5 Zeolites – Zeolites are widely used as catalyst materials. Their well-defined pore size, their performance as solid acids and the ability to ion-exchange different metal ions into the pores have proved very useful in a wide variety of applications ranging from cracking catalysts to shape-selective catalysts. Neyestanaki *et al.*[69] have reported the use of copper-exchanged ZSM-5, ZSM-11 and ZSM-48 as well as Pd-ZSM-5 for propane and natural gas combustion. Although the catalytic activity is good for the zeolite materials, the stability of the materials will probably prohibit the use of zeolites in combustion applications, taking into account the high temperatures and steam which are always present.

3.2.6 Future Promising Materials – Other high temperature stable materials, such as the yttrium–aluminium garnet (YAG) appears to be interesting for its thermal and mechanical properties. During the last 10 years, this class of material has been used for several applications, *e.g.* laser technique, semiconductors, microwave filter. The garnet crystallographic structure enables several substitutions, *e.g.* $Y_3Fe_5O_{12}$[70] and $Y_4Al_2O_9$,[71] and the mechanical strength is remained at temperature above 1400 °C. However, such material has not been reported utilized for catalytic applications so far. Moreover, the Pr_4PdO_7 structure has been patented lately, *cf.* Section 5.5, and could possibly be suitable for stabilizing Pd at high temperatures.

3.2.7 The Platinum Group Metals – Several of the elements of the platinum group metals (PGM) have an outstanding capability in oxidizing hydrocarbons, hydrogen and carbon monoxide. They have found widespread use as active materials in catalysts used at low to moderate temperatures, *e.g.* in volatile organic compound (VOC) catalysts and three-way catalysts. The high activity of PGM implies that they would be an excellent candidate for use in gas turbine applications. Owing to their rarity and volatility at high temperatures, the use is limited to platinum, palladium and rhodium. The activity for PGM in oxidation reactions are generally Ru<Rh<Pd<Os<Ir<Pt.[8]

The choice of metal or metals is dependent on the fuel used. Platinum shows a high activity for oxidizing carbon monoxide and saturated hydrocarbons, while palladium in general is more active when it comes to partial oxidation of olefins, see, for example, Trimm.[11] For a mixture of several types of hydrocarbons, as in diesel or jet fuels, a combination of several different precious metals could be used. Trimm also showed that the combination of, for example, platinum, palladium and rhodium would give a higher activity than the metals alone.

Palladium is the most often used PGM in catalytic combustion applications for natural gas. Palladium exists in the form PdO_x at low temperatures, probably at least on the surface of the particle. As the temperature increases there is a reduction of PdO_x to metallic Pd, around *ca.* 800 °C.[72,73] The reaction is reversible and the Pd metal is re-oxidized back to PdO at 600 °C in air.[73] The Pd metal is less active in methane combustion at temperatures lower than approximately 800 °C. Much discussion has been concerned with which state of PdO_x is the most active, *i.e.* Pd with chemisorbed oxygen, Pd particles with a skin of PdO or bulk PdO. Burch[74] has performed a number of experiments concluding that the bulk PdO seems to be the most active phase. McCarty[75] has shown that the hysteresis in the methane combustion is due to the formation of a stable saturated layer of chemisorbed oxygen on the surface of the Pd particles. Further, Kikuchi *et al.*[76] suggested that the most active species was PdO_x on Pd particles, rather than PdO_x directly on the metal oxide support. Lyubovsky and Pfefferle[77,78] have shown that PdO has a lower activation energy than metallic Pd while Pd has a higher pre-exponential factor than PdO. This results in that, at a certain temperature, metallic Pd will be more active than PdO. Bell and co-workers recently published a study on methane combustion over 9.9 wt% palladium on zirconia.[79,80] It was concluded that palladium oxide forms a shell around a core of palladium metal. Fully reduced palladium metal was inactive at low temperatures (553 K) and the activity increases until a plateau in activity is achieved when 6–7 monolayers of palladium oxide has formed on the surface of the palladium crystallites.

Two of the main problems with the use of platinum group of metals, are sintering and evaporation at high temperatures, *cf.* Section 6. Palladium is more resistant to high temperature sintering than, for example, platinum and the two metals are often combined. It has also been shown that self-oscillation could occur due to the PdO–Pd transition.[81] This self-oscillation can be suppressed by additions of rhodium to the platinum/palladium catalyst system.[82] The PGM loading on the support is also important and has been reviewed by Lee and Trimm[83] concluding that an increase in Pt/Pd loading (between 2.7 and 10 wt%) increased the overall rate of oxidation for methane although the activity per unit metal surface area decreased. The oxidation rate was almost constant for a Pt loading between 0.027–1.4 wt.%, but decreased for a Pt loading above 10 wt%.

The supports on which the PGM are dispersed are also of great importance for the activity of the catalyst. The support could not only decrease the sintering of the PGM particles, but also change the catalytic activity of the catalyst. One example is the spillover effects of Pt/CeO_2, Pd/SnO_2 and Pd/V_2O_5.[11] If a catalytically active support is chosen, the influence of the phase transition between PdO and Pd could be minimized. Sekizawa *et al.* have studied PdO impregnated on different manganese-substituted hexa-aluminates and have shown that the decrease in activity during the transition from PdO_x to Pd is almost gone.[84,85] Sekizawa *et al.* have also investigated the influence of the support for Pd supported on different Al_2O_3–MO_x (where M = Co, Cr, Cu, Fe, Mn, and Ni).[85] It was shown that the catalytic activity was decreased for Co, Cr, Cu, Fe and Mn compared with pure Al_2O_3. This was partly due to a decrease in the surface area of the support. Ni was shown to increase the catalytic activity even if the surface area of the support was decreased.

Kikuchi and co-workers have shown that Pd supported on α-alumina exhibited higher catalytic activity for combustion of methane than Pd supported on γ- or θ-alumina.[76] Pd supported on alumina and the influence of the calcination temperature of alumina from 900 to 1600 °C was studied. The catalytic activity of Pd supported on alumina calcined at 1600 °C was 100 times higher compared to Pd supported on alumina calcined at 1200 °C. Temperature-programmed oxidation studies showed that Pd on alumina calcined up to 1600 °C is oxidized to PdO at temperatures lower than for Pd on alumina calcined up to 1200 °C. Hence, low surface area alumina was able to stabilize a more active Pd-species (PdO_x) than high surface area alumina.

3.2.8 Single-metal Oxides – McCarty and co-workers have compared single supported transition metal oxides to precious metal.[86] They have prepared single phase, high surface area $LaAlO_3$ to be used as a support for single transition metal oxides. They reached a surface area of 16 $m^2\ g^{-1}$ after calcination at 950 °C. Kinetics studies show that supported metal oxides present a high activity in methane combustion. The activity decreased in the following order: $PdO > RuO_2 > CoO_4 > CuO > NiO > Mn_2O_3 > Cr_2O_3$. Ru is more active than Pd at high temperatures.

Marion *et al.* investigated the stability of CuO dispersed on two different types of support, Al_2O_3 and $ZnAl_2O_4$.[87] The thermal stability was examined in CH_4 combustion after an ageing procedure performed at 1030 °C for 24 h under an air/methane stream. Solid-state reaction between CuO and Al_2O_3 led to spinel formation and hence the activity of the catalyst was dramatically reduced. In comparison, $Cu/ZnAl_2O_4$ exhibited a higher thermal resistance and kept its activity after ageing.

$LaAl_{11}O_{18}$- and Al_2O_3-supported Cr_2O_3 have been studied by Zwinkels *et al.*[88] The purpose of this work was to examine the interaction between the support and the active phase after thermal ageing in humid air (10 vol% steam at temperature between 1100 and 1400 °C). It has been suggested that the $LaCrO_3$ perovskite phase was formed by solid-state reactions of La with Cr_2O_3 and then, at higher temperature, aluminium ions were incorporated into the structure, forming the substituted perovskite $LaCr_xAl_{1-x}O_3$. Al_2O_3-supported catalysts were more active than the $LaAl_{11}O_{18}$ ones, both before and after ageing, but did show a decline of activity after ageing. Conversely, the $LaAl_{11}O_{18}$-supported catalyst activity remained the same both before and after ageing.

Single metal oxides have shown that they could be a suitable alternative to precious metals. However, the active phase and the supporting oxide must be carefully selected, to avoid solid-state reactions. In addition, the catalyst temperature should be kept below 1000 °C, to avoid vaporization of the active phase.

4 Mathematical Modelling in Catalytic Combustion

Development work in high temperature catalytic combustion is increasingly supported by theoretical studies. Mathematical models are used for several purposes. For instance, a mathematical model can shed light on problems such as optimal configuration of the catalytic unit. It can also be used to extrapolate laboratory data for scaling up applications, thus minimizing expensive pilot plant tests. Further, mathematical modelling is an important tool for interpreting experimental results, contributing to a deeper interpretation of the complex mechanisms occurring in the catalytic reactor, and interacting with the experimental work at the initial stages of reactor development.

Mathematical models may be very sophisticated and detailed from a mathematical point of view, but may also include many simplifying assumptions. As can be seen in Figure 3, the chemical and physical processes in the catalytic combustor are by far less trivial, including effects such as solid phase conduction, radiation, homogeneous ignition and interactions between the monolith channels. The use of any relevant approximation of the model equations is therefore tractable and usually necessary since sophisticated models of chemical reactors are most often too difficult to solve to be useful. The problem of simplification, that is to what extent mathematical models can be simplified for a given reactor modelling problem, has been devoted to a number of studies for monolith reactors.[45,89–91] In the following text, topics concerning development and simplifying treatment of mathematical models will be described, limiting the discussion to laminar flow conditions since the vast majority of the modelling work is

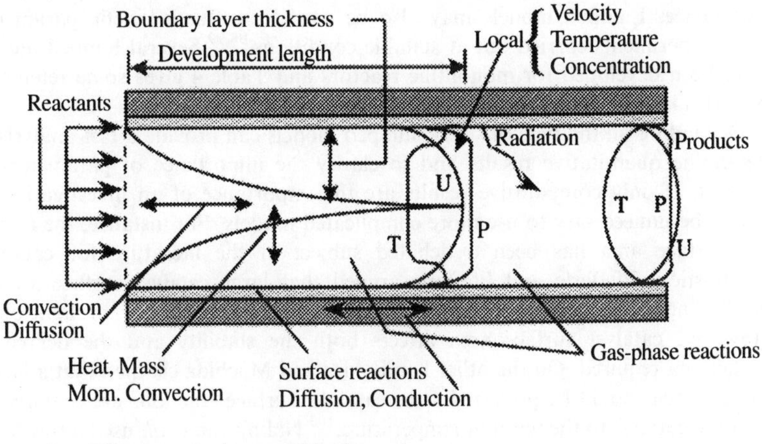

Figure 3 *Flow in a single monolith channel*
(Adapted from Sinha *et al.*[173])

concerned under this type of flow regime. However, it must be emphasized that in stationary gas turbines, working at up to 20 bar with gas velocities between 20 and 40 m s^{-1}, the flow in the monolith channels will be turbulent.[13,18,92,93] In laboratory experiments, on the other hand, the flow is normally laminar due to the low pressure. In some applications, for example the AGATA automotive gas turbine project with an operating pressure of 4 bar, the flow will probably be laminar (see Section 5.3).

In the modelling part, it was felt necessary to review some works earlier than the last five years in order to provide a more complete discussion of the specific problem in focus. Further, no emphasis will be given describing the model equations, and the reader particularly interested in the mathematical formulation is referred to a very recent book on catalytic combustion by Hayes and Kolaczkowski.[18] In addition, the reader can here find valuable references for heterogeneous kinetic rates and a good discussion about the dangers of extrapolating rate models at different operating conditions.

4.1 One-dimensional Models – The models are usually classified into two categories depending on the spatial dimension. The more comprehensive, *distributed models*, account for all necessary spatial variations, while the simpler ones, *lumped models*, are based on average values of the gas-phase properties in the radial and angular directions. These simpler models depend much on proper correlations for heat and mass transfer coefficients in order to account for the resistance between the fluid and the catalytic wall, a topic which will be described in Section 4.2. The advantage with the lumped models is that they offer a great reduction in the computational efforts. This is especially the case when a three-dimensional treatment is required to model square or triangular ducts, or when modelling interaction effects between the monolith channels.[94–97] Further, lumped models are relevant in modelling turbulent flow since distributed models still would require empirical correlations to describe the flow pattern for turbulence. Lumped models may also be used to evaluate kinetic parameters from experiments carried out at suitable conditions.[98,99] Several lumped models have been developed for monolithic reactors and Table 4 gives some references over this kind of work in chronological order.

Depending on the type of study, lumped models can provide a fast analysis to determine quantitative results and to clarify the importance of parameters of interest. If only comparative results are the importance of an investigation, it would be unnecessary to use more complicated models. For instance, the role of the surface area has been a debated subject in the literature for catalytic combustion. Pfefferle and Pfefferle argued that large catalyst surface area is needed neither for combustion at operating temperatures nor at light-off.[10] However, catalyst surface area affects both the stability and the degree of preheating required. On the other hand, Arai and Machida claimed that a larger surface area should be preferred since catalytic surface reactions are required to heat the catalyst to the ignition temperature.[100] Nakhjavan *et al.* used a transient, one-dimensional model to address the significance of the surface area on the catalyst light-off.[101] By varying the interior surface area, washcoat thickness, effective diffusivity and fuel types, results showed that a larger surface area would benefit catalytic ignition. Even when using an extremely fast oxidizing fuel, diffusion limitations did not occur immediately, suggesting that a large surface area would benefit the ignition of the fuel. The reaction could be limited by diffusion only if the initial catalyst temperature was high enough. It was observed that for a fast oxidizing fuel, diffusion limitations occurred above 530 °C.

The role of surface area has also been investigated at steady state conditions.[92] A threshold value was obtained for the surface area. In excess of this value the surface area played a minor role for effective heating, while values lower than the threshold limit the outlet gas temperature decreased to values close to the inlet gas temperature. The threshold value depended on the catalytic activity: the higher the activity the lower would be the specific surface area to achieve a light-off. However, for small enough activities, a higher surface area could not compensate for the lack of effective heating rate.

Groppi *et al.* used a lumped model to investigate the effect of various design parameters such as flow rate, cell configuration and pressure for a catalytic combustor,[102] with a particular focus on the catalytic part of a hybrid reactor. This reactor concept, described in more detail in Section 5.2, divides the reactor into an upstream catalytic section where the fuel is partially oxidized and a downstream section where the homogeneous reactions are sustained. Model results indicated that a catalyst ignition immediately at the inlet of the hybrid reactor is needed. This was important in order to raise the gas exit temperature high enough to sustain homogeneous combustion reactions in a downstream section. At laminar flows, it could be beneficial to use shorter monolith segments resulting in higher gas exit temperatures compared to a single monolith segment of equal length due to the enhanced transport properties. The effect of the gas exit temperatures at different Reynold numbers was also investigated by keeping the inlet velocity constant but varying the operating pressure. For laminar conditions, the outlet gas temperature decreased by increasing the pressure. At the transitional regime the temperature tended to increase again, while at fully turbulent conditions, with high pressures, only small variations in the gas temperature were observed. When laminar conditions prevailed, exit gas temperatures were markedly higher than under turbulent conditions. These results point out the differences that occur from the two flow regions, indicating that it is not straightforward to extrapolate atmospheric tests to full-scale applications without the aid of mathematical tools.

While lumped models provide a quick informative study of the behaviour of the catalytic combustors, this kind of model may fail to predict the wall temperatures and the light-off position accurately in presence of steep axial wall temperatures. In some cases, lumped models may also predict different steady state solutions than distributed models. This kind of different behaviour between lumped and distributed models will be discussed in the next section.

4.2 Comparison of Lumped and Distributed Models at Laminar Flows

4.2.1 Heat and Mass Transfer Coefficients – While one-dimensional models are relatively easy to solve and offer the attraction of a substantial reduction in the

Table 4 *Some examples of references describing modelling in monolith reactors*

Year	*Ref.*	*Model*	*Application/comments*
1970–80	Hegedus[168]	(L), (S)	Solid overtemperature can occur by the nature of the fuel and channel geometry
	Votruba *et al.*[169]	(L), (S)	Investigation of axial heat transfer in monoliths. Conduction can be neglected in ceramics
	Young and Finlayson[89]	(L,D), (S)	Comparison between 1D, 2D, 3D models, influence of backward conduction on multiple steady states
	Heck *et al.*[90]	(L,D), (S,T)	Comparison between 2D and 1D models using variable heat and mass transfer coefficients
	Lee and Aris[107]	(D), (S)	Radiation effects could be approximated by increasing the solid wall conduction
	Stevens and Ziegler[170]	(L), (S)	Effects of friction and momentum convection on velocity change can be neglected
	Sinkule and Hlavacek[171]	(L), (S)	Investigation of radiation effects using different boundary conditions. Existence of five steady state solutions
1980–90	T'ien[172]	(L), (T)	Study of transient response in monolith combustors. Response delay due to substrate inertia
	Sinha *et al.*[173]	(D), (T)	Study of the light-off performance of CO oxidation. Global homogeneous reactions included
	Oh and Cavendish[174]	(L), (T)	Investigation of poison accumulation in washcoats for catalytic mufflers
	Ahn *et al.*[175]	(L), (T)	Parametric investigation of combustor response in operating parameters. Metallic monoliths show faster response time
	Flytzani-Stephanopoulos *et al.*[120]	(MC), (S)	Discrete model for the description of metallic monolith. Model was based on a cooling fin approach
	Kolaczkowski *et al.*[96, 121]	(MC), (S)	Discrete model with a square duct reconfiguration. Comparison with experiments for non-reacting conditions
	Zygourakis[95]	(MC), (T)	Investigation of different flowpatterns at the inlet of a monolith. A continuous approach of multichannel model

1990–98	Hayes *et al.*[108]	(D), (T)	Investigation of axial heat conduction and radiation. Discontinuous values of Nu at transient conditions
	Groppi *et al.*[102]	(L), (S)	Parametric study of the influence of design variables in particular hybrid reactors
	Markatou *et al.*[113]	(D), (S)	Detailed gas-phase chemistry for oxidation of methane. Both catalytic and non-catalytic surfaces were investigated
	Cybulski and Moulijn[123]	(MC), (S)	Multichannel model for the investigation of metallic monoliths. Discrete model based on triangular cells
	Hayes and Kolaczkowski[109]	(D), (S,T)	Study addressing discrepancies between experimentally and numerically determined Nu and Sh numbers
	Groppi *et al.*[91, 94]	(D,L) (S)	Comparison between lumped and distributed models in circular, rectangular and triangular ducts
	Nakhjavan *et al.*[101]	(L), (T)	Investigation of the role of surface area for catalytic ignition
	Groppi *et al.*[45]	(D), (S)	Different levels of model complexity to address simplified treatments in hybrid reactors
	Boehman *et al.*[116]	(D), (T)	Computation of thermal stresses in monolith channels
	Papadias *et al.*[98]	(L), (T,S)	Evaluation of heterogeneous kinetics by model fitting of experimental data
	Leung *et al.*[118]	(D), (S)	Investigation of reaction and diffusion in real washcoat geometries
	Groppi and Tronconi[124]	(MC), (S)	Comparison between discrete and continuous multichannel models

(D) and (L) denote distributed- and lumped-models, respectively; (MC) denotes models considering channel–channel interactions; (S) and (T) address steady state and transient conditions; Nu = Nusselt, Sh = Sherwood.

computational efforts, the predictive results of these models depend strongly on the correlations for heat and mass transfer coefficients that are used. The literature has not been clear over the correct values for Nusselt and Sherwood numbers in monolith channels. Hawthorn[103] derived semi-theoretical correlations for heat and mass transfer coefficients based on earlier work from Kayes and London[104] on laminar flow in ducts. The asymptotic values for fully developed flow and temperature correspond well with the classical Graetz–Nusselt problem for constant wall boundary conditions. However, substantially lower asymptotic values have been reported with correlations derived experimentally by Bennet *et al.*[105] and Ullah *et al.*,[106] performed under reaction conditions being an order of magnitude lower than results by Hawthorn and also results reported by numerical investigations of monolithic reactors.[89,107,108] Hayes and Kolaczkowski[109] investigated by numerical means the reason for the low asymptotic mass transfer coefficients proposed by Ullah *et al.* and Bennet *et al.* In the case of Bennet *et al.*, the mass transfer coefficients were evaluated under a kinetic controlled regime, giving rise to the extremely low coefficients, while the low values reported by Ullah *et al.* could have been caused by diffusion in the washcoat. Asymptotic Nusselt and Sherwood numbers computed with a two-dimensional steady state model were more in agreement with numbers for non-reacting flow.

Using mass and heat transfer correlations adapted for non-reacting conditions with correctly asymptotic behaviour can provide a good agreement between lumped and distributed models at isothermal conditions.[110] However, in highly exothermic reactions such as in the case of high temperature catalytic combustion, the heat and mass transfer correlations fail to predict the dramatic changes that occur in the shift from a kinetic to a diffusion regime. The problem concerning the local variations of heat and mass transfer coefficients in reacting systems has especially been discussed by Young and Finlayson,[89] Heck *et al.*,[90] and for catalytic combustion in particular by Groppi *et al.*,[91] where comparisons between lumped and distributed models have been investigated.

The difference between these two types of model can be illustrated in Figure 4 during a rapid light-off behaviour for the oxidation of carbon monoxide.[111] Prior to and after the ignition of the catalyst, calculated Nusselt numbers using the distributed model show values corresponding to the solutions of the Graetz problem with constant heat flux and constant wall temperature conditions, respectively. However, in the vicinity of the light-off region the distributed model predicts a jump for the Nusselt and Sherwood numbers, giving rise to higher heat and mass transfer coefficients. Higher heat transfer coefficients delay the ignition since more heat is flowing from the catalyst wall to the gas phase.[89,90] The application of heat and mass transfer correlations in lumped models does not account for the strong changes of interface temperature and concentration during the ignition, and the use of the asymptotic values leads to an overoptimistic prediction of the light-off position. Heck *et al.*, however, showed comparable predictions between lumped and distributed models if the heat and mass transfer correlations are allowed to vary with the axial position.[90] Prior to light-off, Nusselt numbers for the analytical solution of constant wall flux were used and after light-off, a new entrance region was assumed with the analytical solution for constant wall temperature. The light-off still occurred earlier with the lumped model, but this study showed the importance of variable heat and mass transfer coefficients for the adequacy of the lumped models.

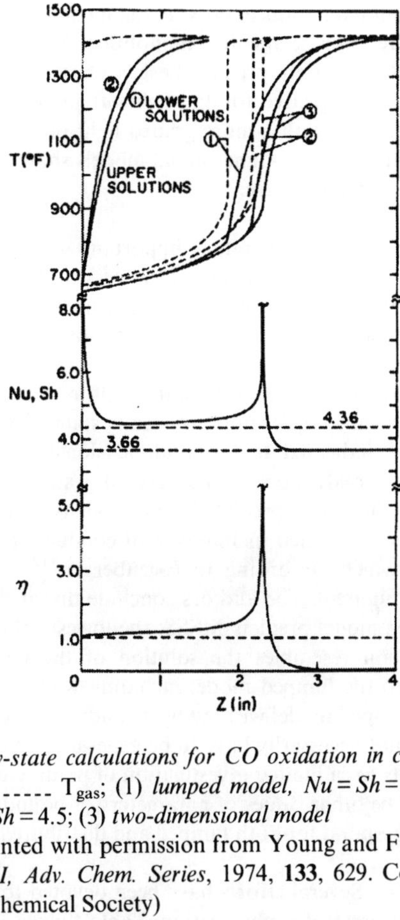

Figure 4 *Steady-state calculations for CO oxidation in circular channels;* —— T_{solid}; ----- T_{gas}; (1) *lumped model,* $Nu = Sh = 3.5$; (2) *lumped model,* $Nu = Sh = 4.5$; (3) *two-dimensional model*
(Reprinted with permission from Young and Finlayson, *Chem. React. Eng.-II, Adv. Chem. Series*, 1974, **133**, 629. Copyright (1974) American Chemical Society)

A comparison between lumped and distributed models for hybrid reactors for catalytic combustion of methane was more recently investigated by Groppi *et al.*[91] The validation of one-dimensional models was made on the basis for different Nusselt correlations and geometrical properties of the monolith channel. In general, the simulations showed a mismatch of the wall temperature profiles predicted with the one-dimensional model. One-dimensional models could show good agreement of the gas exit temperature, which is an important parameter for hybrid reactors as described in the previous section. The catalyst has to ignite in the inlet of the channel in a hybrid reactor and therefore a large part of the channel operates under the diffusion control regime with wall temperatures close to adiabatic conditions. For a long single segment monolith, the authors pointed out that correlations based on interpolating formulas between constant wall flux and temperature boundary conditions, or by using the solution of the Graetz problem for constant wall temperature, provide the most accurate predictions of the exit gas temperature. For segmented monoliths, the lumped model with circular ducts could not accurately predict the gas phase temperature, because this type of model is more sensitive for the light-off prediction in the shorter channels. For square channels with a more graded light-off behaviour and weak maxima in the Nusselt number, one-dimensional models showed good agreement in predicting the gas-phase temperature.

4.2.2 Multiple Steady States – Another important problem in modelling monolith reactions for exothermic reactions is the occurrence of multiple steady state solutions (MSSS). It has been shown, by comparing lumped with distributed models, that a one-dimensional model may give false solutions corresponding to higher multiple steady state solutions.

Young and Finlayson investigated different kinds of models for monolith converters with reactions allowing for multiple steady states.[89] For the conditions investigated, simulations indicated that multidimensional models without the effect of backward heat conduction in the catalyst wall did not predict any MSSS. Lumped models, however, predicted a multiplicity in the solutions over larger parametric values, and when including wall conduction only the highest solution was obtained which, according to Eigenberger,[112] corresponds to an erroneous solution, see Figure 4. The authors conclude that if the heterogeneous kinetics are such that the model predicts MSSS, the lowest solution to a lumped model without conduction resembles the solution of the distributed models. However, the solution to the lumped model including wall conduction and the highest solution to the lumped model without wall conduction is not valid.

The above results do not necessarily have to be general since the study is based on numerical experiments for a limited investigation of parameters and boundary conditions. There might be other values of parameters or boundary conditions in which similar MSSS correspond for both lumped and distributed models.

4.3 Distributed Models – Several efforts have been devoted to model monolith reactors with a various degrees of sophistication. Table 4 gives some examples of references on modelling monolith reactors in which distributed models have been used. The mathematical treatment to include more space dimensions increases the computational cost drastically compared with the lumped models, but in return, a tool with greater resolution capacities is obtained. Lumped models, for instance, become more restricted to use when the influence of gas-phase ignition has to be modelled. Here, distributed models are more suitable in order to investigate the homogeneous–heterogeneous interaction mechanisms occurring in the boundary layer of the catalytic wall.[10,113–115] Distributed models become also more relevant to use when transient conditions prevail,[108] or when thermal stresses in the solid have to be computed under a start-up behaviour.[116] Still, even with today's powerful computer capacities, any degree of reasonable approximation is tractable in order to reduce the complexity of the distributed models.

The degree of complexity of the mathematical treatment can be assessed by comparing simpler models with more comprehensive ones. Such studies have been described in the previous section where lumped models were compared with distributed ones. Recently, Groppi *et al.* investigated the degree of reasonable approximations for modelling hybrid reactors by considering a class of four different two-dimensional models.[45] The effect of hydrodynamical development gave minor differences to the results. Hence, this effect could be neglected for hybrid reactors operating at laminar flow conditions. Including the temperature dependence of the gas properties was important in order to account accurately for the temperature profiles. Since steep gradients prevail in hybrid reactors, the influence of wall conduction was also investigated using a range of conductivities including both ceramic and metallic monoliths. Results for a cordierite showed that the effect of wall conduction was negligible. The high mass flow rates prevailing in a hybrid reactor result in negligible heat transfer along the wall compared with convective contribution. For metallic monoliths, the influence of conduction was important, giving improved light-off performances. The influence of homogeneous reactions was also investigated by including a simplified two-step reaction mechanism. Negligible influence of homogeneous reactions was observed for CH_4 combustion due to bulk temperatures less than 800 °C and short residence times. For other fuels, such as CO- and H_2-rich gasified biomass, the homogeneous reactions start at lower temperatures with a need to include gas-phase chemistry in the models.

4.4 Modelling the Washcoat – When using distributed models, the channel is often modelled in two dimensions. Even if the monolith has a square geometry, the washcoat tends to round the channel, which makes the use of cylindrical co-ordinates a valid approximation. The washcoat, however, is often approximated as being uniformly distributed in the periphery. In practice, there is a tendency for the coating to accumulate in the corners in the honeycomb channel, which can be an order of magnitude thicker compared with the thickness at the side of the channel. Chou and Stewart investigated the accumulation of washcoat in the corners of a monolith and solved numerically the effectiveness factor approximating the coating as a square with rounded corners.[117] The assumption of a perfectly uniform distributed washcoat could overestimate the effectiveness factor markedly. More recently, Leung *et al.* studied the diffusion and reaction in real washcoat geometries using different kinds of rate expressions.[118] The effectiveness factor as a function of Thiele modulus for a first order reaction was compared using a 2D model and a simple 1D approximation of the washcoat. The characteristic length for the 1D model was simply a volume to surface ratio of the washcoat. The 1D approximation did not yield a correct solution. Instead, the value of the one-dimensional Thiele modulus was fitted to the two-dimensional Thiele modulus using an expression of polynomial function. This polynomial expression was used in a reactor model since a fully two-dimensional approach of the washcoat would be too costly to evaluate numerically. It was apparent that when using other kinds of rate expression this would influence the relationship of the polynomial expression. Also, when using the correlations of the polynomial fits to another geometry of a washcoat, the solutions were no longer valid. Hence, a new fit has to be made.

In recent on-going work at our laboratory,[119] a method has been developed to address *a priori* the effectiveness factor in non-uniform washcoats. This method consists of a symbolic sectioning of the washcoat into particles, and remarkably simplifies the analysis of diffusion and reaction in irregular shapes of washcoats. A simple procedure can be used to determine the effectiveness factor instead of using rigorous and time-consuming numerical treatment. This simplified calculation method gives significant advantages in computation time, especially in the case of strong pore diffusion resistance conditions where a fine grid is necessary to obtain good accuracy with the finite element method.

4.5 Modelling Interactions between Channels – Most of the numerical work developed considers the modelling of a single channel as a representative for the whole monolith assuming identical conditions for all the channels. However, thermal interactions between the channels can arise because of poor inlet fuel mixing, flow and heat maldistributions or by intentionally coating sequential channels within the monolith, a configuration that is described in Section 5.2.3. Few models have been developed to account for the radial gradients within the whole monolith matrix. Such investigations have been based by using either discrete or continuous models.

The discrete models reconfigure the monolith matrix into a row of channels in the radial direction. The heat conducted radially in the solid walls is discretized, based on the repetition of the single cell configuration. A one-dimensional model describes the gas-phase equations in each channel. Early attempts using this kind of model have been developed to describe non-reacting conditions. Flytzani-Stephanopoulos *et al.* studied heat transfer in a metallic monolith and compared model results with experimental data giving fairly good agreement with model predictions.[120] However, the physical consistency of the model was too simplified to use for reaction conditions assuming zero thickness of the transverse walls of the channels. Kolaczkowski *et al.* developed a discrete model where the solid heat equations for each channel considered both the radially- and the peripherally-directed walls.[96,121] The model was compared with experiments for non-reacting conditions, and if the model accounted for the inlet radial flow and temperature variations, good predictions of the outlet gas temperature were obtained. Flow maldistributions at the inlet of the monolith gave significant variations in the flow regime of the channels with laminar flow at the outer radius and transition to turbulent flow in the centre of the monolith. More recently, the model was extended to include reaction conditions simulating a hotspot behaviour arising from an inlet fuel maldistribution.[122]

While discrete models have the advantage to describe the performance of the separate channels in the monolith matrix, monoliths containing complex geometries are often reconfigured into simpler ones in order to simplify the mathematical treatment. Such a case concerns the metallic monoliths where the sinusoidal cells are reconfigured into square or triangular ducts.[96,120,123] Moreover, discrete models may become numerically expensive if the monolith contains a large number of channels. Another approach to model the radial gradients in the monolith is to use continuous models where the monolith is described as a homogeneous medium containing the solid and the gas phase. These continuous models do not reconfigure the geometry of the monolith but are strongly dependent of a physical correct radial effective conductivity. Groppi *et al.* derived effective radial conductivities for square and triangular cells and compared discrete and continuous models for a case with a monolith containing square channels.[124] In general, good agreement between these two types of model was obtained. Variations between the two models occurred in the presence of steep temperature gradients. The discrete model with a coarse grid step, based on the cell configuration, was numerically inefficient to describe such steep profiles. In such cases, the homogeneous models can provide a more numerically robust treatment of the model equations.

5 Gas Turbine Applications

During the last five years, several successful pilot- and full-scale demonstrations of catalytic combustors for gas turbine applications have been presented. Here, we have divided these systems into five different classes: the large- and small-size fully catalytic combustor (designs Ia and b) and the hybrid designs with partially inactive catalyst, with secondary fuel and with secondary air (designs IIa, b and c). The first part of this section is devoted to fundamental gas turbine considerations, which will be followed by a summary of demonstrations of catalytic combustors.

5.1 Fundamental Gas Turbine Considerations – A comprehensive description of gas turbine design is beyond the scope of this review. However, some basic considerations regarding the inlet and outlet temperatures of the combustor and catalyst are worth mentioning. The inlet temperature of the gas to the catalyst segment is determined by the pressure ratio (*i.e.* the compression work), gas turbine load and use of pilot-flames. Typically, the inlet air temperature to the combustor is raised to approximately 200 °C for a low pressure turbine and to 400 °C for a high pressure turbine at full load. At idle and partial load conditions, the gas turbines do not reach the same inlet temperature. Therefore, a highly active catalyst should be developed that can ignite the fuel already at the inlet temperature. If this is not possible, pilot flames must be used to heat the gas to the catalyst ignition temperature. These pilot flames must be carefully designed, otherwise they can produce significant amounts of thermal-NO_x.[125,126]

The maximum outlet temperature of the combustor is determined by the amount of cooling air, homogeneity or pattern factor of the fuel–air mixture, temperature losses in the transition duct between combustor and turbine, turbine inlet temperatures and adiabatic temperature rise on combustion of the fuel. Usually, much less cooling air is used in catalytic combustors than conventional ones, since the main purpose with fuel-lean catalytic combustion is to operate in a fuel-lean mode where almost all air is injected to the combustion zone. The air-to-fuel ratio will in fact determine the adiabatic temperature rise, but if the fuel and air mixing is not completely homogeneous, some regions evolve more heat than others. Hence the catalyst must be designed for a higher temperature than the theoretical one. Therefore, development of mixing devices that give a uniform pattern factor is of paramount importance. Finally, it should also be taken into account that gas turbines are designed for a specific turbine inlet temperature, which is not the same as the combustor outlet temperature. Usually, some heat losses must be taken into account in the transition duct between the outlet of the combustor and the inlet of the turbine. In conclusion, if the gas turbine is designed for an inlet temperature of 1300 °C, approximately 100 °C could be lost in cooling air and temperature losses in the transition duct. Further, a 5% inhomogeneity in the air-to-fuel mixing could give a temperature deviation of *ca.* ± 50 °C in adiabatic temperature rise. Therefore, the catalyst material would need to be designed for 1450 °C and not 1300 °C.

The gas turbine combustor can be divided into three different designs; the silo combustor, the canned or cannular combustor and the annular combustor. The silo combustor is mounted on the side of the compressor and gas turbine. This configuration is well suited for catalytic combustor since a change in the combustor geometry can be made without significantly changing the rest of the gas turbine. The canned or cannular design is made of several small combustors, mounted in a ring around the axis of the gas turbine. Here as well, changes in the combustor can be made without any large change in the gas turbine design. The last design, the annular combustor consists of a ring with multiple fuel and air injectors. It will be difficult to design a catalyst shape that would fit the annular geometry. Hence, silo and cannular configuration could be used for retrofitting gas turbine combustors to catalytic combustors, whereas the annular configuration is less suitable.

5.2 Recent Pilot- and Full-scale Demonstrations of Catalytic Combustors – Here, we will report on five different designs for catalytic combustors. The principal schedule is shown in Figure 5 and a summary of results from recent pilot-scale combustion tests is given in Table 5, which concludes some of the information discussed here. Concepts for stationary and mobile applications, as well as for gaseous and liquid fuels, will be discussed. Special attention to the recently finished AGATA-program will be given.

5.2.1 Fully Catalytic Design Ia: Mid- and Large-size Gas Turbines Running on LNG – The first design has been proposed by Sadamori *et al.* and is a co-operation between Osaka Gas, Kobe Steel, and Catalyst and Chemicals Inc.[127] Here, all fuel is mixed together with the preheated air at the inlet of the

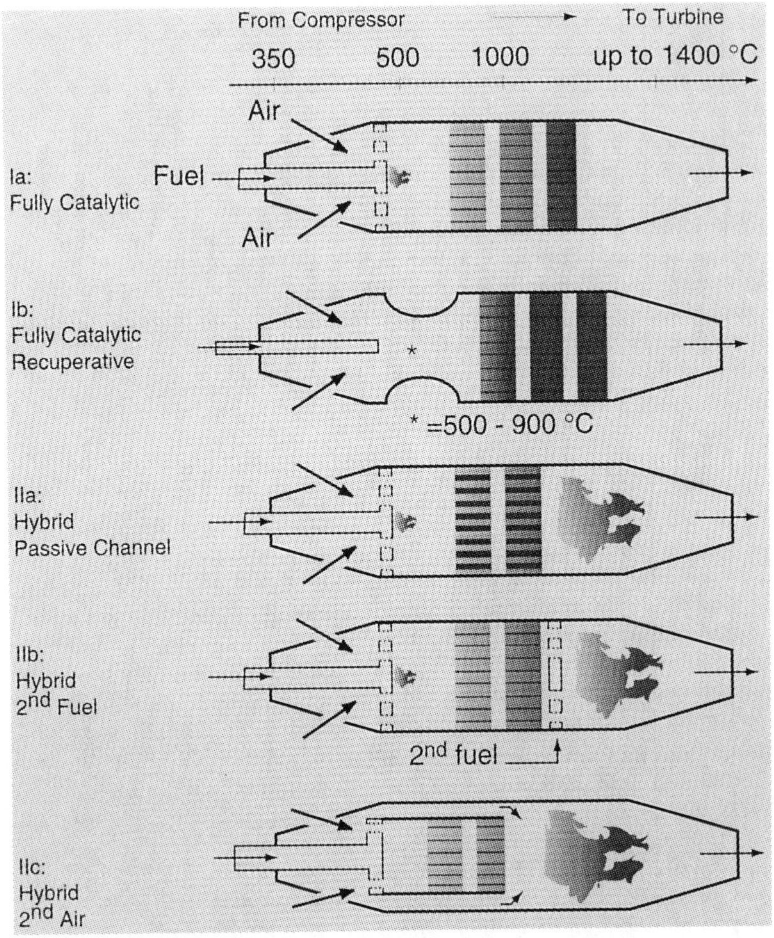

Figure 5 *Designs of catalytic combustion chambers for gas turbine applications*

combustor. Test results from a 160 kW pilot-scale unit showed NO_x emissions between 35 and 60 ppm at an outlet temperature of 1100 °C. However, since a large amount of the NO_x was formed in the pilot-flame, NO_x emissions from the combustor itself were approximately 8 ppm.[126] The catalyst part consists of several segments of honeycomb monoliths with different properties. The first two segments are highly active Pd-based catalysts supported on cordierite at the entrance followed by three less active metal oxide catalysts, namely the extruded hexa-aluminate monoliths developed by Arai and co-workers.[6,14,128] Although this configuration is the least complicated, sufficient thermal stability of this catalyst system has not been verified at 1300 °C for 8000 h. This approach will

Table 5 *Catalytic combustion designs*

Company or organization	*Concept*	*Gas turbine*	*Fuel*	*Catalyst composition*	GHSV[†] [× 1000]	*Potential advantages* (+) *and drawbacks* (−)[#]	*Ref.*
Osaka Gas, Kobe Steel, *etc.*	Ia, fully catalytic	1 MW	Natural gas	3-stages PM/cordierite, 4-stages Mn-hexa-aluminate	380, 10 bar	+ Simplistic configuration − Catalyst Stability	127
AGATA group	Ib, fully catalytic	Recuperative 40–50 kW	Diesel ethanol	Pd on alumina-based washcoat/ ceramic substrate	190, 4 bar	+ Pre-heated air from heat exchanger, no pilot-flame	141, 142
Allied Signal	Ib, fully catalytic	Recuperative single-shaft 50 kW	Diesel or JET	Proprietary washcoat on Corning EX-22 substrate	NR[‡]	+ Pre-heated air from heat exchanger, no pilot-flame, low outlet temperature prevents catalyst overheating	130, 131
Precision Combustion	Ib, fully catalytic	Recuperative small-scale	Gasoline or JET	Advanced Technology Surface Reactor	Atm	+ Ultra-short channels yield low emissions, low pressure drop and good fuel efficiency	19
Catalytica, TKK KK and:	IIa, hybrid, partial inactive catalyst		Natural gas	Detailed composition is proprietary, 2–3 stages FeCrAl-metal monolith washcoated with metal oxide, first stage Pd		+ Prevent catalyst over-temperature, no second mixing − Control of post-catalytic auto-ignition delay time	15, 176, 177
TEPCO/GE	IIa, hybrid	GE MS9001 *ca.* 87 MW	Natural gas	As above	450, 12 bar	+ As above − Pilot-flames	7
Allison	IIa, hybrid and IIc, second fuel and air	4.5–15 MW	Natural gas	As above	NR[‡]	+ Lean premixed post-catalytic homogeneous combustion zone	138
Solar	IIa	Small/mid-size recuperative	Natural gas	As above	NR[‡]	+ No pilot-flames needed in recuperative mode	129, 178

Precision Combustion	IIa, hybrid		Methane	High palladium content metal monolithic	NR‡ atm	+ ultra-short segments prevent too high catalyst temperatures	13
Toshiba/ TEPCO	IIb. Hybrid second fuel	200 MW	Natural gas	Pd/alumina/cordierite	570, 1 bar	+ Prevent catalyst overheating second fuel mixing	136, 179
CRIEPI	IIb. Hybrid second fuel and air	1 combustor in a multi-can 20 MW GT	Natural gas	Pd, Pt, Rh/ZrO_2Cordierite	250, 13 bar	+ Prevent catalyst overheating, pre-mixed second fuel and air Mix between first and second fuel/air	137, 180
NASA Lewis	IIc. Hybrid second air	Aviation	Kerosene or JET	Ceramic	190, 2 bar	+ Reduction of soot, prevent catalyst overheating	139, 140
ULECAT group	NR‡	2.3 MW stationary	Biomass diesel	Ceramic	NR‡	+ Dual fuel, reduction of fuel-NO_x, no net CO_2 from biomass	26, 146

† GHSV calculated as the gas hourly space velocity per reactor volume, based on the overall monolith size and divided by the pressure; # in comparison to other catalytic combustion concepts; ‡ not reported

be successful if catalyst materials can be developed to withstand the highest anticipated temperatures.

5.2.2 Fully Catalytic Design Ib: Small-size Recuperative Gas Turbines – Recuperative gas turbines use a heat exchanger to utilize the turbine exhaust to heat the compressor gases before they are injected into the combustor. This increases the net efficiency for gas turbines that work with a relatively low pressure ratio, typically 4–6 bar. At high pressure ratios, the compressed air will have a too high temperature to make the heat exchanger effective. This design has been studied for mobile applications, such as engines in cars, buses and trucks, but could also be utilized in small-scale domestic applications.

The European AGATA program (Advanced Gas Turbines for Automotive Applications) is one example of this design and is described in more detail at the end of this Section.

Moreover, Dutta *et al.*[129] report on a recuperative design in an on-going co-operation between Solar Turbines and Catalytica in the Advanced Turbine Systems (ATS) Program. Here, Catalytica catalysts are used for a recuperative gas turbine with relatively high combustor inlet temperatures at a low pressure ratio (9:1) and for a medium pressure ratio gas turbine (15:1) fuelled with natural gas. The catalyst is composed of a washcoated two-stage FeCrAl-based metal monolith with Pd as active catalytic phase. The catalyst is reported to ignite at approximately 375 °C, which is lower than the inlet temperature for the recuperative gas turbine above 50% load. Since lower inlet temperatures are achieved at lower load in this turbine and for the non-recuperative medium-size gas turbine, pilot-flames are needed to raise the fuel–air mixture to catalyst ignition temperatures. Tests showed that this catalyst configuration contributed to less than 3 ppm NO_x. Critical design features showed to be the design of the post-catalyst stage with homogeneous combustion to reach low CO emissions, as well as pilot-flame contribution to NO_x and fuel–air mixing.

Several other projects are directed towards the use of catalytic combustors for use in small hybrid vehicles. Allied Signal reports on a compact single-shaft 50 kW gas turbine for hybrid electric vehicles.[130,131] Since the gas turbine is recuperative and works at relatively low turbine inlet temperature (TIT), the catalytic combustor must only raise the gas temperature from 677 to 1010 °C at full load. Combustor tests with diesel fuel No. 2 showed emission levels with $NO_x = 6$, $CO = 97$ and hydrocarbons = 1 ppm, corrected for 15% O_2 at an outlet temperature of 1013 °C. This is significantly lower than ULEV (ultra-low emission vehicle) specifications, recalculated to g km^{-1}. The NO_x emissions were almost strictly a function of the nitrogen content of the fuel, although slightly higher NO_x emissions were reported at outlet temperatures below 950 °C. The catalyst was composed of commercial Corning EX-22 ceramic honeycomb substrates, coated with a proprietary catalyst formulation and arranged in series. The ceramic discs were held in place by an alumina/silica fibre mat.

Further, Kraemer *et al.*[19] describe tests with a catalytic lean-premixed prevaporized gas turbine combustor for recuperative turbogenerators. The catalyst was a combination of noble and base-metals. Tests with Jet-A and gasoline gave single-digit emissions of CO and hydrocarbons, and even less than 1 ppm NO_x.

Although these designs are denoted 'Fully Catalytic', homogeneous reactions are still likely to take place in parallel to the heterogeneously catalysed reaction, at gas and surface temperatures above *ca.* 800 °C. These reactions could be radicals formed at the surface and reacting in the volume, or *vice versa.* Interaction between radicals in the gas phase volume and the catalyst surface at high temperatures are complicated to analyse, both from a modelling and experimental point of view. Hence, the homogeneous–heterogeneous interactions are still not fully understood and have been scarcely reported. Recently, Kiperman discussed gas-phase heterogeneous–homogeneous catalytic reactions which begin on the solid surface of a catalyst and propagate or compete with some or all of the active species escaping to the gas phase.[132]

However, the fully catalytic design (Ia and b) is different compared to a second class of catalytic combustors, denoted hybrid designs (IIa,b and c), where a post-catalytic homogeneous combustion zone is used to complete the combustion and increase the temperature further. Hybrid designs are discussed in the following section.

5.2.3 Hybrid Design IIa: Partially Inactive Catalyst – This design is based on a patent from Westinghouse.[133] A catalyst coating is applied on selected channels in the monolith. Hereby, the substrate is provided with catalytically active and passive channels, where the passive channels cool the monolith. This method has been adopted by Catalytica Combustion Systems and Tanaka Kikinzoku Kogyo (TKK KK). Recent results from the joint project of Catalytica, General Electric (GE) and Tokyo Electric Power Company (TEPCO) are reported by Schlatter *et al.*[7] All fuel is mixed with the combustion air at the combustor inlet. Since the catalysts are composed of active and inactive monolith channels, only part of the fuel is combusted over the catalysts. During manufacture of the monolith, one folded and one straight metal sheet are rolled together. Only one side of the folded metal is washcoated and this gives the finished monolith its unique characteristic; 50% of the channels will be without catalyst and therefore remain passive.[133,134] It should be pointed out that the ratio of washcoated and non-washcoated channels could be varied by applying washcoat on both sides of the folded metal. This makes it possible to fine-tune the conversion of fuel and the catalyst temperature. Hence, fuel and air passing in the passive channels will not react and the exothermic heat release will be limited. The metal monolith will also work as a heat exchanger between channels with and without heat production. In the post-catalytic zone, unburned fuel and combustion gases will ignite homogeneously to reach the final turbine inlet temperature. Hence, a large part of the fuel is combusted downstream of the catalytic reactor. This design makes it possible to reduce catalyst temperature to around 900 °C, while combustor outlet temperatures can be raised to 1300–1400 °C. The catalyst is a FeCrAl-based metal monolith, washcoated with a proprietary catalyst formulation. The active component is PdO, and the PdO to Pd transition is used to find a smooth operating window. Full-scale tests simulating a GE MS9001E gas turbine showed emission levels for $NO_x = 3.3$ ppm, $CO = 2$ ppm and hydrocarbons = 0 ppm at a combustor exit temperature of 1192 °C and a pressure of 1.25 MPa.

Catalytica has recently announced promising results from tests with a small-scale Kawasaki engine equipped with a silo-combustor running for 1000 h at an outlet temperature of 1300 °C. A one-year test started in 1998.

In addition, Precision Combustion has separately developed an approach with only partial conversion of the fuel mixture over the monolith, as reported by Karim *et al.*[135] A catalyst with a high palladium loading was used. Short length catalytic monoliths were chosed to ensure that catalyst temperature remains well below adiabatic flame temperature.

5.2.4 Hybrid Design IIb: Secondary Fuel – The secondary fuel design has been suggested by Toshiba,[136] and with slight modifications, by Ozawa and co-workers.[137] Here, the catalyst temperature is limited by staged combustion, *i.e.* secondary fuel is introduced after the catalyst segment. The Osaka gas configuration introduces pure fuel after the catalyst, whereas the design proposed by CRIEPI premixes secondary fuel and air before the mixing point.

In the design proposed by Toshiba and Tokyo Electric Co., pure secondary fuel is injected after a Pd/alumina/cordierite catalyst segment.[136] Although this system has shown low NO_x emissions, it will be difficult to achieve fuel–air homogeneity in the post-catalyst combustion zone.

Ozawa *et al.* report recent results from testing of a 10 MW-class gas turbine combustor running on liquefied natural gas (LNG).[137] The combustor has been designed to prevent catalyst temperatures above 1000 °C. This is achieved by feeding primary and secondary premixed fuel and air before and after the catalyst section. Catalyst inlet ignition temperatures are 320 and 360 °C for a fresh and aged catalyst, respectively. Outlet temperature from the catalyst is lower than 1000 °C and outlet temperature from the combustor after injection of premixed secondary fuel and air is between 1350 and 1390 °C at 13.3 MPa inlet air pressure. The catalyst is a 60 cpsi cordierite honeycomb monolith, washcoated with a Pd/Pt/Rh impregnated ZrO_2-stabilized Al_2O_3 (Zr:Al 1:5 mass ratio). The metal loadings of Pd, Pt and Rh are 20, 5 and 2 g l^{-1} of monolith, respectively. Rh was added to the catalyst formulation, to decrease temperature and combustion oscillations. The catalyst segments are shaped as six pieces of cake. Secondary fuel and air is feed from a specially designed nozzle, mounted directly behind and between the six catalyst segments. Test results from four different tests showed that for an adiabatic combustion temperature between 1350 and 1390 °C, NO_x emissions were between 3.7 and 4.4 ppm and the combustion efficiency was 99.99% or above. The total pressure drop was between 3.5 and 4.0% and the pattern factor varied between 2.4 and 3.7%. The results showed that it was possible to reduce NO_x formation by applying a thermal barrier coating to the afterburner, *i.e.* a ceramic liner. Hereby the maximum temperature in the combustor could be increased and the amount of cooling air lowered. Hence, some of the cooling air could be used to dilute the fuel–air mixture, making it even leaner. However, the stability of the catalytic combustion was lowered and a higher catalyst bed temperature was needed to stabilize the combustion. This could be prevented by increasing the amount of catalyst, *i.e.* decreasing the gas hourly space velocity over the catalyst.

Recent results from a co-operation between Allison and Catalytica are described by Smith *et al.*[138] Here, the combustor consists of three parts: first, a catalytic reactor, second, a post-catalytic homogeneous combustion zone, and third, a lean-premixed (LPM) flame region downstream of the catalytic and post-catalytic zone. The fuel and air are premixed prior to entering the catalytic reactor, where part of the fuel is converted, maintaining the catalyst temperature below 660 °C. After the catalytic reactor, the remaining fuel is combusted homogeneously. The lean-premixed flame region is used to anchor the homogeneous post-catalytic flame over a wide range of pressure, temperature and stoichiometric conditions, from idle to full load conditions. The operation procedure of the combustor is to combust all fuel in the LPM-module from idle up to 50% load. At 50% and higher load, the inlet temperature is high enough for ignition of the fuel over the catalyst. Therefore, parts of the fuel and air are fed over the catalyst. The fuel fed to the catalyst zone is only partly combusted; hereby, the catalyst outlet temperature is held almost constantly at 660 °C. The post-catalytic lean-premixed module burns the rest of the unburned fuel from the catalyst zone and secondary fuel to reach the desired load. Bench-scale testing of the catalytic segment showed emissions below 3 ppm and CO less than 2 ppm. Rig tests showed that NO_x and CO emissions could be kept low over a wide range of test conditions, with less than 9 ppm NO_x. Ignition delay time and design of the post-catalytic combustor design were important to maintain low NO_x and CO emissions and prevent flashback.

5.2.5 Hybrid Design IIc: Secondary Air – Researchers at NASA Lewis have studied fuel-rich catalytic combustion for kerosene-fuelled gas turbines,[139,140] probably intended for supersonic aircraft running on jet-fuels. The catalyst temperature is limited to less than 1000 °C by introducing secondary air after the catalytic combustor. The catalyst works in the fuel-rich zone, where fuel and air are mixed at an equivalence ratio of 5.0 to 7.6 ($\lambda = 0.20$ to 0.13). Hence, the fuel is only partially oxidized to a mixture of H_2O, CO_2, CO, H_2 and C_1–C_4 hydrocarbons as well as N_2 originating from the air. Since the long chain hydrocarbons have been converted in the low temperature fuel-rich zone before they enter the hotter fuel-lean zone, they have been withdrawn from the soot formation mechanism that starts to get significant at a threshold temperature of 1080 °C. NO_x formation is also lowered, since a better fuel and air mixing can be achieved. First, evaporation and mixing of the liquid fuel and primary air occur prior to the low temperature catalytic zone. After the catalytic zone, secondary air and partly-converted fuel are mixed. By this design, peak temperatures with higher formation of thermal-NO_x can be avoided. Results show that no soot formation could be detected and NO_x levels were very low. Pt and Pd as well as Ni-based catalysts supported on ceramic discs (honeycomb substrates or foams) were tested.

5.3 Case Study: The European AGATA Project – The European EUREKA project EU 209, otherwise known as AGATA (Advanced Gas Turbine for

Automobiles) commenced in early 1993 and has occupied a five year period until March 1998.[141] The programme was dedicated to the development of three critical ceramic components – a catalytic combustor, a radial turbine wheel and a static heat exchanger – for a 60 kW turbogenerator in a hybrid electric vehicle. These three components, which are of critical importance to the achievement of low emissions and high efficiency, have been designed, developed, manufactured and tested as part of a full-scale feasibility study. AGATA is a joint project conducted by the following partners: AC Cerama, Volvo Aero Corporation and Volvo Aero Turbines in Sweden, Aerospatiale, Allied Signal Catal. Environ., Allied Signal Turbo SA, CEA/CEREM, CEA/LETI, Céramiques & Composites, GRETH, IFP, ONERA, Peugeot SA and Renault in France. The project is partly funded by the Swedish National Energy Administration and French Ministries of Research and Industry.

5.3.1 Technical Specification – Technical specifications for the gas turbine, excluding electric generator, are as follows:

- mechanical output: 60 kW;
- specific fuel consumption: 200 g kWh^{-1};
- turbine inlet temperature: 1623 K;
- max. pollution emission: ULEV or similar European standards;
- fuel: diesel or ethanol.

As described by Gabrielsson and Holmqvist,[142] steady state conditions at 298 K and 101.3 kPa inlet conditions have been calculated according to Table 6.

Table 6 *AGATA engine steady state calculations at 298 K and 101.3 kPa*

Load (%)	*SFC** (g kWh^{-1})	*Thermal efficiency* (%)	*Combustor inlet temperature* (K)	*Turbine inlet temperature* (K)
100	200	42	1208	1623
50	211	39	1208	1518

* Specific fuel consumption.

The critical condition for the ceramic parts is the cold start. Therefore, a special cold-start schedule has been chosen in order to separate thermal and centrifugal mechanical stress. This is achieved by a start procedure where the turbine inlet temperature (TIT) is increased to the maximum 1623 K and the gas generator speed is accelerated to 80%. The gas generator speed is held at 80% for 60 s, allowing temperature gradients to even out. During hot start conditions there is, however, no limitation on the start cycle. In practice, the driving cycle for a recuperated gas turbine would normally only include one cold start per day because of heat storage in the recuperator. During the typical hot start, the acceleration to 100% speed and maximum power takes 9 s.

5.3.2 Catalytic Combustor – The catalytic combustor section is shown in Figure 6. At full load conditions, the temperature in the catalytic section is increased from inlet 1208 K to the exhaust 1623 K. Therefore, structural components in the hot section are made of ceramic materials. The catalytic combustor includes the following components:

- Preheater, to be operated only during the start-up period of 60 s and then by-passed for all other engine conditions.
- Premix duct based on a lean premixed prevaporised (LPP) concept with a venturi configuration, which has been optimized in order to obtain a good velocity profile at the inlet to the catalytic section. Fuel is injected upstream of the venturi minimum section with injectors angled relative to the airflow. Mixing is achieved with an inlet radial/axial turbulence generator in the venturi inlet.
- Catalytic reactor, Figure 7, with catalyst substrates made of a ceramic honeycomb of an extruded oxide ceramic material inside the catalytic section envelope made of silicon nitride material.
- Afterburner cone, Figure 7, for exhaust gases, designed for high strength and thermal shock resistance and for operations at 1623 K is made of silicon nitride material.

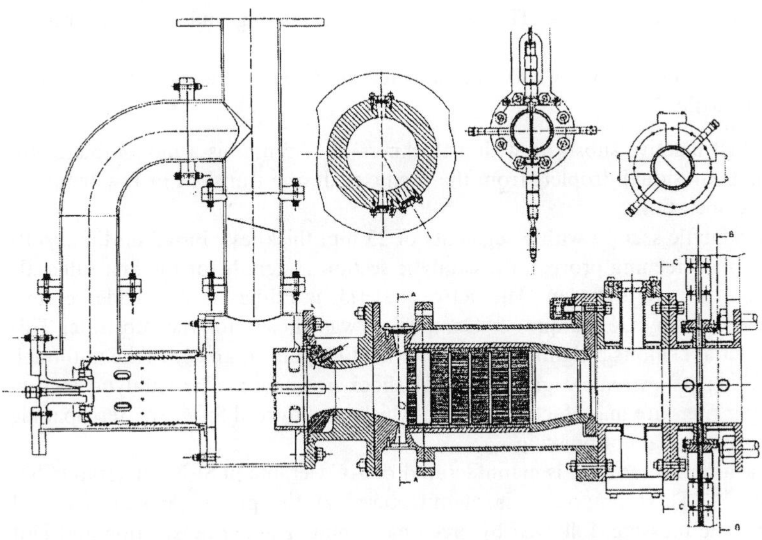

Figure 6 *Catalytic combustor with exhaust measurement section*

The combustor validation during steady state and transient conditions is performed in a test rig at ONERA, France, where the high pressure/temperature conditions can be achieved.[143] As was pointed out earlier,[142] in order to obtain high hetero/homogenous combustion efficiency with low emission values it is important to create an even velocity profile at the catalytic section inlet. Extensive

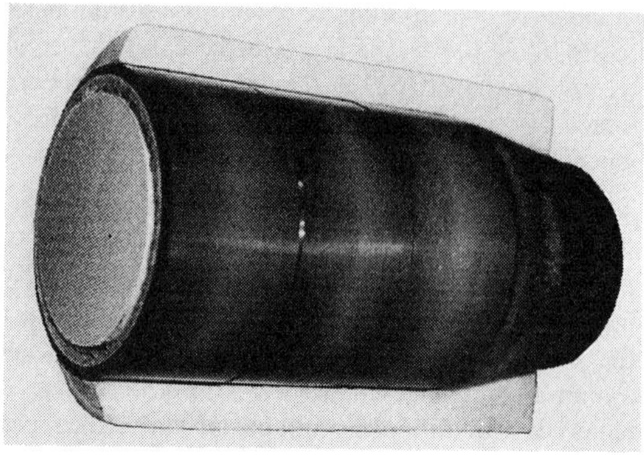

Figure 7 *AGATA catalytic and afterburner section*

Laser Dopple Velocity (LDV) measurements in the venturi premix duct outlet therefore were carried out. These tests, which were run at cold conditions at the design point with the same Reynolds number, show a very symmetric and even velocity profile at all circumferential angles in the duct outlet. The measured RMS values, 30–35% of the mean velocity, also showed that a high turbulence level exists in the venturi. This will enhance the mixing and fuel vaporization process.

The catalytic section of 100 mm diameter, as shown in Figure 7, is built up in sections with:

- A first inactive substrate with a thickness of 12.5 mm as a monolithic guard for the residual droplets from the premixer during transient or low temperature operations.
- A catalytic section with 5 segments of 25 mm thickness. Based on the results from a screening process, the catalytic section materials for the first full-scale prototype were chosen. The active material is Pd-based. An oxide ceramic developed by Céramiques & Composites was chosen for the substrate. This material exhibits a significantly higher melting point than cordierite, although lacking its extremely good thermal shock resistance. The catalytic section outer rings are manufactured in SiC-whisker reinforced Si_3N_4 by Aerospatiale using uniaxial hot pressing.
- The afterburner cone is manufactured by AC Cerama in Si_3N_4 material (CSN 101).[144] This component is manufactured at the prototype scale by cold isostatic pressing, followed by green machining, glass encapsulation and Hot Isostatic Pressing (HIP). In a future larger production volume, such a piece could be formed by injection moulding. The outer and inner surfaces of this component will be used with as-HIPed surface finish. Only the contact surfaces are machined after HIP.

Laser diagnostic measurements were conducted by Lund Institute of Technology at Volvo Aero Turbines, Sweden. Temperature and oxygen profiles downstream of the catalytic reactor were measured. These tests indicated cold streaks at the periphery, which meant that it will be possible to reduce the CO and HC emissions further near the periphery of the combustor.

The catalytic combustor was run on diesel fuel during the complete test period. A test campaign comparing exhaust emissions when running on ethanol fuel was performed at Volvo Aero Turbines. These results showed that the catalyst reaction rate and CO/HC/NO_x emissions were similar. This means that the chosen catalytic combustor can be used as a duel fuel combustor diesel/ethanol.

The final test campaign at Onera in France was run up to temperatures slightly above the specified maximum design temperatures: the inlet temperature 1235 K (design 1208 K) and the exhaust temperature 1635 (design 1623 K). These tests showed that NO_x emission levels below 4 ppm at 15% O_2 were obtained when low CO and HC emissions levels were measured at full load conditions.

5.4 Use of Different Fuels and Their Characteristics – The best commercial, dry, low NO_x combustors today are optimized for clean-burning natural gas. However, with raised natural gas prices over the next decade, power plants may be forced to burn low-heating value fuel gas, products of gasification or low quality residual fuels. As the combustion becomes more complicated, *e.g.* lean-premixed combustors, to handle NO_x emissions from different fuels is bound to become more complex, too.[145] Hence, development of the catalytic combustor must also be directed towards fuels other than natural gas. These fuels could be other hydrocarbon feedstock, *e.g.* diesel fuels,[146] which are more available than natural gas in some parts of the world, and kerosene,[139,140] which is used for jet-turbines in aeroplanes. An increased use of renewable fuels, such as methanol, ethanol and low-heating value fuels derived from biomass or waste will also lead to a demand to put these fuels to use in gas turbines.[26,58,147]

In a recent study at our laboratory, some of these alternative fuels have been compared over $LaMnAl_{11}O_{19}$-washcoated cordierite monoliths. The typical light-off temperature, at a gas hourly space velocity (GHSV) of 50 000 h^{-1} and a λ-value around 3 is given in Table 7. Clearly, comparing ignition temperatures for these different fuels, they typically ignite at lower temperatures than methane. Hence catalytic ignition is not a problem here. However, other challenges are to be solved for these fuels, *e.g.* evaporation and homogenous mixing of liquid diesel and air, fuel-NO_x conversion for LHV-fuels such as

Table 7 *Typical light-off temperature for various fuels*

Fuel	*Typical light-off** (°C)	Remark
Diesel	400	
Methane	550–600	
LHV-fuel	350–550	For CO–H_2
	550–600	For CH_4
Ethanol	400	

* At GHSV = 50 000 h^{-1}, $\lambda = 3$, reported as the inlet temperature for 50% conversion.

gasified biomass and deactivation from sulfur species in both diesel and LHV-fuels. The conversion of fuel-nitrogen is presently studied at our laboratory within the framework of a European project.[26,146] Corroborating results from studies in the early 1980s,[8] it has been shown that formation of fuel-NO_x can be reduced by catalytic combustion. Factors that will influence the formation of fuel-NO_x are catalyst composition, temperature and the stoichiometry of the fuel–air mixture.

5.5 Recent Patents – A selection of patents published during 1993–1998 describing new high temperature catalytic materials or catalytic combustion applications in gas turbines is given in Table 8. Title, affiliation or inventor, patent number and a short description are summarized in the table, but are not given in the reference section. As shown, several patents are similar and are therefore grouped together, *e.g.* for patent applications in several countries or for similar patents from the same company. Although the table is not complete, it can be concluded that surprisingly many of the 34 patents are describing the use of passive channels, *i.e.* design IIa originally proposed by Westinghouse. Catalytica, Mitsubishi, Grace, Toshiba, W.C. Pfefferle and ICI have all proposed the approach with active and passive channels for catalytic combustors recently. Moreover, Precision Combustion suggest a similar hybrid approach where the catalyst segment is so short that part of the fuel exits the catalyst without being burnt. Several of the patents are describing the use of precious metals, *e.g.* PdO/Pd for the first catalytic segment. New patents on other interesting materials cover hafnia-modified zirconia, stannic oxide, pyrochlore compounds and catalysts having the formula Pr_4PdO_7.

6 Deactivation of Combustion Catalysts

One of the main hurdles for combustion catalysts is to maintain the catalytic activity, during prolonged operation. Typically, gas turbine manufacturers demand at least one-year-operation, *e.g.* 8000 h for stationary gas turbines and approximately the same number of hours for mobile applications. Combustion catalysts will be exposed to severe conditions, *i.e.* gas velocities around 10–25 $m\ s^{-1}$, peak temperatures between 1000 and 1400 °C depending on the system design, transients with temperature drops around 500 °C s^{-1} or higher, and fuel mixtures that can contain known catalyst poisons. Hence, the catalysts could be deactivated by:

- sintering, *i.e.* loss of active surface area, and vaporization of active material, due to long-term exposure to heat and steam;
- thermal stresses, which could lead to fracture of the catalyst;
- reversible and irreversible poisoning of active sites by: (1) sulfur compounds in the fuel, such as thiophenes, H_2S or SO_2, (2) CO from the fuel, and (3) alkali salts from combustion air at offshore or coastal installations or from the fuel, *e.g.* gasified biomass.

Table 8 *Examples of recent patents on applications of catalytic combustion in gas turbines*

Approximate title sorted chronologically	*Patent no. and holder*	*Brief remark*
1998		
Catalytic combustors containing palladium and manganese for gas turbines	JP 10047610 Mitsubishi Heavy Industries Ltd.	Pd-containing catalyst as an initial stage followed by Mn-containing catalyst with magnetoplumbite structure
Catalytic method	US 5720605, 5720609 W.C. Pfefferle	Lean hybrid combustion with very short channels (*e.g.* Microlith™)
1997		
Catalytic combustion of combustible gases *and* combustion method for combustible gases	JP 09262472, JP 09049609 *and* JP 09049608 Mitsubishi Heavy Industries Ltd.	Catalyst system with first catalyst consisting of PdO–PGM (except for Pd) and second one of PdO *or* La-perovskite *or* selected from Al_2O_3, SiO_2, TiO_2 and ZrO_2
Palladium oxide combustion catalysts for high temperature uses and their preparation	JP 09285730 Petroleum Energy Centre	Catalyst system based on PdO supported on δ-alumina
Combustion catalyst structure for gas turbine combustor	JP 09192501 Mitsubishi Heavy Industries Ltd.	Catalyst consisting of support of sheets with 'bumps' acting as spacers between the sheets
Combustion method	US 5634784 Precision Combustion Inc.	Hybrid combustion with catalyst heated with hot recirculated partially reacted combustion gases
Catalyst support for high temperature applications and catalysts	WO 9705057 Catalytica Inc., Tanaka K.K.	HfO_2 based catalyst support material
1996		
Palladium oxide catalysts for high temperature combustion and their manufacture	JP 08309184 Petrol. Energy Centre Foundation	Catalyst consisting of PdO on a zirconia-based support and its manufacturing
Combustion catalysts for fuel gases and catalytic combustors for gas turbines	JP 08252463 Tokyo Shibaura Electric Co.	PGM-containing catalyst as an initial stage followed by alkaline- or rare earth metal oxide based catalyst

Continued

Table 8 *Continued*

Approximate title sorted chronologically	*Patent no. and holder*	*Brief remark*
Catalytic combustion in gas turbine	WO 9325852, EP 646228, etc. ICI Plc.	Combustor with honeycomb catalyst with passive and active channels and secondary downstream air/fuel injection
Method for efficient combustion of fuel air mixtures in catalytic gas turbines	EP 677707, CA 2147024, *etc.* Precision Combustion Inc.	Hybrid combustor type II b
Apparatus to secure and support a monolithic catalyst structure	EP 733188 Catalytica Inc., Tanaka K.K.	An apparatus for secure and support the catalyst structure to the combustor wall
Stabilised catalyst carrier member for catalytic combustion of gaseous carbonaceous fuels	US 5509796 Engelhard Corp.	A silica–alumina–magnesia material for support coated with an alumina washcoat
Partial combustion process operating below auto-combustive temperature	WO 9209848, *etc.* Catalytica Inc., Tanaka K.K.	Partial combustion process using metal monolith partly deposited with Pd, preferably on a zirconia support
1995		
Sequentially staged combustion using a catalyst, *e.g.* for gas turbines	EP 686813, AU 9520099, *etc.* Westinghouse Electric Corp.	Hybrid combustor type IIb
Catalytic combustion of methane using stannic oxide	WO 9531675, AU 9524145 Johnson Matthey PLC	Catalyst system based on SnO_2 and optional Pt, Pd, Rh, Ru
Catalytic stabilized gas phase combustion system	EP 668471 A2 W.C. Pfefferle.	Hybrid combustor type IIa
Ceramic catalyst burner for catalytic combustion	JP 07119927 Mitsubishi Jukogoyo KK	A SiC mesh based burner, with an internal heating system.
Catalytic combustion device	JP 07229623 Toshiba KK	Catalyst with passive channels based on Pd/PdO
Improved catalytic combustion system for gaseous hydrocarbons	WO 9420789, *etc.* Engelhard Corp.	Design of catalyst system for natural gas combustion using a supported palladium oxide catalyst
Praseodymium–palladium binary oxide	US 5378142 Engelhard Corp.	Catalyst having the formula Pr_4PdO_7
Combustion fuel mixtures	WO 9209849, *etc.* Catalytica Inc., Tanaka K.K.	Hybrid combustor with catalytic part consisting of three catalyst segments using Pd and Pt based catalysts

Catalytic combustion process for gas turbine	EP 491481, *etc.* ICI Plc.	Hybrid combustor with part of the air/fuel being combusted over the catalyst
1994		
Combustion device for high temperature environment	US 5346389 Grace & Co.	Multi-staged combustor comprising of, on one side, coated corrugated metal strip
Combustion catalyst for catalytic combustor for gas turbine	JP 06082013 Mitsubishi Jukogoyo KK	Honeycomb with passive and active channels
Catalytic combustion chamber for gas turbine	GB 2269764 Rolls-Royce Plc.	Catalytic combustor with heat conductor that transfer heat from downstream part to upstream end
Gas turbine combustion chamber	EP 576697 ABB Res. Ltd.	The combustor consists of pre-mixed burners and catalytic burners which allow very lean operation
Gas turbine engine catalytic combustion chamber	GB 2268694 Rolls-Royce Plc.	Catalytic combustor with electrical preheater for start up and additions of hydrogen to heat the catalyst.
Oxidation catalyst for gaseous fuel combustion	EP 620654 ICI Plc.	Catalyst based on an intimate mixture of Pr and group IIIA elements and optional zirconium oxides.
Catalyst structure with integral heat exchange surfaces	EP 559844, JP 06506290 Catalytica Inc., Tanaka K.K.	Metallic monolith with parts of the channels coated with catalyst
1993		
Catalytic combustion of a carbonaceous fuel, *e.g.* natural gas *and* catalytic combustion process	US 5214912 *and* WO 9318347, *etc.* Engelhard Corp.	Combustor system using the PdO to Pd transition to control the temperature. Use of metal oxide (*e.g.* ceria, titania) supported PdO and restoring of catalyst activity
Two-stage process for combusting fuel mixtures *and* multistage process for combusting fuel mixtures using oxide catalysts in hot stages	US 5183401 *and* US 5232357 Catalytica Inc., Tanaka K.K.	Catalytic combustor with two or more catalytic stages using metallic monoliths with parts of the channels inactive and Pd-based catalyst in the first stage *or* using oxide catalyst in the hot stage
Graded palladium-containing partial combustion catalyst and a process for using it	US 5248251 Catalytica Inc	A palladium catalyst with higher activity in the beginning of the catalyst.
Catalytic combustion of methane	JP 05272719 Eniricerche Spa, Snamprogetti Spa.	$X_aY_bZ_cSn_2O_7$ Pyrochlore-type catalyst for catalytic combustion of methane

These three mechanisms for deactivation will be briefly discussed in the following section. In addition, coking could occur for fuels with longer hydrocarbon chains or large amounts of CO, in applications using fuel-rich catalytic combustion. Also blocking by dust particles could be a problem. However, coking and blocking have not been reported to be a significant problem so far.

6.1 Sintering and Vaporization – The severe conditions, especially heat generation at the catalyst surface, will cause sintering of the catalyst, *i.e.* decrease the surface area of the support, as well as causing migration and encapsulation of the active species on the washcoat. For example, high surface area γ-Al_2O_3 is well known to transform to low surface area α-Al_2O_3 around 1000 °C, *i.e.* a change in surface area from 150 to 2 $m^2\ g^{-1}$. In addition, vaporization of catalyst material is a very significant process at temperatures above 1000 °C. Dalla Betta reports vapour pressure of several active and support components in an equimolar H_2O/O_2 mixture at 1 atm.[28] The vapour pressures at 1000 °C ranged from 10^{-4} to 10^{-9} atm. Cu, Co, Ba and Cr showed the highest vapour pressures, whereas Pd and Al showed the lowest. The vapour pressure increases with approximately two orders of magnitude, as the temperature is increased to 1400 °C. The loss of Pt from a monolithic support, predicted by a diffusion-controlled model,[28] showed that approximately 70% of the original Pt loading would be lost after 15 h on stream at 1000 °C. Usually, if the material is stabilized by reactions with other materials, the resulting compound would have a lower activity. The rate for sintering and vaporization is also accelerated in the presence of steam and formation of surface oxides or hydroxides. Solid phase reactions are also likely to take place, such as the spinel formation,[32] between active phase and washcoat along with phase transformation of the oxide materials.

6.2 Thermal Stresses and Fracture of the Catalyst – Boehman *et al.*[116] studied dynamic stress formation in catalytic combustion of methane over ceramic-supported PdO catalysts by a transient 2D model. The simulation illustrated that thermal stress during warm-up and cool-down could form fracture stress above the fracture strength of typical monolith materials, such as cordierite. In fact, warm-up during uncontrolled light-off created the highest stresses along with ramping of the inlet fuel concentration. The formation of fracture stresses could be decreased by several innovative designs; *e.g.* using higher inlet temperatures, such as with a recuperative design, *cf.* Section 5.2.2, or dividing the ceramic substrate into several cake-like pieces, *cf.* Section 5.2.4, or using passive channels that will cool the active channels, *cf.* Section 5.2.3. The uses of secondary fuel, see Section 5.2, will also reduce the heat generation in the catalytic part of the combustor. However, the best alternative would be to find a novel material that can stand these fracture stresses.

For catalyst systems used below 1000 °C, the substrate is washcoated. Here, the washcoat–substrate adhesion and the ratio of the washcoat–substrate thermal expansion are of primary importance.

6.3 Poisoning – Sulfur species are known catalyst poisons and have been studied extensively. They can be present in natural gas, diesel oil or low-heating value fuels originating from gasification of biomass or coal. Adsorption of sulfur on most catalysts causes rearrangement of the surface structure and, hence, affect the catalytic properties.[86,148] It has been shown that base metals are more likely to suffer from sulfur poisoning than precious metals.[149] However, deactivation by sulfur species is not so severe at the higher operating temperatures in catalytic combustion as for other catalytic applications operating at lower temperatures.[8] Sulfur exists in different chemical states depending on the reaction conditions. For the oxidative conditions applicable to catalytic combustion, sulfur species usually react to form sulfur dioxide or trioxide and sulfate species. Hoyos *et al.* studied combustion of methane over Pd on alumina or silica catalysts.[150] Sulfur in the inlet gas readily poisoned the catalyst. Sulfate groups bound to the Pd particles were found to be responsible for the deactivation and there is a direct relationship between the catalytic activity and the amount of sulfates formed. The sulfates could be thermally destroyed at higher temperatures. However, the stability of sulfates varies.[151] Sulfates such as $MnSO_4$ can decompose easily at 755 °C, whereas others such as $SrSO_4$ can be stable up to 1374 °C. Even though sulfur poisoning seems to deactivate more strongly at lower temperatures,[8] *e.g.* during start-up, the catalyst material must be selected to avoid formation of stable sulfates.

Ignition of CO- and H_2-rich fuels, such as gasified biomass, is controlled by the ignition of CO, reported by Zwinkels *et al.*[152] and Johansson *et al.*[58] CO adsorbs strongly on some catalyst systems and retards adsorption of other fuel species and oxygen and hence ignition at low temperatures. Moreover, steam has been reported to inhibit the rate for total oxidation of methane. These results are summarized in a recent PhD Thesis by van Giezen.[153]

Alkali salts from combustion air at coastal installations or low-heating value fuels originating from gasification of biomass may be present in the fuel or combustion air. Salt deposition could be a problem at coastal or offshore installations.[15] Chlorine deactivation was studied by Agarwal and co-workers,[154] with respect to catalytic combustion of mixed chlorohydrocarbons over chrome-alumina catalysts. A decline in catalyst activity downstream of the catalyst bed was observed after 153 days-on-stream.

In conclusion, deactivation or poisoning by the fuel or combustion air compounds must be considered, especially if present levels in fuels are lower than gas turbine specifications, but which could still deactivate the catalysts significantly. This could be solved either by the modification of the catalyst or by pre-treatment of the fuel or combustion air.

7 Spin-off Benefits for Other Areas in Combustion

Though most recent studies on catalytic combustion are focused towards either low temperatures (for elimination of VOCs) or high temperatures (gas turbines), fall-off or spin-off benefits from such studies are being harvested in several other areas of combustion. Typical examples for these are household appliances and utensils ranging from cooking stoves and local/central heating to gas-heated hair-curlers and soldering torches. Large-scale applications as in industrial ovens and furnaces are being constantly upgraded for both energy saving and reduced pollution. Only one typical example for this is illustrated in the next Section.

7.1 Eco-friendly Coke Combustion in FCC[155] – The fluid catalytic cracking (FCC) plant or catcracker is the heart of a modern petroleum refinery. The FCC process converts gas oil and heavier/resid feedstocks, mainly in the 315–650 °C boiling range, into lighter transportation fuels such as gasoline and diesel, cycle oil and other middle distillates and C_3–C_4 gases for alkylation, *etc.* Tightening environmental regulations are forcing changes in the design and operation of the FCC process. Up to 7 wt% of the FCC feedstock is converted to coke on catalyst; this coke has to be burned off continuously to regenerate the catalyst and generate the heat necessary for the endothermic cracking. In the United States alone, over 1 million tonnes of gas oil are cracked every day. That means combustion of 50 000–70 000 tonnes of coke per day. NO_x levels in regenerator flue gas are in the range 50–200 ppm, which are increasingly coming under local and regional legislative pressures. No NO_x control technology, however, can be allowed to interfere with the smooth operation of the catcracker, nor should it have any adverse impact on the emissions of other catcracker pollutants (CO, CO_2, SO_x and VOCs).

The chemistry of the conversion of coke-bound nitrogen on the FCC catalyst into NO_x, though not yet very clear, is analogous to the three modes of NO_x formation in ordinary combustion,[156] namely thermal-NO_x, prompt NO_x and fuel-NO_x. The conversion of fuel/coke nitrogen first to HCN and NH_3 and then to NO_x is schematically shown in Figure 8.

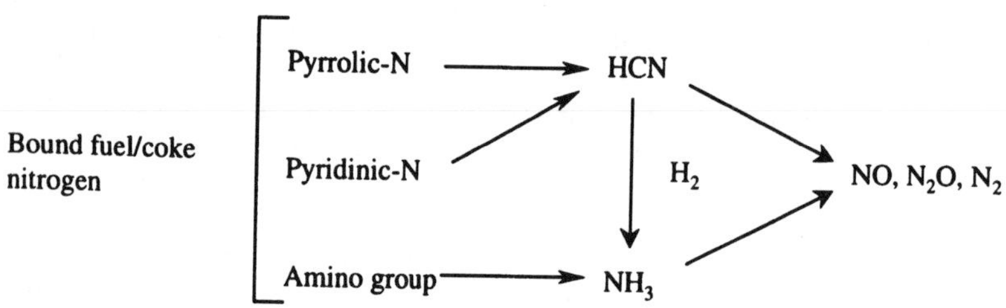

Figure 8 *Conversion of fuel/coke nitrogen*

Peters *et al.* attempted to obtain a nitrogen balance around the FCC unit.[157] The results showed that about 8% of the feed nitrogen is converted to ammonia and 30–40% into molecular nitrogen. Only about 10–15% of the nitrogen in the coke was converted to NO, the rest ended up as molecular nitrogen. In commercial FCC regenerators operating at 650–750 °C, the nitrogen released as NO varies from 3 to 25%, depending on the regenerator design and the FCC operating conditions.[158] For instance, selective noncatalytic reduction of NO_x to N_2 can occur by interaction with the CO present in the regenerator flue gas. However, to enhance CO combustion to produce more heat in the regenerator and to minimize CO emissions into the atmosphere, a CO-combustion catalyst component (*e.g.* ppm levels of Pt) is often added as an integral part of modern FCC catalysts. This catalytic combustion additive in FCC catalysts reduces CO, and consequently increases NO_x, in the regenerator flue gas. Older type FCC units have often a separate CO boiler, instead of the combustion catalyst additive; but then NO_x emission from the CO boiler itself becomes a separate issue. In the high-efficiency regenerator, developed by Universal Oil Products (UOP), full CO combustion is achieved by using a riser tube only and not using any Pt combustion promoter; here the NO_x emission can again be high since there is no CO for the selective non-catalytic reduction of NO_x.

Direct additives into the FCC catalysts to reduce NO_x emissions are of the spinel/perovskite types, just as for high temperature catalytic combustion also. Thus, Cu/Mn perovskites and $MgAl_2O_4$ spinel have been claimed to reduce NO_x to N_2, in the presence of SO_2 and oxygen.[159,160] Once again these additives are most efficient when there is enough CO in the flue gas, which means that little or no Pt-based CO-combustion component is added to the FCC catalyst blend. These are very conflicting demands indeed. Other NO_x-reducing additives, claimed in the patent literature, are: oxides of La or Y, lanthanum titanate,[161] Zn-based compounds,[162] ceria,[163] and ZSM-5 in a matrix containing titania and zirconia, impregnated with Cu and rare-earth cations.[164]

The simultaneous reduction of both SO_x and NO_x emissions continues to be a challenge in FCC, just as in other areas of catalytic combustion also. The process requirements are diametrically opposite: NO_x control needs reducing agents, whereas SO_x additives all work in an oxidizing environment. Additives for the direct decomposition of NO_x in the presence of oxygen still remain as a challenge for future research. Perhaps at a still earlier stage, a closer understanding has to emerge of the relationship between fuel/coke combustion chemistry and the formation and destruction of NO_x. The know-how and experience emerging from studies of high temperature catalytic combustion as for gas turbines have thus a widespread beneficial effect for other areas in combustion as well. The case of FCC, cited here, is only one example for it.

8 Summary

As a preventive solution to the problem of NO_x emissions, catalytic combustion has come to the forefront during the last two decades. The focus of major interest here is its application in gas turbines for power generation and for transportation by road, water or air. Any catalyst for catalytic combustion, however, has to face extreme demands: above 1000 °C in the presence of oxygen and steam, in uninterrupted operation for at least one year; also, resistance to impurities or poisons in the fuel and to severe thermal and mechanical shocks. Promising developments in catalytic supports, washcoats and active materials are reviewed briefly.

Mathematical modelling is proving to bc a powerful tool to probe the complexities of catalytic combustion in monolithic honeycomb catalyst geometry, to guide experimental research, and to analyse the experimental results. The various models proposed have been surveyed critically.

Since material science and catalysis cannot as yet meet the exacting demands in catalytic combustion, elegant systems engineering is being applied to overcome the problems and to develop practical solutions for the time being. Several approaches of this type developed in Japan and the United States and in the European AGATA project are presented briefly. Finally, a survey of recent patents (from 1993) indicates the emerging tendencies in this field.

Acknowledgments

Funding from the European Union, contract JOR3-CT96-0071, the Swedish National Energy Administration (formerly NUTEK – the Swedish National Board for Industrial and Technical Development), contracts P910, P4469, P7057, and P10547, and SSF-Cecost, contract PROJ9701-SOT, is gratefully acknowledged by KTH. Likewise, Volvo Aero Corporation is indebted to the Swedish National Energy Administration and French Ministries of Research and Industry for funding to the AGATA-project.

References

1 W.C. Pfefferle, Belgian Patent 814,752, 1974.
2 W.C. Pfefferle, *J. Energy*, 1978, **2**, 142.
3 S.T. Gulati, in *Structured Catalysts and Reactors*, ed. A. Cybulski and J.A. Moulijn, Marcel Dekker, New York, 1998.
4 R.N. Carter, P. Menacherry, W.C. Pfefferle, G. Muench and S. Roychoudhary, SAE Paper 980672, 1998.
5 M.V. Twigg and D.E. Webster, in *Structured Catalysts and Reactors*, ed. A. Cybulski and J.A. Moulijn, Marcel Dekker, New York, 1998.
6 M. Machida, K. Eguchi and H. Arai, *J. Catal.*, 1989, **120**, 377.
7 J.C. Schlatter, R.A. Dalla Betta, S.G. Nickolas, M.B. Cutrone, K.W. Beebe and T. Tsuchiya, ASME Paper 97-GT-57, 1997.
8 R. Prasad, L.A. Kennedy and E. Ruckenstein, *Catal. Rev.-Sci. Eng.*, 1984, **26**, 1.
9 J.P. Kesselring, in *Advanced Combustion Methods*, ed. F.J. Weinberg, Academic Press, London, 1986.
10 L.D. Pfefferle and W.C. Pfefferle, *Catal. Rev.-Sci. Eng.*, 1987, **29**, 219.
11 D.L. Trimm, in Chemistry of the Platinum Group Metals – Recent Developments, ed. F.R. Hartley, Elsevier, Amsterdam, 1991.
12 M.F.M. Zwinkels, S.G. Järås, P.G. Menon and T.A. Griffin, *Catal. Rev.-Sci. Eng.*, 1993, **35**, 319.
13 S.T. Kolaczkowski, *Trans. IChemE.*, 1995, **73A**, 168.
14 K. Eguchi and H. Arai, *Catal. Today*, 1996, **29**, 379.
15 R.L. Garten, R.A. Dalla Betta and J.C. Schlatter, in *Handbook of Heterogeneous Catalysis*, ed. G. Ertl, H. Knözinger and J. Weitkamp, Wiley-VCH, 1997.
16 M.F.M. Zwinkels, S.G. Järås and P.G. Menon, in *Structured Catalysts and Reactors*, ed. A. Cybulski and J.A. Moulijn, Marcel Dekker, New York, 1998.

17 P.G. Menon, M.F.M. Zwinkels and S.G. Järås, in *Studies in Surface Science and Catalysis. Recent Advances in Basic and Applied Aspects of Industrial Catalysis*, ed. T.S.R.P. Rao and G.M. Dhar, Elsevier Science B.V., 1998.

18 R.E. Hayes and S.T. Kolaczkowski, *Introduction to Catalytic Combustion*, Gordon and Breach Science Publishers, Amsterdam, 1997.

19 G. Kraemer, T. Strickland, W.C. Pfefferle, J. Ritter, EC-Vol. 5 Proc. Int. Joint Power Gen. Conf., ed. A. Sanyal, A. Gupta and J. Veilleux, Book no. G01702, 1997.

20 J. Zeldovich, *Acta Physicochimica URSS*, 1946, **21**, 578.

21 D.L. Baulch, A.M. Cox, T. Just, J.A. Kerr, M. Pilling, J. Troe, R.W. Walker and J. Warnatz, *J. Phys. Chem., Ref. Data*, 21:3, 1991.

22 S.M. Correa, *Combust. Sci. Technol.*, 1992, **87**, 329.

23 C.P. Fenimore, *13th Symp. (Int.) Combustion*, Combust. Inst., Pittsburgh, PA, 1970.

24 I. Glassman, *Combustion*, Academic Press, San Diego, CA, 1996.

25 J.A. Miller and C.T. Bowman, *Prog. Energy Combust. Sci.*, 1989, **15**, 287.

26 E.M. Johansson, K.M.J. Danielsson, A.G. Ersson and S.G. Järås, ASME Paper 98-GT-338, 1998.

27 I. Stambler, *Gas Turbine World*, 1993, **23**, 32.

28 R.A. Dalla Betta, *Catal. Today*, 1997, **35**, 129.

29 K.C. Taylor, in *Catalysis – Science and Technology*, ed. J.R. Anderson and M. Boudart, Springer-Verlag, Berlin, 1984.

30 R.M. Heck and R.J. Farrauto, *Catalytic Air Pollution Control. Commercial Technology*, van Nostrand Reinhold, New York, 1995.

31 T. Ishikawa, R. Ohashi, H. Nakabayashi, N. Kakuta, A. Ueno and A. Furuta, *J. Catal.*, 1992, **134**, 87.

32 P.H. Bolt, M.E. v. Ipenburg, J.W. Geus and F.H.P.M. Habraken, *MRS Symp. Proc, Mat. Res. Soc.*, 1994.

33 J.A. Cusumano, R.A. Dalla Betta and R.B. Levy, *Catalysis in Coal Conversion*, Academic Press, New York, 1978.

34 H. Arai and M. Machida, *Appl. Catal., A*, 1996, **138**, 161.

35 Y. Hashimoto, O. Takayasu and I. Matsuura, *Chem. Express*, 1991, **6**, 81.

36 K. Matsuura, Y. Hashimoto, O. Takayasu, K. Nitta and Y. Yoshida, *Appl. Catal.*, 1991, **74**, 273.

37 M. Berg and S. Järås, *Catal. Today*, 1995, **26**, 223.

38 R. Wang, WO Patent 97/05057, 1996.

39 M. Ledoux, S. Hantzer, C.P. Huu, J. Guille and M.P. Desaneaux, *J. Catal.*, 1988, **114**, 176.

40 H. Inoue, K. Sekizawa, K. Eguchi and H. Arai, *J. Am. Ceram. Soc.*, 1997, **80**, 584.

41 H. Inoue, K. Sekizawa, K. Eguchi and H. Arai, *J. Mater. Sci.*, 1997, **32**, 4627.

42 D.K. Agrawal, C.-Y. Huang and H.A. McKinstry, *Int. J. Thermophys.*, 1991, **12**, 697.

43 S.Y. Limaye, D.K. Agrawal and H.A. McKinstry, *J. Am. Chem. Soc.*, 1987, **70**, C232.

44 S.L. Andersson and N.-H. Schöön, *Ind. Eng. Chem. Res.*, 1993, **32**, 1081.

45 C. Groppi, A. Belloli, E. Tronconi and P. Forzatti, *AIChE J.*, 1995, **41**, 2250.

46 G.L. Vaneman, *Catalysis and Automotive Pollution Control II*, Elsevier Sci. Publ. B.V., Amsterdam, 1991.

47 M.F.M. Zwinkels, S.G. Järås and P.G. Menon, in *Preparation of Catalysts VI*, ed. G. Poncelet, P.A. Jacobs, P. Grange and B. Delmon, Elsevier Science B.V., Amsterdam, 1994.

48 G. Groppi, C. Cristiani and P. Forzatti, *Catalysis*, Specialist Periodical Reports, vol. 13, The Royal Society of Chemistry, Cambridge, UK, 1997, p. 85.
49 K. Eguchi, H. Takahara, H. Inoue, K. Sekizawa and H. Arai, 'Symp. Catalytic Combustion, San Francisco', Am. Chem. Soc. Div. Petrol. Chem., **42**, 1997.
50 M. Machida, A. Sato, T. Kijima, H. Inoue, K. Eguchi and H. Arai, *Catal. Today*, 1995, **26**, 239.
51 H. Inoue, K. Sekizawa, K. Eguchi and H. Arai, *J. Solid State Chem.*, 1996, **121**, 190.
52 K.S. Ramesh, J.J. Kingsley, T.L. Hubler and D.E. McCready, in *Ceramic Transactions*, ed. K.S. Ramesh, M. Misono and P. Gai, Am. Ceram. Soc., Westerville, 1997.
53 M. Machida, K. Eguchi and H. Arai, *J. Catal.*, 1990, **123**, 477.
54 J.J. Spivey, B.W.-L. Jang, G. Marcelin, M. Öcal, R. Oukaci and R.V. Serauskas, 'Int. Gas Research Conf', 1995.
55 A.G. Ersson, E.M. Johansson and S.G. Järås, in *Preparation of Catalysts VII*, ed. G. Poncelet, B. Delmon, P.A. Jacobs, R. Maggi, J.A. Martens and P. Grange, Elsevier B.V., Amsterdam, 1998, pp. 601–608.
56 A.G. Ersson, E.M. Johansson and S.G. Järås, *Book of abstracts, 8th Nordic Symp. on Catalysis, Oslo, Norway*, Norwegian Chem. Soc., 1998.
57 G. Groppi, A. Belloli, E. Tronconi and P. Forzatti, *Catal. Today*, 1996, **29**, 403.
58 E.M. Johansson and S.G. Järås, 'Symp. Catalytic Combustion, San Francisco', Am. Chem. Soc., Div. Petrol. Chem., **42**, 1997.
59 L.G. Tejuca, *J. Less-common Metals*, 1989, **146**, 251.
60 L.G. Tejuca, *J. Less-common Metals*, 1989, **146**, 261.
61 K.S. Ramesh, J.L. Cox and W.P.J. Parks, in *Proc. 2nd Int. Workshop Catal. Comb.*, ed. H. Arai, Catal. Soc. Japan, Tokyo, 1994.
62 M.F.M. Zwinkels, S. Druesne, E. Björnbom, P.G. Menon and S.G. Järås, *Ind. Eng. Chem. Res.*, 1998, **37**, 391.
63 D.M. Lowe, M.I. Gusman and J.G. McCarty, in *Preparation of Catalysts VI*, ed. G. Poncelet, P.A. Jacobs, P. Grange and B. Delmon, Elsevier Science B.V., Amsterdam, 1994.
64 C. Cristiani, G. Groppi, P. Forzatti, E. Tronconi, G. Busca and M. Daturi, in *11th Int. Cong. Catalysis. Studies in Surface Science and Catalysis*, ed. J.W. Hightower, W.N. Delgass, E. Iglesia and A.T. Bell, Elsevier Science B.V., Amsterdam, 1996.
65 P.H. Bolt, F.H.P.M. Habraken and J.W. Geus, *J. Catal.*, 1995, **151**, 300.
66 P.E. Marti, M. Maciejewski and A. Baiker, *Appl. Catal., B:*, 1994, **4**, 225.
67 M.A. Subramanian, *Mat. Res. Bull.*, 1990, **25**, 107.
68 A. Castro, I. Rasines and X.M. Turrillas, *J. Solid State Chem.*, 1989, **80**, 227.
69 A.K. Neyestanaki, N. Kumar and L.-E. Lindfors, *Appl. Catal., B:*, 1995, **7**, 95.
70 Y.S. Cho, V.L. Burdick and R.W. Amarakoon, *J. Am. Ceram. Soc.*, 1997, **80**, 1605.
71 S. Liu and Q. Su, *J. Alloys Comp.*, 1997, **255**, 102.
72 R.J. Farrauto, M. Larkin, J. Fu and J. Feeley, *Mat. Res. Soc. Symp. Proc.*, 1994, **344**, 101.
73 L.M. Quick and S. Kamitomai, *Catal. Today*, 1995, **26**, 303.
74 R. Burch, *Pure Appl. Chem.*, 1996, **68**, 377.
75 J.G. McCarty, *Catal. Today*, 1995, **26**, 283.
76 E. Kikuchi, T. Matsuda and N. Takahashi, 'Symp. Catalytic Combustion, San Francisco', Am. Chem. Soc., Div. Petrol. Chem., **42**, 1997.
77 M. Lyubovsky and L. Pfefferle, *Catal. Today*, 1999, **47**, 29.
78 M. Lyubovsky and L. Pfefferle, *Appl. Catal., A*, in press.
79 S. C. Su, J. N. Carstens and A.T. Bell, *J. Catal.*, 1998, **176**, 125.

80 J. N. Carstens, S. C. Su and A.T. Bell, *J. Catal.*, 1998, **176**, 136.
81 T. Furuya, K. Sasaki, Y. Hanakata, T. Ohhashi, M. Yamada, T. Tsuchiya and Y. Furuse, *Catal. Today*, 1995, **26**, 345.
82 Y. Ozawa, T. Fujii, M. Sato, T. Kanazawa and H. Inoue, *Catal. Today*, 1999, **47**, 399.
83 J.H. Lee and D.L. Trimm, *Fuel Proc. Technol.*, 1995, **42**, 339.
84 K. Sekizawa, M. Machida, K. Eguchi and H. Arai, *J. Catal.*, 1993, **142**, 655.
85 K. Sekizawa, K. Eguchi, H. Widjaja, M. Machida and H. Arai, *Catal. Today*, 1996, **28**, 245.
86 J.G. McCarty, Y.-F. Chang, V.L. Wong and E.M. Johansson, 'Symp. Catalytic Combustion, San Francisco', Am. Chem. Soc., Div. Petrol. Chem., **42**, 1997.
87 M.C. Marion, E. Garbowski and M. Primet, *J. Chem. Soc., Faraday Trans.*, 1991, **87**, 1795.
88 M.F.M. Zwinkels, O. Haussner, P.G. Menon and S.G. Järås, *Catal. Today*, 1999, **47**, 73.
89 L.C. Young and B.A. Finlayson, *AIChE J.*, 1976, **22**, 343.
90 R.H. Heck, J. Wei and J.R. Kratzer, *AIChE J.*, 1976, **22**, 477.
91 G. Groppi, A. Belloli, E. Tronconi and P. Forzatti, *Chem. Eng. Sci.*, 1995, **50**, 2705.
92 G. Groppi, E. Tronconi and P. Forzatti, *Appl. Catal., A*, 1996, **138**, 177.
93 S.T. Kolaczkowski and S. Serbetcioglou, *Appl. Catal., A*, 1996, **138**, 199.
94 G. Groppi and E. Tronconi, *Chem. Eng. Sci.*, 1997, **52**, 3521.
95 K. Zygourakis, *Chem. Eng. Sci.*, 1989, **44**, 2075.
96 S.T. Kolaczkowski, P. Crumpton and A. Spence, *Chem. Eng. Sci.*, 1988, **43**, 227.
97 D.J. Worth, S.T. Kolaczkowski and A. Spence, *Trans. IChemE*, 1993, **71**, 331.
98 D. Papadias, M.F.M. Zwinkels, L. Edsberg and P. Björnbom, *Catal. Today*, 1999, **47**, 315.
99 D. Papadias, E.M. Johansson, L. Edsberg and P. Björnbom, 'Symp. Catalytic Combustion, San Francisco', Am. Chem. Soc., Div. Petrol. Chem., **42**, 1997.
100 H. Arai and M. Machida, *Catal. Today*, 1991, **10**, 81.
101 A. Nakhjavan, P. Björnbom, M. Zwinkels and S. Järås, *Chem. Eng. Sci.*, 1995, **50**, 2255.
102 G. Groppi, E. Tronconi and P. Forzatti, *Catal. Today*, 1993, **17**, 237.
103 R.D. Hawthorn, *AIChE Symp. Ser.*, 1974, **70**, 428.
104 W.M. Kays and A.L. London, *Compact Heat Exchangers*, McGraw-Hill, New York, 1964.
105 C.J. Bennet, S.T. Kolaczkowski and W.J. Thomas, *Trans. Instn. Chem. Engrs., Part B*, 1991, **69**, 209.
106 U. Ullah, S.P. Waldram, C.J. Bennet and T. Truex, *Chem. Eng. Sci.*, 1992, **47**, 2413.
107 S. Lee and R. Aris, *Chem. Eng. Sci.*, 1977, **32**, 827.
108 R.E. Hayes, S.T. Kolaczkowski and J.W. Thomas, *Comput. Chem. Eng.*, 1992, **16**, 645.
109 R.E. Hayes and S.T. Kolaczkowski, *Chem. Eng. Sci.*, 1994, **49**, 3587.
110 E. Tronconi and P. Forzatti, *AIChE J.*, 1992, **38**, 201.
111 L.C. Young and B.A. Finlayson, *Chem. React. Eng.-II, Adv. Chem. Series*, 1974, **133**, 629.
112 G. Eigenberger, *Chem. Eng. Sci.*, 1972, **27**, 1909.
113 P. Markatou, L.D. Pfefferle and M.D. Smooke, *Combust. Flame*, 1993, **93**, 185.
114 D.G. Vlachos, L.D. Schmidt and R. Aris, *AIChE J.*, 1994, **40**, 1005.
115 L. Pfefferle, *Catal. Today*, 1995, **26**, 255.

116 A.L. Boehman, J.W. Simons, S. Niksa and J.G. McCarty, *Combust. Sci. Technol.*, 1997, **122**, 257.
117 C.T. Chou and W.E. Stewart, *Chem. Eng. Sci.*, 1986, **41**, 202.
118 D. Leung, R.E. Hayes and S.T. Kolaczkowski, *Can. J. Chem. Eng.*, 1996, **74**, 94.
119 D. Papadias and P. Björnbom, submitted to *Chem. Eng. Sci.*
120 M. Flytzani-Stephanopoulos, G.E. Voecks and T. Charng, *Chem. Eng. Sci.*, 1986, **41**, 1203.
121 S.T. Kolaczkowski, P. Crumpton, R.P.J. Lee and A. Spence, *Chem. Eng. J.*, 1989, **42**, 167.
122 S.T. Kolaczkowski and D.J. Worth, *Catal. Today*, 1995, **26**, 275.
123 A. Cybulski and J.A. Moulijn, *Chem. Eng. Sci.*, 1994, **49**, 19.
124 G. Groppi and E. Tronconi, *AIChE J.*, 1996, **42**, 2382.
125 R.A. Dalla Betta, D.G. Löffler and S. Magno, Symp. Catalytic Combustion, San Francisco', Am. Chem. Soc., Div. Petrol. Chem., **42**, 1997.
126 H. Sadamori, T. Tanioka and T. Matsuhisa, *Catal. Today*, 1995, **26**, 337.
127 H. Sadamori, T. Nishida, T. Yamashita, A. Furuya and T. Matsuhisa, '95-Yokohama-IGTC-140', 1995.
128 H. Arai and H. Fukuzawa, *Catal. Today*, 1995, **26**, 217.
129 P. Dutta, D.K. Yee and R.A. Dalla Betta, ASME Paper 97-GT-497, 1997.
130 P. O'Brien, ASME Paper 98-GT-400, 1998.
131 J.J. Lipinski, P.R. Brine, R.D. Buch and G.R. Lester, ASME Paper 98-GT-390, 1998.
132 S.L. Kiperman, *Kinet. Catal. (Engl.)*, 1994, **35**, 37.
133 W.E. Young and D.E. Carl, US Patent 4,870,824, 1989.
134 R.A. Dalla Betta, N. Ezawa, K. Tsumuri, J.C. Sclatter and S.G. Nicklolas, US Patent 5,183,401, 1993.
135 H. Karim, S. Etemad, L.L. Smith and W.C. Pfefferle, *Catal. Today*, 1999, **47**, 305.
136 T. Furuya, K. Sasaki, Y. Hanakata, T. Ohhashi, M. Yamada, T. Tsuchiya and Y. Furuse, *Catal. Today*, 1995, **26**, 345.
137 Y. Ozawa, Y. Tochihara, N. Mori, I. Yuri, T. Kanazawa and K. Sagimori, ASME Paper 98-GT-381, 1998.
138 D.A. Smith, S.F. Frey, D.M. Stansei and M.K. Razdan, ASME Paper 97-GT-311, 1998.
139 T.A. Brabbs and S.A. Merrit, NASA Technical Paper 3281, 1993.
140 J. Rollbuhler, '27th Joint Propulsion Conf.', Sacramento, California, 1991.
141 R. Gabrielsson, R. Lundberg and P. Avran, ASME Paper 98-GT-392, 1998.
142 R. Gabrielsson and G. Holmqvist, ASME Paper 96-GT-362, 1996.
143 P. Avran, A. Leclair and G. Payen, ASME Paper 97-GT-69, 1997.
144 R. Lundberg and M. Moret, ASME Paper 98-GT-566, 1998.
145 R. Swanekamp, *Power*, 1995, June, 15.
146 J.-H. Le Gal, G. Martin and D. Durand, ASME Paper 98-GT-294, 1998.
147 E.M. Johansson and S.G. Järås, *Catal. Today*, 1999, **47**, 359.
148 J. Oudar, *Catal. Rev.-Sci. Eng.*, 1980, **22**, 171.
149 M. Shelef, K. Otto and N.C. Otto, *Adv. Catal.*, 1978, **27**, 311.
150 L.J. Hoyos, H. Praliaud and M. Primet, *Appl. Catal., A*, 1993, **98**, 125.
151 A.G. Ostroff and R.T. Sanderson, *J. Inorg. Nucl. Chem.*, 1959, **9**, 45.
152 M.F.M. Zwinkels, G.M.E. Heginuz, B.H. Gregertsen, K. Sjöström and S.G. Järås, *Appl. Catal., A*, 1997, **148**, 325.
153 J. C. van Giezen, PhD Thesis, University Utrecht, The Netherlands, 1997.
154 S.K. Agarwal, J.J. Spivey and J.B. Butt, *Appl. Catal., A*, 1992, **81**, 239.

155 W.C. Cheng, G. Kim, A.W. Peters, X. Zhao, K. Rajagopalan, M.S. Siebarth and J.C. Periera, *Catal. Rev.-Sci. Eng.*, 1998, **40**, 39.
156 J.S. Magee and J.J. Blazek, in *Zeolite Chemistry and Catalysis*, ed. J.A. Rabo, Am. Chem. Soc., Washington, DC, 1976.
157 A.W. Peters, G.D. Weatherbee and X. Zhao, *Fuel Reformul.*, 1995, **5**, 45.
158 NPRA Q & A Session on Refining and Petrochemical Technology, Washington, DC, 1994.
159 J.S. Yoo, C.A. Radlowski and J.A. Karch, in 'Preprints of 207th ACS National Meeting', Am. Chem. Soc., 1994.
160 G.H. Dieckmann and E.Q. Labrador, US Patent 5,364,517, 1994.
161 A.A. Chin, US Patent 5,021,146, 1991.
162 A.A. Chin, US Patent 5,002,654, 1991.
163 G.J. Green, D.S. Shihabi and T.Y. Yan, US Patent 4,973,399, 1990.
164 C.J. Periera and M.D. Amiridis, in *Reduction of Nitrogen Oxide Emissions, ACS Symp. Ser. 587*, ed. U. Ozkan *et al.*, Am. Chem. Soc., Washington, DC, 1998.
165 D.L. Trimm, *Appl. Catal.*, 1983, **7**, 249.
166 A. Cybulski and J.A. Moulijn, *Catal. Rev.-Sci. Eng.*, 1994, **36**, 179.
167 K. Masuda, T. Sano, F. Mizukami, T. Takezaki and K. Kuno, *Appl. Catal., B*, 1994, **4**, 187.
168 L.L. Hegedus, *AIChE J.*, 1975, **21**, 849.
169 J. Votruba, J. Sinkule, V. Hlavcek and J. Skrivnek, *Chem. Eng. Sci.*, 1975, **30**, 117.
170 J.G. Stevens and E.N. Ziegler, *Chem. Eng. Sci.*, 1977, **32**, 385.
171 J. Sinkule and V. Hlavacek, *Chem. Eng. Sci.*, 1977, **33**, 839.
172 J.S. T'ien, *Comb. Sci. Technol.*, 1981, **26**, 65.
173 N. Sinha, C. Bruno and F.V. Bracco, *Physicochemical Hydrodynamics*, 1985, **6**, 373.
174 S.H. Oh and J.C. Cavendish, *Ind. Eng. Chem. Res.*, 1983, **22**, 509.
175 T. Ahn, W.V. Pinczewski and D.L. Trimm, *Chem. Eng. Sci.*, 1986, **41**, 55.
176 R.A. Dalla Betta, J.C. Schlatter, D.K. Yee, D.G. Loffler and T. Shoji, *Catal. Today*, 1995, **26**, 329.
177 R.A. Dalla Betta, N. Ezawa, K. Tsurumi, J.C. Schlatter and S.G. Nickolas, US Patent 5,183,401, 1993.
178 P. Dutta, L.H. Cowell, D.K. Yee and R.A. Dalla Betta, ASME Paper 97-GT-292, 1998.
179 T. Yoshine, T. Furuya, Y. Hara and T. Tsuchiya, 'Proc. Yokohama Int. Gas Turbine Cong.', 1991.
180 Y. Ozawa, J. Hirano, M. Sato, M. Saiga and S. Watanabe, *J. Eng. Gas Turbines Power, Trans. ASME*, 1994, **116**, 511.

7
The Oligomerization of Alkenes by Heterogeneous Catalysts

BY MEHRI SANATI, CHRISTINA HÖRNELL AND SVEN G. JÄRÅS

1 Introduction

The acid-catalysed oligomerization of propylene and butylene was introduced in 1930 as the first commercial catalytic process of the petroleum industry.[1] The product was gasoline-range iso-olefins (C_6–C_{10}) using phosphoric acid impregnated on silica clay.[2] The mechanism of branched alkene oligomerization was studied in the presence of sulfuric acid.[3,4] Ethylene and propene were instead best oligomerized in the presence of phosphoric acid, kieselguhr with 60% P_2O_5 became an item of commerce. Because of corrosion problems, other acidic solid catalysts like zeolites were then studied to replace the phosphoric acid-containing catalysts. Also the shape-selective zeolites (where the shape of the product molecules is governed primarily by the pore structure of the zeolite catalyst) have been investigated as catalysts for improving liquid product selectivities in oligomerization of low molecular weight olefins into liquid products. This resulted in a process, MOGD (Mobil olefin to gasoline and distillate), which uses the ZSM-5 class of zeolites to convert light olefins to higher molecular weight gasoline and diesel fuel.[5,6]

The oligomerization of olefins is an important industrial reaction and represents a route to the production of gasoline and diesel fuel-blending stocks, solvent and lubricant stock, plasticizers, dyes, detergents and additives. Oligomerization of olefins aims to yield substances composed of only a few monomer units, as opposed to polymerization which implies the production of high molecular weight products. The literature data are divided into two parts; low molecular weight alkene (ethylene, propene and butene) oligomerization, and higher molecular weight alkene oligomerization. This oligomerization reaction is carried out in the presence of a catalyst, which comprises either an acid catalyst, particularly modified or unmodified zeolites with various metal ions, or supported phosphoric acid and solid nickel catalysts. Then the olefins combine with each other forming dimers, trimers, tetramers and even higher oligomers. The product spectrum is influenced by both the nature of the catalyst and the reaction conditions. The nature of the acid sites in acidic solid catalysts and their physicochemical characterization have been extensively reviewed by Corma[7] and Minachev.[8] At high temperature (>350 °C) depending on catalyst type, besides

Catalysis, Volume 14

oligomerization the olefins also undergo isomerization, disproportionation, and cracking, a sequence of aromatization and hydrogen-transfer reactions that yields a broad spectrum of molecular types.

Oligomerization of low molecular weight alkenes leads to the formation of liquid fuel and synthetic lube oils, whereas the high molecular weight alkene oligomerization results particularly in different synthetic lube oils. The unsaturated oligomer products are then hydrogenated to improve their oxidative stability, or alternatively, are used as feedstocks for various chemical processes using heavy olefins.

Most of the available literature on oligomerization concerns some model reaction using only one reactant, although the situation is different in practice when multi-component feeds are less expensive and widely available. Still, when working on the laboratory scale, obtained data should be interpretable, which is why it is desirable to work only with one component in a model reaction. The main product of propene oligomerization over acid catalysts is trimers, whereas the main product of butene oligomerization is dimers. Still, these main reactions of a single olefinic component may not correspond to the whole product spectrum, *e.g.* gasoline, distillate or α-olefin base-stock. Usually, a refinery fraction, containing many components, is used as feed.

O'Connor has written surveys of both homogeneous and heterogeneous catalysis[9–11] while the reviews of Skupinska[12] and Al-Jarallah *et al.*[13] focus particularly on homogeneous catalysis. Recently, Minachev *et al.* have written about their work on the catalytic and physicochemical properties of the zeolites based on pentasils for oligomerizing lower olefins and paraffins into a mixture of aliphatic hydrocarbons of the composition C_6–C_{10} or into a concentrate of aromatic hydrocarbons, depending on the reaction conditions.[14]

The present review focuses on oligomerization in heterogeneous systems. It provides a comprehensive review and a supplement to earlier surveys of alkene oligomerization by solid catalysts. Related publications and patents to date have been reviewed and evaluated in terms of catalytic systems used, reaction conditions and product selectivity.

2 Principles of Oligomerization Reactions

In particular, oligomerization refers to the preparation of molecules consisting of only a relatively few monomers, in contrast to polymerization which implies the production of high molecular weight products. The chain growth depends on the number of reacting molecules, and is called dimerization when $n = 2$, oligomerization when $2 > n > 100$, and polymerization when $n > 100$. Oligomerization occurs in the presence of catalysts, and consists of two steps, propagation (chain growth), and elimination (hydrogen elimination from a β carbon to the catalytic centre).[12] The terms k_p and k_e denote the rate constants for the propagation and elimination reactions, respectively. If $k_p >> k_e$, polyolefin will be formed, when $k_p << k_e$, dimers are obtained, and finally when $k_p \approx k_e$ oligomers are produced.

The oligomerization reactions are not elementary and are often accompanied by various parallel reactions. In addition to oligomers, aromatics, coke and saturates are often formed, which is indicative of the occurrence of cracking, isomerization, dehydrocyclization, methyl and hydrogen transfer even at low temperatures.[5,9,15–18] Figure 1 shows a simplified scheme for oligomerization reactions.

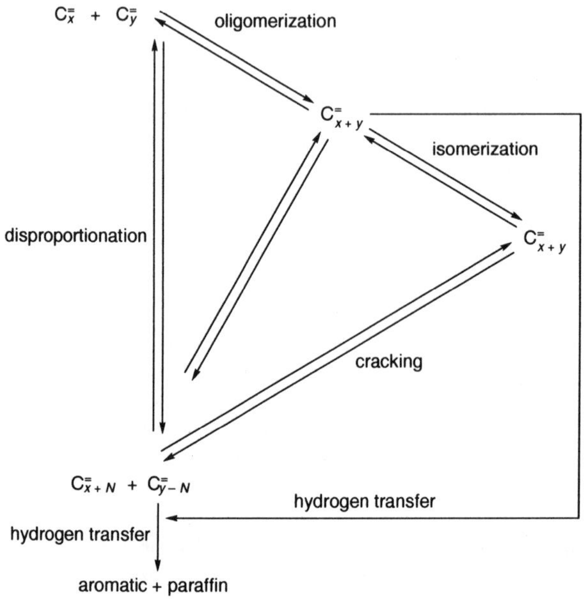

Figure 1 *Olefin oligomerization reaction scheme*

As shown in Figure 1, there are several distinct reactions occurring. First, two olefins $C_x^=$ and $C_y^=$, will oligomerize to a single higher olefin $C_{x+y}^=$ where x and y refer to olefin carbon numbers. Starting with a propene feed, for example, this reaction would yield the oligomers $C_6^=$, $C_9^=$, $C_{12}^=$, *etc.* Depending on catalyst nature, double bond and skeletal isomerization have been reported in previous literature. In the case of the oligomerization of alkenes on aluminosilicates and the NiO/SiO_2 system, the carbocation mechanism tends to give skeletal isomerization.[19] On the other hand, at higher temperatures the oligomers undergo a carbenium ion decomposition.[19] The carbenium ion mechanism leading to double bond isomerization was also proposed by Kazansky *et al.*[20] Skeletal isomerization over high-silica zeolites was observed by Miller using ^{13}C NMR.[21] In the study of light olefin oligomerization over a ZSM-5 catalyst and analysis by ^1H NMR, double bond isomerization at high space velocity (5 WHSV) and skeletal isomerization at low space velocity (1 WHSV) were suggested.[22] Additionally, any two olefins may react to disproportionate, *via* a carbenium ion intermediate, to two olefins of different carbon numbers. Olefin cracking may also occur simultaneously with oligomerization and disproportionation. Hydrogen transfer, cyclization, and aromatization may also take place. In practice, the kinetics of the oligomerization, disproportionation, and cracking reactions determine the olefin product distribution under process conditions.[22] Also, oligomerization, isomerization, cracking, and interpolymerization have been suggested for the upgrading of the lower olefins to lubricant-range hydrocarbons over HZSM-22.[23]

However, it has been postulated that the hydrogen-transfer activity of zeolites may be undesirable due to rapid and early chain termination in oligomerization reactions, which leads to production of aromatics and coke species.[9]

In the reviews by both Skupinska, and Al-Jarallah *et al.*, it is suggested that the activities and selectivities of the catalysts are affected by strong adsorption of products, by isomerization or by a tendency to copolymerize into branched products.[12,13]

3 Mechanistic Investigation and Kinetics of Alkene Oligomerization

On the basis of the previous work, several mechanistic manners for the oligomerization of olefinic feedstocks have been suggested:

(1) the carbenium ion species intermediate;
(2) surface ethoxy structure intermediate;
(3) cationic intermediate;
(4) cyclic derivatives intermediate;
(5) an ionic mechanism.

The proposed mechanisms (1)–(3), which can describe the oligomerization of olefins over zeolites, such as ZSM-5, Y, *etc.*, are also called hydride intermediate mechanisms. Simplified schemes for the adsorption site for these three suggested mechanisms could be illustrated as in Figure 2.

Evidence for these intermediates was revealed by adsorption measurements of the oligomerization reaction in an infrared spectrometer cell under static conditions. The infrared spectroscopic data showed that the cationic oligomerization [Figure 2(c)] was initiated by Lewis acid sites of the dehydroxylated zeolite. The obtained products were mainly branched oligomers.[24] On the other hand, the spectroscopic results of olefin oligomerization over HZSM-5, which contained Brønsted acid sites, showed that the product was composed of linear oligomers, and a mechanism similar to the one shown in Figure 2(a) was proposed.[25] Two reaction mechanisms for alkene oligomerization in the presence of Brønsted acid sites are possible, as shown in Figures 2(a) and 2(b). However, a carbenium ion mechanism which leads to the formation of C_2–C_7 branched oligomers has also been suggested.[26] On the other hand, it was reported that the oligomer-chain branching depends on the zeolite porosity, and long-chain olefins in good yields were obtained using zeolites with large pore size.[27] Experimentally, it was found that the degree of chain branching decreases with pore size in the following order: omega > HY > mordenite > ZSM-5 > offretite > boralite.[28,29] Zeolite omega, having the largest pore opening, gives the most branched oligomers. From previous spectroscopic and reactor studies in the oligomerization of different

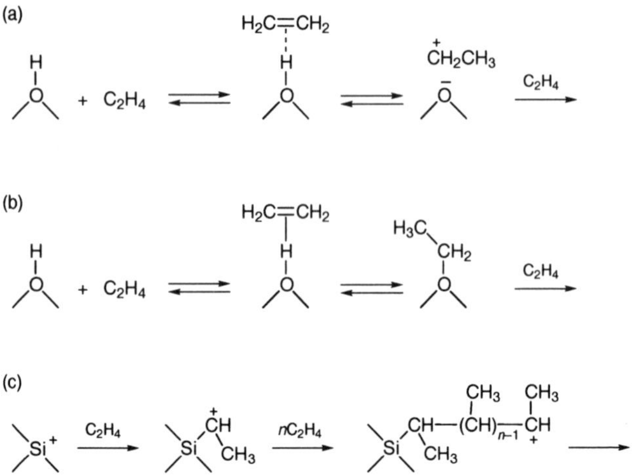

Figure 2 *Adsorption mechanisms on Brønsted and Lewis acid sites;* (a) *carbenium ion mechanism (adsorption on Brønsted acid sites),* (b) *ethoxy structure mechanism (adsorption on Brønsted acid sites),* (c) *cationic mechanism (adsorption on Lewis acid sites)*

olefinic feedstocks at constant reaction conditions, identical product distributions for different feedstocks were obtained, suggesting that the reactants go through a common intermediate stage.[9,18,26,30] The common intermediate stage was also explained from the expected thermodynamic, equilibrium considerations. All of the mechanistic approaches, mentioned above, seem to be in accordance with the evidence of a common intermediate stage. The oligomerization of even higher olefins over outer surface-deactivated HZSM-5 (Si/Al = 70) results in nearly linear products.[31] When HZSM-5 without modification was used as catalyst it resulted in a complex structure of products. The authors' interpretation was that for an outer surface-deactivated zeolite where the reaction took place inside the zeolite channels and resulted in linear products. However, it has been claimed that at Si/Al ratios above *ca.* 35, it is unlikely that there will be much difference in the strength of acid sites since the framework aluminium atoms will probably not have any next nearest neighbours to influence their acidity by withdrawing electrons.[9]

On the basis of the accompaniment of the isomerization reaction, in the study of α-olefin oligomerization to lubricating base oil over sulfated zirconia catalyst, the carbocation mechanism was also suggested.[32]

The fourth mechanistic approach was suggested by various investigators, and it explains the linear structure of the products obtained by the olefinic oligomerization reaction.[31,33,34] The authors have proposed a scheme for an oligomerization reaction in the propene HZSM-5 system, proceeding *via* protonated cyclopropyl intermediates as shown in Figure 3.[34]

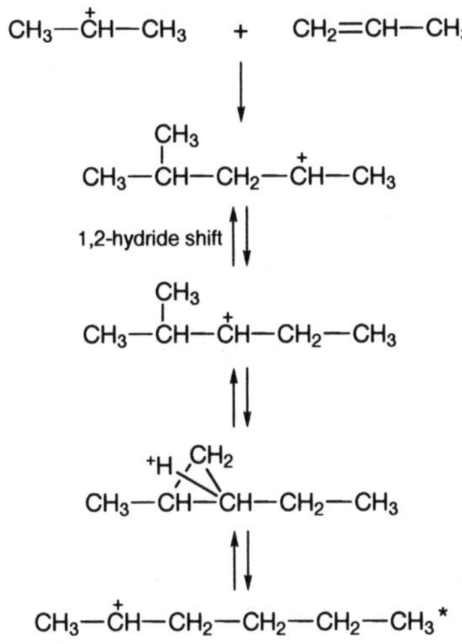

Figure 3 *The cyclic intermediate mechanism (from refs. 5, 31)*

A similar manner as in the above mechanism, losing methyl branches through the protonated cyclopropyl intermediate, was suggested for obtaining linear products from dec-1-ene and isobutene.[31] However, linear products were obtained over modified zeolites, and branched products over unmodified zeolite. This would seem to imply that it is the channel in the zeolite which governs product structure rather than a cyclic intermediate as was proposed by Chen and Bridger.[31] It should be noted that Van den Berg *et al.*[34] worked with unmodified zeolite. A cyclic derivative is assumed to be the intermediate in order to explain the primary product distribution in oligomerization reactions over NiO on silica–alumina catalysts.[33]

The suggested ionic mechanism was used to explain the activity and selectivity behaviour in the oligomerization of olefins over alkali metal ion-exchanged zeolites. In olefin oligomerization, the activity pattern for the alkali metal ion forms in zeolites is inversely proportional to the ionic radius, *e.g.* $Li^+ > Na^+ > K^+ > Rb^+$; thus, the carbenium mechanism does not explain this behaviour. For an ionic mechanism, the spatial constraints within the zeolite determine the selectivity routes towards the oligomerization reaction. According to this approach, the entry of further olefin molecules, which would lead to polymerization, is impeded.[5]

4 Ethylene Oligomerization

The most important product in ethylene oligomerization is α-olefins. Oligomerization of ethylene is at present the primary source of linear olefins and is likely to be the basis for such production for many decades. Current commercial processes for making α-olefins are based on homogeneous catalysts, trialkylammonium (Chevron and Ethyl) and nickel complexes (Shell). The product spectrum for heterogeneously-oligomerized ethylene is more diversified than for homogeneously-oligomerized ethylene; products tend to include both paraffins and aromatics. Still, there is a variety of solid catalytic systems for oligomerization of ethylene to higher molecular weight olefins. Heterogeneous catalysts studied in the literature may be divided into zeolites, solid-acid and nickel-supported catalysts.[26,35–37] The upgrading of light olefins to higher molecular weight distillate fuels, as well as gasoline and lube oil-range products, depends upon both reaction conditions and catalyst structure. Since ethylene has a relatively low reactivity in comparison with other light hydrocarbons, and needs a higher reaction temperature, methods have been developed to counteract the low reactivity, especially when the feed consists of a mixture of C_2–C_5 hydrocarbons. For example, multistep reactors with different temperatures, pressures and catalysts have been studied. In a mixed feed, conversion of input ethylene is about 10–30% in a multistep reactor, while conversion of the remaining feed components approaches total conversion. Some of the catalytic systems in ethylene oligomerization are referred to in the text, but Table 1 (page 244) contains the remaining systems found in the literature.

4.1 Zeolites – The oligomerization of ethylene to synthetic fuel and lubricating oil base-stock over various zeolites has been the subject of many investigations.[20,38–40] The catalytic performance of this oligomerization is related to the nature and number of acidic sites present in the zeolite. It has generally been found that the oligomerization rate of ethylene over ZSM-5 is rather slow compared with that of other olefins. Also, ethylene oligomerization begins at higher temperature.

IR spectroscopic studies of ethylene oligomerization revealed that the degree of product branching was not restricted by the size of the cavities and channels in the zeolites,[20] contrary to the findings of Miller.[16] According to Kustov *et al.*,[20] linear oligomers are only obtained using high-silica zeolites possessing strong Brønsted acid sites. In this case it was assumed that ethoxy groups were the intermediates in ethylene oligomerization.

Oligomerization of ethylene carried out on HZSM-48, which has a different pore structure from HZSM-5 and HZSM-11, yielded a product with shorter oligomer chains and gave lower ethylene conversion.[40] An increase of diesel-range products was obtained for ethylene oligomerization carried out on ion-exchanged NaY with Ni.[39] In the same investigation, ethylene conversion as a function of temperature showed two temperature regions corresponding to maximum activity for ethylene oligomerization at 3.5 MPa, one near 120 °C and the other above 300 °C. At 120 °C, the product followed Schulz–Flory statistics, and at 300 °C acid sites contributed to the product spectrum. The catalyst deactivated rapidly under conditions which produced high-quality distillate.

TPD–TGA experiments on ethylene and propene oligomerization on HZSM-5 gave similar weight increases, and product distribution for both feeds was also similar. The reason for the similarity was that, in both cases, reaction stopped after the zeolite pore volume surrounding the acid sites had been practically filled. The only difference was that ethylene required a higher temperature to initiate the reaction.[26] Kofke and Gorte[26] attributed the higher temperature requirement to the ethylene intermediate formed being less energetically favoured, and less accessible for participation in the reaction.

Because of the low severity distillate mode used in the oligomerization of lower olefinic feedstocks, C_2–C_6, most of the ethylene present over acid crystalline zeolite is not converted. The extent of conversion for propene, butene and others in distillate mode is maybe 50–90%, while only about 10–30% of the ethylene is converted. Garwood *et al.*[41] demonstrated a multistage catalytic technique with a bifunctional metallic zeolite in the presence of a reducing component such as hydrogen in the oligomerization of an ethylene-rich feed to distillate-range products. To some extent, the multistage technique reduced the problematic temperature effect which is of importance in the oligomerization of ethylene. Co-feeding of water was used to promote the reaction.

There is plenty of research concerning the activity properties of the external and/or internal surface sites in zeolites, related to the catalytic performance in the olefin oligomerization reactions.[42,43] Several techniques may be used to vary the relative ratio of intra-crystalline acid sites to external surface sites due to the geometric relationship between volume and superficial surface area, *e.g.* varying crystal size. The effect of crystal size on the catalytic properties of ZSM-5 has been investigated by Schwartz *et al.*[44] in the oligomerization of propene. The crystal size was related to varying Si/Al ratios. A decrease in Si/Al ratio was reported to increase the oligomerization of olefins.

It is reported that with zeolites having an internal/external surface area ratio of greater than 300, the contribution of external active sites to the reaction is assumed to be negligible, whereas for small zeolite crystallites (<100 nm in diameter), the effects of the external activity can become important, particularly for very rapid reactions or highly diffusion-limited reactions. ZSM-5, synthesized to give small size crystals, was used in the oligomerization of ethylene.[38] The catalytic lifetime was related to the relative rate of the coke formation on the external and internal surfaces of the small crystal-size zeolite. Coke formation was less favoured on the external surface area in comparison with the internal one. It was concluded that the largest external surface area results in the longest lifetime. The shape selectivity was diminished with large external surface area and more C_9+ aromatic hydrocarbons were formed

4.2 Other Solid-acid Catalysts – Chromium on an aluminophosphate support, which is supposed to be a polymerization catalyst, was used in the oligomerization of light olefins and ethylene.[36] For ethylene oligomerization, the catalyst exhibited activity for dimerization, with around 60% conversion to C_4 fractions. In another patent[45] the support and the active phase were the same as in the previous patent, except that the support had been treated with a solution of triethylaluminium in toluene before being contacted with the chromium compound. The catalyst system showed selectivity towards trimerization. When

Table 1 *Oligomerization of ethylene*

Year	*Catalysts*	*Reaction conditions/remarks*	*Products*	*Ref(s)*
1997	Ni/TiO_2	EPR study at 0.027 MPa room temperature; coordinately unsaturated Ni(I) ions on the surface of support are active for oligomerization of olefin	Hexene, butene and propene	49
1997	γ-Alumina support (BET area 102 $m^2 g^{-1}$), and impregnation or ion-exchange of the non-porous material with Ni(II) ions, further sulfation with $(NH_4)_2SO_4$ solution on impregnated or ion-exchanged catalysts	(1) Gas–solid batch microreactor, initial outgassing of the catalyst for 2 h at 500 °C under vacuum; (2) gas–liquid–solid slurry reactor, *n*-heptane liquid phase, 6 °C, 100 % conversion of ethylene, run time 34 h	But-1-ene and hex-1-ene	35
1994	ZSM 5 of $SiO_2/Al_2O_3 = 25$, $OH/SiO_2 = 0.2–0.3$, $H_2O/SiO_2 = 50$, and tetrapropylammoniumbromide $(TPaBr)/SiO_2 = 0.01–0.1$, with a small crystal size (30–50 nm) and a large external surface area	W/F 1.5 gh/ $mole^{-1}$, 400 °C, 0.1 MPa, and $C_2H_4/N_2 = 50/50$ the external surface area was 3 times greater than that of other zeolites, which indicates that the lifetime of zeolite in oligomerization depends on the external surface area or crystal size; consequently the coke selectivity was low on the zeolite with large external surface area	Gasoline-range hydrocarbons	38
1993	Primary oligomerization over a surface-deactivated (by a sterically hindered trialkylpyridine), medium pore zeolite (ZSM-5 or ZSM-23) and second oligomerization over a supported group VI(B) metal, SiO_2-supported reduced Cr oxide	Multistep process: WHSV 0.21, 6.21 MPa, 200 °C, resulting in 91% propylene conversion and yielding 78.6 % product: lubricating oil 54.1, diesel fuel 20.5 and kerosene/jet fuel 15.0 wt%	Lubricating oil (bp >343 °C, viscosity 50 cSt at 100 °C) and high cetane number (cetane no. 50–75 after hydrogenation) fuels (bp 177–343 °C)	50[a]
1993	ZSM-5 of medium pore size, *e.g.* 5–7 Å and having a SiO_2/Al_2O_3 ratio of about 20 : 1 or greater	Fluidized bed, WHSV >0.1, 0.14–1.5 MPa, 342–482 °C, sufficient for concurrent oligomerization of light olefins while cracking *n*-paraffins	C_5+ gasoline and dewaxed distillate	51[a]

1992	Cr(II) on an aluminosilicate support	0.126 l min^{-1} (STP), 1.03 MPa, 250 °C; after 1 h no liquid was observed	Gas product: but-1-ene 64, branched hexene 1.3, hex-1-ene 18.6, oct-1-ene 2.3, dec-1-ene 0.1, C_4–C_5 4.3, double bond isomers $C_6 = 1.7$, unknown C_6–C_8 3.0, and unknown C_8–C_9 4.3%	36
1991	Zr $(OEt)_2Cl_2$–Et_2AlCl supported on SiO_2 or Al_2O_3	Selectivity to linear olefins was enhanced compared with homogenous catalyst	C_{4-10} olefins	52
1991	amorphous SiO_2–Al_2O_3 (Si/Al = 72 : 1), impregnated or ion-exchanged with Ni	Ion-exchanged catalyst is better for ethene oligomerization to diesel fuel, C_{10}+ fraction in product was 41% *vs.* 23% for impregnated catalyst	Diesel fuel	53
1991	HZSM-5, HY, and H-mordenite	NMR investigation	The degree of branching of oligomers was increased with increasing pore dimension, alkoxide species were observed as intermediates	54
1991	HZSM-5	Fluidized catalyst bed, 5.5 MPa, 300–650 °C, conversion 50 wt%	Fuel, C_5–C_{10} hydrocarbons	55[a]
1990	ZSM-5 with a SiO_2/Al_2O_3 molar ratio of 70 : 1	Multi-stage reactor: first reactor for contacting feed with a medium pore shape-selective zeolite catalyst for converting the C_3-olefinics to liquid hydrocarbons comprising C_5+, in second stage the process conditions should be effective to convert a major amount of ethene	C_5+ gasoline	41
1989	HZSM 5 with SiO_2/Al_2O_3 ratio of 70	Temperature-programmed desorption (TPD) study *in vacuo*; at >97 °C oligomerization of C_2H_4 occurred. When a coverage of 7.8 g/100 g was reached on the zeolite, the reaction stopped, suggesting that reaction ceased only when blocking prevented reactants from approaching the acid sites	Similar oligomer chains: a mixture of C_{2-7} olefin products from C_2H_4, C_3H_6 and 2-methyl-2-propanol on the same catalyst	26

Continued

Table 1 *Continued*

Year	*Catalysts*	*Reaction conditions/remarks*	*Products*	*Ref(s)*
1989	An inner core portion, *e.g.* ZSM-5 or ZSM-23, and an outer portion comprising a fluoride-containing crystal shell		High viscosity lubricating oils, and gasoline	56[a]
1989	ZSM-5 and metal oxides of Co, Ru, Cu, Zn, Cr and Al	CH_4 is first oxidatively coupled to form C_2H_4, CO and H_2, the mixture is reacted over a dual catalyst, a metal oxide for hydrogenation of CO and ZSM-5 for the oligomerization of C_2H_4	Oligomers	57
1988	ZSM-5 3, 7.8 wt% WO_3 on SiO_2 1, and MgO 3 parts by weight	Reaction time 5 h under oligomerization conditions >99 % C_2H_4 conversion and 67–75 % selectivity	C_5+ gasoline	58
1988	HZSM-5, and ion-exchanged acid ZSM-5 zeolite	Continuous two-stage process by using HZSM-5 as the primary-stage catalyst, and ion-exchanged acid ZSM-5 zeolite as the second catalyst	In a second-stage product with yields of C_2 olefins of 5.5 wt% and C_5 olefins 67.5 wt% and an unstabilized liquid product with research octane no. 92.2	59[b]
1988	Ni-exchanged/Al_2O_3–SiO_2 (amorphous solid, containing 0.73 wt% Ni), and Ni-impregnated catalyst	2 WHSV (fixed bed micoreactor), 3.5 MPa, 115 °C, 99.3% conversion *vs.* a conversion of 97.7% for Ni-impregnated catalyst	Butene 19.1, hexene 14.4, octene 23.3, and C_{10-20} oligomers 41.4 wt%, *vs.* butene 35.6, hexene 24.7, octene 16, and C_{10}–C_{18} 22.6 wt% for Ni-impregnated catalyst	37
1988	Ni-exchanged ZSM-5	WHSV1, 1.83 MPa, 376 °C, 7.5 days, 89.0 wt% conversion, the selectivity to $\geq C_5$ fraction 82.3 wt%	Gasoline with octane no. (R + O) 93.1	60[b]
1988	Ni-exchanged zeolite Y (10–40 %)	3.5 MPa, at 100–150 °C or at 300–330 °C, at low temperature the oligomerization proceeds by a selective growth reaction, 26% C_{10}+ product, at high temperature, about 300 °C, a purely acid-catalysed oligomerization	Diesel-range products	39

1987	Ni-exchanged amorphous silica–aluminas		Oligomerization activity and molecular size of product increases with increasing acid strength of support	61
1986	HZSM-5, HZSM-11, and HZSM-48		Oligomers with shorter chains are formed on HZSM-48 in comparison with HZSM-5 and HZSM-11	40
1985	ZSM-5	39.2 WHSV and 0.45 s contact time, 0.41 MPa, 350–528 °C, 95.1% C_2H_4 conversion and 68.1% selectivity to C_5+ hydrocarbons	C_5+	62
1985	ZSM-5	Four fixed bed reactors	Gasoline and jet fuels	63[a]
1984	ZSM-5	Fixed-bed, staged oligomerization; zeolites partially deactivated in a primary stage are used before regeneration in a secondary stage at high temperature	Gasoline	64[b]
1984	ZSM-5 type medium pore zeolites, with a SiO_2/Al_2O_3 molar ratio of 70 and extruded with 35 % alumin	(1) In distillate mode (fixed bed, WHSV 0.5–1, 4–10 MPa, 190–310 °C; (2) in gasoline mode (0.4–3 MPa, 285–385 °C)	(1) 80 % of the product was distillate fuels C_{11}–C_{20} with a cetane number after hydrogenation of over 50; (2) C_5+ olefinic gasoline with octane no. 92	65[a]
1983	HZSM-5	TPA/TPD: 27–327 °C; at 27 °C only the stronger part of the Bronsted acid sites are active in ethene oligomerization, at increased temperature more sites become active, in propene and isobutene oligomerization all sites are active at 27 °C	Analogous products from ethene, propene, and iso-butene above 100 °C. Increased branching at higher reaction temperatures; cracking starts at 127 °C, at 217 °C cracking rate equals oligomerization rate	34[a]
1979	Spinel-coated support (Al_2O_3 coated with $MgAl_2O_4$) and impregnated with Mg, Cr and Pt	704 °C	Mixtures of ethene and benzene	66
1972	Al_2O_3 and 15% alkali metal	WHSV 5.1, 8.83 MPa, 120 °C	4.24% C_{6-8}, 94.76% C_{10-16}, 1% C_{18-20}	67

[a] Uses also propylene and/or butene as feed; [b] uses also propene as input feed.

ethylene was the reactant, the observed product spectrum over a catalyst treated in this way was 85% hexene, 11% decene and 2% tetradecene.

The most investigated Ni catalysts in the oligomerization of ethylene deal with homogeneous systems, which have been reviewed extensively.[13] The majority of the investigated heterogeneous Ni catalyst systems have deactivated rapidly and have been incapable of achieving steady state reaction performances.[46,47] Ni-exchanged zeolites exhibit activity for ethylene dimerization,[2–5] but it was demonstrated that a Ni-exchanged NaY-zeolite catalysed ethylene oligomerization at low temperature to produce diesel-range hydrocarbons.[39]

The homogeneous η^5-cyclopentadienyl nickel [η^5-$C_5H_5Ni(C_8H_{13})$] derivatives resulted in high activity for oligomerizing ethylene and propene.[48] It was reported that heterogenization of the η^5-cyclopentadienyl nickel by styrene/divinylbenzene gave HDPE from ethylene instead of oligomers.

The Ni salt was supported on non-porous alumina either by impregnation or ion exchange.[35] Increased activity and selectivity in ethylene oligomerization was observed for the ion-exchanged catalyst in comparison with the impregnated catalyst.

Two catalytic sites on a Ni-supported TiO_2 catalyst were identified using EPR.[49] Low-valent nickel (Ni^+ ions) was exposed on the surface, while high-valent nickel (Ni^{3+}), dominated in the bulk and near the interface of the catalyst. The activity in the ethylene oligomerization was attributed to the nickel surface sites.

5 Propene Oligomerization

The main conversion products in the oligomerization of propene are the C_5+ gasoline range and diesel fuels, but also lube stocks. Either of the three main products are obtainable by varying the reaction conditions with the same catalyst. At moderate temperature and relatively high pressure, conversion conditions favour products with a normal boiling point of at least 165 °C. Gasoline (C_5–C_{10}) is readily formed at elevated temperatures and moderate total pressure, from ambient to about 5.5 MPa. At moderate temperature and relatively high pressure, the conversion conditions favour C_{20}+ aliphatic products.

Previous work of O'Connor[68] on high pressure oligomerization of propene over different zeolites, heterogeneous phosphoric acid, incorporated acids in clay, and nickel on different supports revealed that the major product in all cases was trimers. The catalyst systems referred to in the text may also be found in Table 2 (page 252).

5.1 Zeolites – A number of zeolites with or without different modifications are reported in the literature as being active for alkene oligomerization. For example, in an earlier report,[69] ion-exchanged zeolites of types X and Y were used for propene oligomerization. With the exception of NiX, which showed high selectivity to dimerizing, all other zeolites showed the same spectrum of products, namely saturated C_2, C_4, C_5 and C_7. The interpretation of the appearance of saturated products was that it was indicative of cracking reactions as well as hydrogen transfer.

Studies for correlating liquid product selectivities took place with different zeolites (ZSM-5, boralite, offretite, HY, mordenite and omega, all extruded with 20 wt% sepiolite) for high pressure propene oligomerization.[28] Although the ZSM-5 had a silica–alumina ratio of 78, at WHSV 1 and 4.8 MPa the catalyst was unusually unreactive and showed poor activity below 300 °C. Under similar conditions, boralite was more active, achieving complete conversion. The activity of H-omega and H-offretite increased as the reaction temperature was increased from 200 to 350 °C, and from 150 to 250 °C, respectively. H-mordenite converted little propene below 250 °C; deactivation due to strong adsorption of reactants and/or products at low temperature was supposed. HY was active at 40 °C. The degree of branching of the liquid products, as calculated from ^{1}H and ^{13}C NMR spectra, increased in the order boralite < offretite < ZSM-5 ~ HM < HY < omega.[29]

At present, ZSM-5 is known to be the most promising catalyst for the conversion of light olefins to higher molecular weight distillate fuels and/or gasoline, and is by far the most utilized zeolite since it is employed in the MOGD process.[22,70,71]

Since the HZSM-5 zeolite has a strong activity for hydrogen transfer, olefins in the reactants are partially hydrogenated to the corresponding paraffins by the hydrogen released during simultaneous aromatization, giving a decrease in selectivity to gasoline-range hydrocarbons.[9] However, the medium-pore, shape-selective HZSM-5 zeolite used in the oligomerization of propene produced oligomers with a significantly less complex structure compared with those from conventional acid catalysts.[18,30,71,72]

It is believed that two modes of olefin oligomerization can take place over shape-selective acidic zeolites such as HZSM-5. One reaction sequence takes place at Brønsted acid sites inside the channels, producing essentially linear materials. The other reaction sequence occurs on the outer surface, producing more branched material. By decreasing the external surface acidity of such zeolites, reduced methyl branching occurs, resulting in products of higher viscosity index and higher cetane number.[31,72,73] On the other hand, ESR and UV–Vis–IR spectroscopy with propene as probe molecule found that low temperature catalytic oligomerization takes place on the outer surface of HZSM-5 crystals, whereas the high temperature formation of aromatics takes place inside the zeolite channel.[74]

The ratio of intra-crystalline acid sites to external surface active sites in zeolites tends to increase with crystal size due to the geometric relation between volume and external surface area. Also, how varying the Si–Al ratio in the ZSM-5 catalyst affects crystal size in the oligomerization of propene has been investigated.[44] The optimum Si–Al ratio was about 20, which yielded the best lifetime and activity.

Certain basic materials have been used to deactivate surface acidity. By modifying the surface acidity of HZSM-5 with 4-methylquinoline or hexamethyldisilazane[72] more linear products have been obtained in propene oligomerization, and, by modifying the surface acidity of HZSM-5 with 2,6-di-*tert*-butylpyridine, relatively linear oligomers have been obtained from propene and other olefins.[31] The oils obtained using unmodified HZSM-5 have virtually every possible type of branching. By modification of the surface acidity of medium-pore zeolite HZSM-23 with 2,4,6-collidine,[73] substantially linear hydrocarbons were obtained in the oligomerization of propene. The surface acidity of ZSM-23 was reduced with dicarboxylic acid; when the catalyst was used in propene oligomerization, the product was almost linear and useful as an alkylating agent.[75] The surface of ZSM-23 was inactivated by coking with an olefin at high pressure, the oligomers produced were almost linear.[76] Confirmation of the shape selectivity with surface modified zeolite was examined in isobutene and propene oligomerization over the same catalyst at 230 °C. The products were almost identical and the viscosity indices for the obtained oils were the same.[72]

Crystal size and purity of HZSM-22 were controlled by the addition of seed during crystallization. The authors used the pure HZSM-22 in oligomerization of propene and, according to their activity measurements, the active sites for oligomerization were located near the zeolite external surface.[77]

In the oligomerization of propene over aluminosilicate HZSM-22 and ZSM-5 catalysts, the yields of lubricant-quality fraction were 21 and 18 wt%, respectively.[23] In propene oligomerization with NH_4Cl-exchanged ZSM-22 catalyst the conversion increased from 53 to 79% by increasing amount of added water from 0.02 to 0.15 mole%.[78]

When ZSM-5 was employed as catalyst, the oligomer product contained a low portion of dimers, but with ZSM-22 the major product was dimers[79] which is contrary to the findings of O'Connor.[68] Also, an admixture or extrudate of ZSM-5 and ZSM-22 was examined in propene oligomerization to gain maximum selectivity, *e.g.* for nonenes.

Experimental evidence showed that crystalline MCM-49 in propene oligomerization to components boiling in the gasoline, distillate and/or lube-range was more active than other zeolites such as ZSM-5, ZSM-23, ZSM-35 and MCM-22.[80] The product contained much methyl branching (methyls/C_{12} >2.1) with more than 50% of the product in the gasoline-range.

Catalytic oligomerization of propene was carried out at low temperature, 121 °C and 7 MPa pressure using a large pore size metallosilicate, MCM-41.[81] The catalyst showed higher activity and selectivity for the formation of trimers and tetramers in comparison with ZSM-5 and ZSM-23. By altering operating conditions over MCM-41, it was easy to vary the degree of branching of the oligomers.

The selectivity of the different zeolite systems to produce the lube oil fraction was examined in oligomerization of propene. The lube fraction of the product from oligomerizing propene with a ZSM-5 type catalyst generally has the feature of a low pour point (−50 °C) and low viscosity index (79).[82] The lubricant-range hydrocarbons produced from the oligomerization of propene over HZSM-22 at about 240 °C had a viscosity index of 85.[23]

5.2 Other Solid-acid Catalysts – $NiSO_4/\gamma\text{-}Al_2O_3$ catalyst was used in oligomerization of C_2–C_4 alkenes.[83] It was presumed that two different characteristic active sites in the $NiSO_4/\gamma\text{-}Al_2O_3$ catalyst accounted for the olefinic reactions.

The coordination sites contributed to the dimerization of ethylene, while oligomerization of propene proceeded both *via* coordination and acidic sites. However, but-1-ene only oligomerized *via* acidic sites.

It is believed that the available hydroxyl groups in ZrO_2 react with tungsten to form an acidic catalyst. The coprecipitated or impregnated form of WO_x/ZrO_2 was used in the oligomerization of propene.[84] Under identical conditions the coprecipitated catalyst was more active, resulting in higher propene conversion and higher yields of diesel and lube.

Solid phosphoric acid is prepared by mixing 85% orthophosphoric acid with kieselguhr. The product of propene oligomerization over solid phosphoric acid at 200 °C and 3 MPa contains about 50% trimers, 25% tetramers and less than 5% dimers.[9] Figure 4 illustrates the mechanism usually proposed for this reaction.

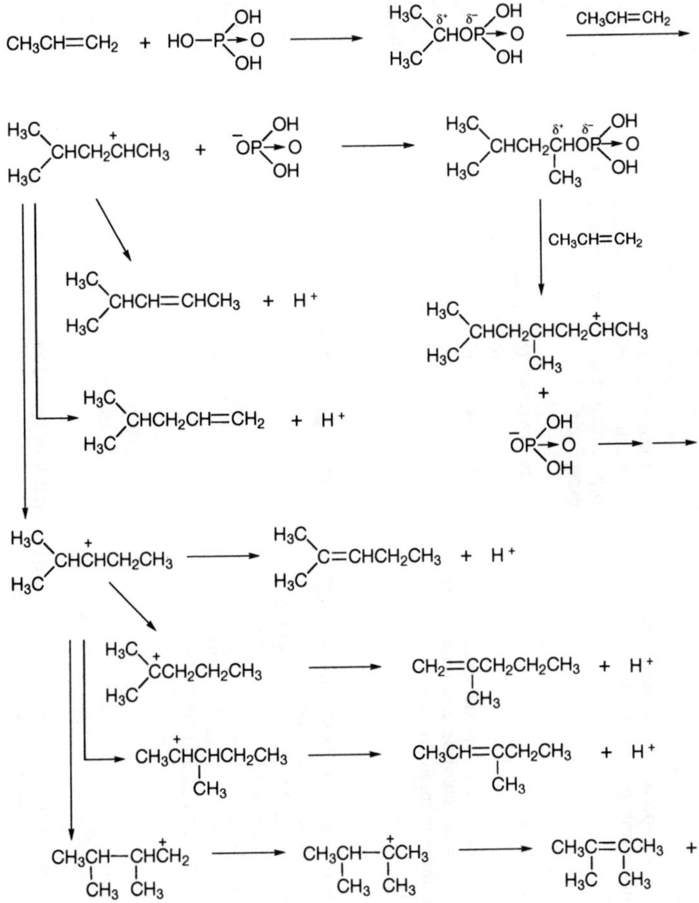

Figure 4 *Mechanism of propene oligomerization over solid phosphoric acid (from refs. 9, 85)*

Table 2 *Oligomerization of propene*

Year	*Catalysts*	*Reaction conditions/remarks*	*Products*	*Ref(s)*
1997	ZrO_2-supported WO_3 using impregnation and co-precipitation methods	Fixed-bed tubular reactor, 7 ml h^{-1}, 2.76 MPa, 160 °C; the conversion was 94.4 and 69.8 % for co-precipitated and impregnated catalyst, respectively	For co-precipitated catalyst, the C_5+ distribution was 29.5 gasoline, 61.6 diesel and 8.9 % lube; for impregnated catalyst, the C_5+ distribution was 38.5 gasoline, 56 diesel and 5.5 % lube	84
1997	HZSM-22	7 MPa, 180–250 °C, and addition of small amount of water to increase the activity	Oligomers	77
1996	Molecular sieve ion-exchanged with ammonium chloride, ammonium nitrate or ammonium hydroxide	WHSV 0.75–3.5, 6–8 MPa, 180–260 °C	An olefinic oligomer with a reduced degree of branching; these less highly-branched oligomers are used in alkylation of aromatic hydrocarbons in the manufacturing of surfactants and polyolefin stabilizers	90[b]
1995	An admixture of ZSM-5 and ZSM-22 in the form of an extrudate: (A) 100% ZSM-22 and 0 % ZSM-5; (B) 75% ZSM-22 and 25% ZSM-5 (C) 50% ZSM-22 and 50% ZSM-5 (D) 25% ZSM-22 and 75% ZSM-5 (E) 0% ZSM-22 and 100% ZSM-5	WHSV(feed) 2, 7 MPa, 195–205 °C	A. C_6 59, C_9 26, C_{12} 8, and C_{15}+ 3%, degree of branching 1.75, and conversion 92.4%; (B) C_6 34, C_9 43, C_{12} 14, and C_{15}+ 5%, degree of branching 1.84, and conversion 88.0%; (C) C_6 13, C_9 51, C_{12} 18, and C_{15}+ 9%, degree of branching 1.93, and conversion 89.0%; (D) C_6 8, C_9 51, C_{12} 20, and C_{15}+ 10%, degree of branching 1.93, and conversion 91.8%; (E) C_6 4, C_9 36, C_{12} 23, and C_{15}+ 16%, degree of branching 2.01, and conversion 91.5%	79
1995	(1) Fe-silicate, with a Si/Fe atomic ratio of 100; (2) Pt/H-Fe-silicate for oligomerization and hydrogenation (0.01 wt% Pt)	GHSV of 1000–5000, 290 °C, tubular reactor	Gasoline synthesis from light olefins	91[a]

1995	Amorphous micro/mesoporous silica–alumina gel having surface area of 500 $m^2 g^{-1}$ with a molar ratio of $SiO_2/Al_2O_3 \geq 30:1$	WHSV: 4 g propene/g active phase/h, 3.8 MPa, 120–180 °C, and reaction time about 240 h	Gasoline and jet fuel	92
1993	Aluminosilicate HZSM-22, with a Si/Al molar ratio ≥ 20	WHSV 0.5, 10.34 MPa, 200–235 °C; liquid products 96%	21% lubricant base-stock fraction	23[a]
1993	MCM-49, $SiO_2/Al_2O_3 = 34–15$, $OH^-/SiO_2 = 0.15–0.25$, $Na^+/SiO_2 = 0.20–0.45$, $HMI/SiO_2 = 0.35$, $H_2O/SiO_2 = 18–19$	WHSV $0.50\,h^{-1}$, 6.89 MPa, 65–105 °C; in comparison with ZSM-5 and MCM-22, MCM-49 has higher methyl branching for propene oligomerization with more than 50% of the products in the C_4–C_{10} gasoline range	At 105 °C the product distribution is: C_4–C_7 5.39, C_7–C_{10} 48.02, C_{10}–C_{13} 21.19, C_{13}–C_{16} 14.89, C_{16}–C_{19} 6.23, C_{19}–C_{22} 2.49, C_{22}–C_{25} 0.98, C_{24}+ 0.82 wt%, propylene conversion 99.54 wt%	80
1993	Coke-selectivated ZSM-23 prepared by extrusion of ZSM-23 and Al_2O_3 and conversion to H form by ammonium exchange and calcination; the catalyst was then coke-selectivated by processing in 30 mol% ethylene in N_2 at 5.52 MPa and 350 °C, for 1 day	For propylene/propane feed and coke-selectivated ZSM 23, WHSV about 1, 6.89 MPa, 150–200 °C	For propylene/propane feed and coke-selectivated ZSM-23 and 175 °C the product distribution was: C_4–C_7 45.86, C_7–C_{10} 37.87, C_{10}–C_{13} 8.05, C_{13}–C_{16} 3.79, C_{16}–C_{19} 1.54, C_{19}–C_{22} 0.89, C_{22}–C_{25} 1.15, C_{24}+ 0.85% and propylene conversion was 50.23 wt%	76
1993	Silica-alumina-nickel oxide gel; SiO_2/Al_2O_3 molar ratio of 30–500 : 1 and NiO/SiO_2 molar ratio from 0.001–0.01 : 1	Dimerization of isobutene with a catalyst of SiO_2/Al_2O_3 molar ratio of 102, NiO/SiO_2 molar ratio 0.021 in autoclave at 60 °C, 15 min and conversion 44%; oligomerization of propylene with the above formula for catalyst, space velocity $2.87\,h^{-1}$, 3.08 MPa, 94 °C, conversion 79%	Isobutene dimerization: product, C_8 78, C_{12} 19, and C_{16}+ 5% and dimer composition was 86% alpha-, and 14% beta-isobutene; propylene product: hexene 45%, nonenes 40% and oligomers higher than C_9 15%; increasing temperature increases the amount of $>C_9$ fraction	93[b]
1993	Medium pore, shape-selective zeolite	WHSV about 1, 6 MPa, 130–250 °C	Gasoline and diesel hydrocarbons	94[a]

Continued

Table 2 *Continued*

Year	*Catalysts*	*Reaction conditions/remarks*	*Products*	*Ref(s)*
1992	Cr(II) on an aluminosilicate support	0.126 l min^{-1} (STP), 1.03 MPa, 250 °C; after 1 h no liquid was observed	Gas product: but-1-ene 64, branched hexene 1.3, hex-1-ene 18.6, oct-1-ene 2.3, dec-1-ene 0.1, C_4–C_5 4.3, double bond isomers C_6 = 1.7, unknown C_6–C_8 3.0, and unknown C_8–C_9 4.3%	36
1991	Reduced Cr [Cr (II)] on silica	Fixed bed down-flow reactor: (1) conditions for C_3H_6 as feed 1.52 MPa, 190 °C; (2) for C_3H_6 and C_2H_4 ($C_3/C_2 = 3$) as feed, 1.38–2.07 MPa, 190 °C, 15 h, and yield 75%	(1) Propylene as feed gives : C_6–C_{30}, C_6–C_{12} after hydrogenation was used as gasoline; (2) C_3H_6 and C_2H_4 as feed give lube oil product with viscosity 46.03 cSt at at 100 °C, 703.25 cSt at 40 °C and viscosity index (VI) 112	86
1991	Silica–alumina co-gel dispersed in alumina matrix		Gasoline and diesel fuel	95
1990	Modified natural zeolite such as clinoptilolite with alkali and acid washing and subsequent calcining	Space velocity 3 h^{-1}, at 5 MPA and 250–350 °C	85% selectivity to C_6 (34 : 62 : 4) *n*-hexenes, methylpentenes, dimethylbutenes; 13% C_9 alkenes and 2% alkenes C_{12}+	96
1989	ZSM-5 with various ratios of Si/Al 20–500 : 1	Space velocity 12.0 ± 1 h^{-1}, 5 MPa, 220–300 °C. A decrease in Si/Al ratio increased the activity of ZSM-5 for propene oligomerization, and but-1-ene was more reactive than propene, which in turn was more reactive than hex-1-ene as an oligomerization feed for ZSM-5	In all cases C_{12} products the major oligomers formed	44[a]
1989	HZSM-5 with SiO_2/Al_2O_3 ratio of 70	Temperature-programmed desorption (TPD) study *in vacuo*, propene oligomerized rapidly inside the zeolite at 22 °C. When a coverage of	Similar oligomer chains produced, a mixture of C_{2-7} olefin products from different feeds, *i.e.* C_2H_4, C_3H_6 and 2-	26

		7.8 g/100 g was reached on the zeolite the reaction stopped, suggesting that reaction ceased only when blocking prevented reactants from approaching the acid sites	methyl-2-propanol, on the same catalyst	
1989	Mixture of pillared clay (*e.g.*, bentonite) and WO_3/SiO_2	The catalyst was more active for conversion of C_3H_6 to liquid fuel than a mixture of unpillared clay and WO_3/SiO_2	C_5+ hydrocarbons (liquid fuel)	97
1989	HZSM-5	Low-temperature oligomerization took place on the outer surface of HZSM-5; high temperature formation of aromatics took place inside its channels	Low temperature oligomers, high temperature aromatics	74[b]
1989	Ni supported on silica–alumina using three different techniques: homogeneous deposition precipitation, impregnation and coprecipitation	Homogeneously-deposited Ni catalyst was more active than impregnated or coprecipitated Ni. Ni content gave no marked variation in product distribution for propene oligomerization at 1 and 4.1 MPa and 80 °C	Product containing dimers, trimers, tetramers and small amount of pentamers	87
1988	SiO_2/Al_2O_3 161, SiO_2/TiO_2 46	0.6 h^{-1}, 260 °C, and 98% conversion	11% C_4, 17% C_5, 17% C_6 and 54% C_7+oligomers	98
1988	SiO_2/Fe_2O_3 573, SiO_2/TiO_2 46	0.6 h^{-1}, 310 °C	C_4, C_5, C_6 and C_6+ oligomers, 21, 25, 18 and 35%, respectively	99
1988	Surface-deactivated ZSM-5 containing NH_4F	Oligomerization conditions	Liquid hydrocarbons and a bottom lube fraction of 13 wt% obtained, with a VI of 111.9	100
1988	EDTA-extracted ZSM-5	Four days under oligomerization conditions yielded 42 wt%	343 °C lubricant having a VI >80	82
1988	Zeolite of Si, Ti, and Ga oxides, formula pHGaO$_2$.qTiO$_2$.SiO$_2$ (p = <0–<0.5; q = >0–<0.25), SiO_2/Ga_2O_3 = 195, SiO_2/TiO_2 = 54	0.6 WHSV, 0.1 MPa, 240 °C and 45% conversion	Selectivity to C_2 0.2, C_3H_8 trace, C_4 19, C_5 21, C_6 22, and C ≥7 37.8%	101

Continued

Table 2 *Continued*

Year	*Catalysts*	*Reaction conditions/remarks*	*Products*	*Ref(s)*
1988	Medium pore, shape-selective acid ZSM-5 ($SiO_2/Al_2O_3 = 70$)	0.75 WHSV (fixed bed reactor), 6.9 MPa, 117 °C	329 °C lubricating oil, viscosity index 125	102
1988	Ni-exchanged pillared silicotitanate (surface area 299 $m^2 g^{-1}$, containing TiO_2 70.2, SiO_2 21.7, and Na 3.3 wt%)	10.03 MPa, 150 °C for 24 h	Liquid product gasoline, containing C_6 16.6, C_9 56.7, C_{12} 17.7, C_{15} 6.7, C_{18} 2, and C $\geq$21 0.4 wt%	103
1987	Surface-inactive HZSM-5 whose surface Brønsted acid sites have been inactivated with a sterically-hindered basic organophosphorus compound, the organophosphorus compound cross section is larger than zeolite pore (*e.g.*, treated with 10% Ph_4PBr)	Pressure reactor, 230 °C, 19% yield after 8 days	Lubricating oil with a viscosity index of 132, and bp >343 °C	104
1986	ZSM-5 zeolite, HZSM-5 of 1.22 wt% alumina, ZSM-5 poisoned with a large excess of hexamethyldisilazane in order to replace the acidic surface hydroxyl groups by trimethylsiloxyl groups, addition of 4-methylquinoline to the feed supposed to neutralize the sites due to alumina at external surface of zeolite	Fixed bed, 2.4 MPa, 189–359 °C. Low temperature gives dimer, trimer, tetramer, *etc.* without shape selectivity; reaction occurs on the external surface of the zeolite rather than within the channel. At higher temperature the external surface is poisoned and shape-selectivity is apparent	Diesel fuel	72
1986	Hydrogen form zeolite, HY, H-ZSM, H-mordenite	FTIR study from ambient temperature to 300 °C	At ambient temperature oligomer formation occurs slowly on HY, at moderate speed on HZSM-5 and on HM, the oligomers formed on HZSM-5 and HM are unbranched and on HY highly branched	105

1986	HZSM-5	3 catalyst beds, 0.6–1 VHSV, 4.2–10.4 MPa, 205–290 °C	C_{9-10} fraction with bp 165–290 °C, diesel fuel and/or jet aircraft fuels	106[b]
1986	Ni–Zn-HZSM-5	Two-stage oligomerization: first stage 0.5 -1 LHSV for 40 h, 11.0 MPa, 54 °C, to produce 98 wt% conversion (71 wt% C_3H_6 dimer), in second stage 0.5–1 LHSV, 5.52 MPa, 149 °C to convert >60% of the feed to the fraction with bp >177 °C, of which approx. 60 wt% was C_3H_6 tetramer	C_3H_6 tetramer	107
1986	ZSM-5 (SiO_2/Al_2O_3 26 000 : 1) impregnated with Na-aluminate	0.38 LHSV, 10.3 MPa, 110 °C and conversion 83.5%	46% with bp −1 to 170 °C, 51% 166–343 °C and 3% >343 °C	108
1986	ZSM-5	Oligomerization conditions as function of temperature, pressure and space velocity	Gasoline (C_{5-10}), and diesel (C_{10-20})	71[b]
1985	Ferrosilicate zeolite (ZSM-5-type)	10.4 MPa, 110 °C	C_{1-5} hydrocarbons 1.1%, C_6-fraction 22.7% (bp 166 °C), 64.9% with bp 166–343 °C, and 11.3 wt% with bp > 343 °C	109
1985	ZSM-5	0.4 LHSV, 10.4 MPa, 247 °C, the total conversion was 98.74 wt%	The product contained 45.39 wt% lubricating oil fraction (bp >343 °C)	110[b]
1985	Zeolite	WHSV 0.1–5, 0.69–2.07 MPa, 177–316 °C	Olefinic gasoline stream and a distillate stream, product is hydrotreated with hydrogen to recover lubricating oils	111
1985	ZSM-5	Approx. 5.8 MPa, 315 °C, conversion about 95%	Gasoline and diesel fuels	112[b]
1984	HZSM-5 grain size <0.5 μm or >0.5 μm	(a) WHSV 1, 0.34 MPa, 260 °C over large crystal catalyst to give 98% distillate; (b) the distillate was then heated at 210 °C for 48 h in presence of small crystal zeolite	40% yield of lubricating oil bp >316 °C with viscosity index 100	113[b]
1983	Al_2O_3-free silicate molecular sieve (SiO_2/Al_2O_3 >500 : 1, preferably >1000 : 1), contained Cd or Zn or their compounds	0.1 MPa at 316 °C	60% $\geq C_5$ hydrocarbons	114
1983	ZSM-5	Fluidized bed, 316 °C	70% C_4+ and 61% C_5+	115

Continued

Table 2 *Continued*

Year	*Catalysts*	*Reaction conditions/remarks*	*Products*	*Ref(s)*
1983	Solid phosphoric acid	Feed first dehydrogenated over Pt, Sn and alkali metal on Al_2O_3, then passed through catalytic condensation zone	Dimers and trimers	116[b]
1983	Al_2O_3-free silicate molecular sieve	Oligomerization conditions	Liquid fuel-range products	117[a]
1983	CZM or zeolite impregnated with Cd and/or Zn	Oligomerization conditions	Liquid fuel-range products	118[a]
1982	ZSM-5	0.6 WHSV, 0.12 MPa, 277 °C	95% C_{3-11} olefins	30
1981	Zeolites	Oligomerization conditions	Liquid fuel	119
1981	HZSM-12	4.14 MPa at 47 °C, conversion 48.1 wt%		6
1979	SiO_2–Al_2O_3 or fluorinated SiO_2–Al_2O_3	1.4 MPa, 100–300 °C	Hydraulic fluids, gasoline with high octane no., and jet fuels or white spirit-type solvent	120
1977	$TiCl_4$ supported on Group III(A) metal	2 h at 200 °C	C_{6-7} hydrocarbon fraction	121
1977	γ-Al_2O_3 cold rolled with H_2TiF_6	2 h at 200 °C	15.5% dimethylbutene, 41.5% methylpentenes, 9.2% hex-1-ene + 2-methylpentene, 5% hexene+ethylbutene, and 4.2% *trans*-2-hexene	122
1977	LaX, CeX, MgX, NiY, CoY, AlY, MgX, MnY, NiX, CoX, and CaX	Fixed bed, 0.1 MPa, at 190 °C, with the exception of NiX; all zeolites tested showed 95.5% selectivity to dimer formation	The catalyst is more focused on dimerization	69
1973	Al_2O_3/an organoaluminum halide, Ni(II) or Co(II)	660 v/v h, 201 °C	72% dimers (of which 16.5% was *n*-hexene), 20% trimers, and 8% higher oligomers	123
1971	Silicon oxide–zinc phosphate		Heptenes	124[b]

1971	$Ni(CO)_2(Ph_3P)_2$ on Al_2O_3–SiO_2	0.7–1.7 MPa, 60–71 °C; 87% conversion	Mixture of 15% 2,3-dimethylbutenes, 53% 2-methylpentenes, 16% *n*-hexenes, and 15% high boiling products, mostly C_9 olefins	125
1970	Mixture of SiO_2 and polyphosphoric acid		Gasoline with high octane no.	126
1970	Cobalt-activated coal	Continuous flow reactor, an activity of 8.4 g oligomer/g catalyst/h, 2.07 MPa, 20 °C	83% hexenes (approx. 70% straight chain) and 17% nonenes	127
1968	Cobalt–carbon catalysts by adding transition metal, Cu, Ni, or Cr	WHSV 10, 2.07 MPa, 25 °C,the Co catalyst showed no activity, but the Co-Cu had an activity of 4.16 g oligomer/g catalyst/h	5% nonenes and 95% hexane, of which 51% straight chain hexenes	128
1967	Cobalt-activated coal	WHSV 24, 5.52 MPa, 25 °C		129

[a] Uses also butene and/or higher olefins as feed; [b] uses also butene as feed.

Cr/SiO_2, as described,[86] was used for oligomerization of propene alone or co-fed with ethylene to produce HVI-PAO (high viscosity index-poly alpha-olefin). This lubricant material had a structure with a CH_3/CH_2 ratio of less than 0.19, compared with a ratio greater than 0.20 for conventional PAO.

Nickel incorporated onto silica–alumina using various procedures: impregnation, coprecipitation and homogeneous deposition precipitation, was tested in propene oligomerization.[87] Homogeneous deposition precipitation was found to be the most active. At a conversion of around 90% of the propene, the products were mainly dimers (*ca.* 60%) and trimers (*ca.* 25%). The product spectrum was unaffected by pressure changes, while higher temperatures resulted in more heavy products.

Catalysts containing nickel supported by or incorporated into clay structures have been used for oligomerization. Bercik *et al.*[88] have shown that nickel-substituted micamontmorillonite is a very active catalyst for oligomerizing propene and butene. The incorporation of nickel, zinc and cobalt into micamontmorillonite by ion exchange had very little effect on selectivity for propene oligomerization; mainly trimers and tetramers were formed.[89] However, with this catalyst an improved lifetime was obtained.

6 Butene Oligomerization

Lower olefins are produced in large quantities by cracking processes like steam or fluid catalytic cracking. C_2–C_4 olefins are the most important base chemicals. For many years, the usage of the C_4-olefins has been limited to mainly alkylate production, gasoline blending or simply burning.

The elimination of lead in gasoline initiated the development of methyl-*tert*-butylether (MTBE) production from isobutene and methanol. Nowadays, MTBE is an integral component of the gasoline pool, and still rapidly increasing demand will result in a more or less quantitative consumption of the isobutene contained in C_4-cuts. Further, sometimes there is insufficient isobutene available in a refinery to permit all olefins from the different cuts to be catalytically alkylated to MTBE.

Development in zeolite catalysis promoted the MOGD oligomerization to convert the lower olefins into gasoline and distillate.[71] This process has significance as a safe, environmentally acceptable technique for utilization of lower olefins. Most of the available literature concerning oligomerization of the C_4-fraction treats the production of liquid fuel or lube oil base-stock. Table 3 (page 262) contains a summary of solid catalyst systems used in the oligomerization of butene.

6.1 Zeolites – The zeolites, and in particular pentasil zeolite ZSM-5, have been intensively investigated in butene oligomerization for various purposes. The performance of this catalyst for alkene oligomerization was associated both with the catalyst synthesis, pre-treatment, and the reaction conditions.[70] The product spectra were reported in the oligomerization of butene over mordenite and solid phosphoric acid at 5 MPa and 250 and 200 °C, respectively.[9] The mordenite catalyst gave mainly dimers and trimers with a minor fraction consisting of tetramers and pentamers. The solid phosphoric acid catalyst products were essentially dimers.

Synthetic oils with high viscosity indices were produced in the oligomerization of isobutene using HZSM-5 catalyst, surface-deactivated with 2,6-di-*tert*-butyl-pyridine.[31] Chen and Bridger[31] proposed a protonated cyclopropyl intermediate for the isobutene–HZSM-5-system in the formation of linear oligomers.

Olefinic oligomers with a reduced degree of branching were produced with a silica-rich ZSM-22 outer layer catalyst in the oligomerization of a feed containing mixed butenes, diluted with butanes, with the approximate proportions 65% olefins and 35% saturates.[90] It was reported that a high viscosity index lubricant was produced by oligomerization of lower olefins, essentially the C_3–C_4 fractions, with a medium-pore, shape-selective aluminosilicate HZSM-22 catalyst.[23]

The effect of varying the NH_4/Na content in NH_4NaY on the oligomerization of butenes at 5 MPa has been examined.[130] The activity decreased with increasing Na content but the product spectrum was independent of percent exchange. Varying the NH_4/Na content was also investigated on the oligomerization of butenes over NH_4-mordenite at 200 °C and 5 MPa,[131] the mordenite catalyst behaved like the Y zeolite. NaY, partly exchanged with corresponding nitrates and subsequently impregnated with NH_4Cl, $(NH_4)_2SO_4$, $(NH_4)_2HPO_4$, $NaHSO_4$, H_3PO_4, and H_3BO_3, was examined for isobutylene di- and trimerization at 120–200 °C at atmospheric pressure. It was found that zeolites modified with Cl^- and SO_4^{2-} facilitated isomerization and cracking of initially formed dimers and trimers.[132]

The reaction of butenes over hydroxylated Y zeolites at atmospheric pressure has been investigated using infrared spectroscopy.[19] The results showed a reactivity pattern of the reactants in the order of *trans*-but-2-ene < *cis*-but-2-ene < but-1-ene.

6.2 Other Solid-acid Catalysts – γ-Al_2O_3 supported with different amounts of sulfate ion, and other carriers such as iron oxide, zinc oxide, silica, and silica–alumina impregnated with sulfate ion were used in the oligomerization reaction of a C_4 olefin-containing material.[133] Conversion of butene passed through a maximum at 80% when the catalyst sulfate ion content was around 20% by weight. Butene conversion was low for the different supports except for silica–alumina-supported sulfate ion, for which both the conversion and yield of trimer and tetramer were improved.

The product of butene oligomerization over solid phosphoric acid at 200 °C and 3 MPa was predominantly dimers.[9] As the H_3PO_4 concentration was reduced, the rate of butene formation declined linearly, with almost no change being observed in product selectivity.

It was described that nickel supported on H-mordenite was improved in stability and less regeneration was needed when the catalyst was used in oligomerization of butene.[134]

A high yield of linear dimers and trimers was reported in the oligomerization

Table 3 *Oligomerization of butene*

Year	*Catalysts*	*Reaction conditions/remarks*	*Products*	*Ref(s)*
1997	Heterogeneous acid catalysts	Catalytic distillation	Oligomers (mostly dimers)	136
1996	HZSM-5 impregnated with Cr salts	Oligomerization conditions	Feedstocks for diesel fuel and lubricant oil	137
1996	Resin with cation exchange, functionalized with sulfonic groups, Amberlyst 15	Feeding space velocity of less than 20, 5 MPa, 30–100 °C gave joint production of ethers and oligomers with a high octane number	Ethers, dimers, and trimers; if the reaction carried out without primary alcohols the selectivity to dimers is reduced	138
1996	Sulfate ion supported on iron oxide, zinc oxide, γ- Al_2O_3, silica and a silica–alumina carrier	A fixed bed flow type reactor LHSV of 1, 4.8 MPa, 50–150 °C, 24 h from the initiation of the reaction; at 150 °C with 15.3 wt% sulfate ion, butene conversion was 100%, selectivity to dimer 20% and to trimer and tetramer 64%	With C_4 olefin-containing material as a feed, the products are dimers, trimers and tetramers; with C_8 olefin-containing material as feed the products are dimers and trimers, the oligomers have a low degree of branching, which is useful as a material for a plasticizer	133[b]
1991	Monolayer of SiO_2 on Al_2O_3, coated with a Ti-containing precursor solution, catalyst containing 9% TiO_2 and 10% SiO_2 on Al_2O_3	125 °C, under autogenous pressure	C_8, C_{12} and C_{16} olefins with selectivity of 59.7, 17.23, and 6.7%, respectively	139
1991	Ni [(a) 11.5 or (b) 10.3 wt%] on mordenite in hydrogen form	WHSV 1, 3 MPa, 215 °C. After 170 h conversion of butene 71 wt% over (a), and after 210 h, 76 wt% over (b)	Liquid hydrocarbons	134
1990	Alumina spheres impregnated with Ni 4.77, Sn 2.75 and Cl 1.0%	4.83 MPa, 70 °C	*n*-Octene 16, methylheptene 66, and dimethylhexene 18%	140

1990	Silica spheres, large pore diameter, impregnated and reimpregnated with orthophosphoric acid	WHSV 0.543 g cm^{-3}, 0.1 MPa, 160 °C, conversion 98 wt% of isobutene, degree of conversion affected after 4th impregnation	Liquid products containing 79% dimer, 18% trimer and 3% tetramer	141
1989	HZSM-5 with SiO_2/Al_2O_3 ratio of 70 (520)	Temperature-programmed desorption (TPD) study *in vacuo*, oligomerization occurred at 22 °C. When a coverage of 7.8 g/100 g was reached on the zeolite the reaction stopped, suggesting that reaction ceased only when blocking prevented reactants from approaching the acid sites	Similar oligomer chains produced, a mixture of C_{2-7} olefin products from different feeds, *i.e.* C_2H_4, C_3H_6 and 2-methyl-2-propanol, on the same catalyst. For SiO_2/Al_2O_3 ratio 520 this oligomerization with 2-methyl-2-propanol could be minimized and methylpropane was a major product	26
1989	Ni/Ba/CSZ-1 (0.33 Cs_2O : 0.58 BaO : NiO : Al_2O_3 : 5.86 SiO_2)	82.5 g but-1-ene and 12.05 g catalyst, 15 min 110 °C, 50.7% conversion	Dimer 72.4, trimer 21.6, and tetramer 5.6%	142
1989	Alumina impregnated with $Ni(NO_3)_2 \cdot 6H_2O$ and $SnCl_4 \cdot 5H_2O$ then activated with $AlCl_3$ and Et_2AlCl	50 °C in the presence of H_2 gives slower deactivation of catalyst for oligomerization	Oligomers	143
1989	Fluorinated γ-alumina containing 3.8% Ni	70 °C, and 70% conversion	C_8 dimers and C_{12} trimers with 84.6 and 14.6% selectivity, respectively	144
1989	SiO_2, Al_2O_3, MgO, or TiO_2/H_3PO_4	– 103 to 20 °C	Oligomers C_{20-50}	145
1989	HZSM-5, Cr(V), V(IV), or Zn(II)	320 °C	Oligomers C_{6-10} (60%)	146
1988	Silica, alumina, titania, and zirconia impregnated with H_3PO_4	FTIR study: H_3PO_4 on silica is the strongest Brønsted acidic catalyst; the other supports generate Brønsted acidity and modify Lewis acidity and basicity. Isobutene and but-1-ene oligomerize on all catalysts at 0.027 MPa, 20 °C. Ti-support was most active.	Oligomers	147
1988	Al_2O_3 impregnated with Ni nitrate hexahydrate and $SnCl_4 \cdot 5H_2O$	Space velocity 2.0 h^{-1}, 70 °C, 4.83 MPa, 12 h of operation, butene conversion 71%, C_8 selectivity 84.4%	C_8 isomer distribution: 1–10.4, methylheptene 64.7 and dimethylhexene 24.9%	148

Continued

Table 3 *Continued*

Year	*Catalysts*	*Reaction conditions/remarks*	*Products*	*Ref(s)*
1988	Al_2O_3 impregnated with Ni nitrate hexahydrate	Space velocity 2.0 h^{-1}, 70 °C, 4.83 MPa, 18 h of operation, and 40% butene conversion	C_8 isomers with distribution octene 10, methylheptene 72, and dimethylhexene 28 wt%	149
1988	Varying NH_4/Na content on NH_4Na-mordenite	5 MPa, 200 °C		131
1988	Ni mordenite catalyst (Si–Al 9 : 1, Ni–Al 1 : 1)	Two-stage process; mixture first passed over catalyst under non-oligomerization conditions at butene space velocity 2 h^{-1}, 1.6 MPa, 42 °C; effluent was passed over the same catalyst in a second stage oligomerization at space velocity 2 h^{-1}, 1.6 MPa, 215 °C	Liquid hydrocarbons	150
1987	L and Y zeolites: mordenite, erionite and pentasil	50–200 °C	Oligomers C_{8-12}	151
1987	NaHY-zeolites, degree of Na exchange 5–100%	WHSV ≅ 1.0, 100–400 °C, degree of exchange has no marked effect on the extent of catalyst deactivation, and zeolite-Y is not suitable for synthesizing long straight chain hydrocarbons		130
1986	Ni/HZSM-5	120–230 °C	Oligomers C_{16}	152
1985	Acidic Al_2O_3	0.1 MPa, 310 °C	15.4 wt% C_5+ fraction (at 84.8% conversion) and motor octane no. 94 and 81, respectively	153
1984	TaCl supported on SiO_2	WHSV 22, 3.45 MPa, 38 °C	100% conversion to products, which were used as a blending agent for gasoline and gave an increased octane no.	154
1984	BF_3-containing Al_2O_3 catalyst treated with dimethyl ether	0.69 MPa, 150 °C	Methylheptene 8%, dimethylhexene 91.5% and trimethylpentene 0.2 wt%, a blending stock for gasoline	155

1981	NaY zeolite (SiO_2/Al_2O_3 5.22)	20 °C, reactivity: *trans*-but-2-ene < *cis*-but-2-ene < but-1-ene		19
1979	Amberlyst 15	0.7 MPa, 60 °C, yield 57.1%		156
1978	Perfluorosulfonic acid resin	145 °C, 1000 h	C_8 50.8%, C_{12} 47.5%, and C_{16} 1.8%	157
1976	74:26 P_2O_5/silica gel	Space velocity 1.5h^{-1}, 3 MPa, 190 °C	3.9% C_{5-7} olefins, 67.5% octenes, 3.7% C_{9-11} olefins, 22.1% dodecenes, 2.8% higher olefins	158
1973	Ta_2O_5–alumina	175 °C	7% C_8 and C_{12} olefins with 90% selectivity to C_8 olefins	159
1972	X, Y, HX, HY, and alkali metal (Li^+, Na^+, K^+, and Rb^+) ion-containing zeolites	X-type zeolites are more active than zeolite Y, the activity order for alkali metal ion forms is inversely proportional to the ionic radius in general, the catalysts are not stable due to the build-up of coke		160–163

[a] Uses also higher olefins as input feed.

of a mixture of butane/butene with NiO supported on different carriers such as TiO_2, SiO_2 and/or a mixture of TiO_2 and SiO_2 as catalysts.[135]

7 Long-chain Alkene Oligomerization

Conversion of light olefins to gasoline and/or distillate products is well documented in the literature by various investigators. The light olefinic feedstocks alone or in admixture with paraffins are converted into liquid fuels by contacting the feedstocks with a catalyst containing an acidic solid. Light olefins can also be oligomerized to α-olefins (RCH = CH, where R is an alkyl radical of 8–22 C atoms) or internal olefins (R′CH = CHR″, where R′ and R″ are the same or different alkyl radicals of 1–21 C atoms). The synthetic lubricants are produced by the oligomerization of alpha or internal, linear or branched alkenes or 1-alkenes. Higher molecular weight olefin oligomerization is known to be catalysed homogeneously, *e.g.* boron trifluoride plus promoters, as well as zeolites, acidified montmorillonite clays, transition metal salts and complexes.

Still there are efforts to improve the performance of natural mineral oil-based lubricants by the synthesis of oligomeric hydrocarbons, which has been the subject of important research and development in the petroleum industry for many years and has led to commercialization of a number of synthetic lubricants. These materials are based on the oligomerization of α-olefins such as C_6–C_{20} olefins. Industrial research effort on synthetic lubricants has generally focused on improved viscosity index, thermal and oxidative stability, and a pour point equal to or better than that of the corresponding mineral oil lubricants.

It is common practice to classify the base-stocks by their viscosities, measured in centistokes (cSt) at 100 °C. Those with viscosities less than or equal to about 4 cSt are commonly referred to as low viscosity, whereas those with a viscosity in the range of about 40–100 cSt are referred to as high viscosity base-stocks. Base-stocks having a viscosity of 4–8 cSt are called medium viscosity base-stocks. The low viscosity materials are recommended for low temperature applications. Motor oil, turbine lubricants and other industrial lubricants generally require medium viscosity base-stocks. High viscosity base-stocks are used in gear oils and as blending stocks. The viscosity of the base-stocks is determined by the length of the oligomer molecules formed during the oligomerization reaction.

The degree of oligomerization is affected by the catalyst and reaction conditions, as well as the length of the carbon chain of the monomer used as starting material. Fluids prepared from short chain monomers have low pour points and low viscosity indices, whereas fluids prepared from long-chain monomers have moderately low pour points and higher viscosity indices. Oligomers prepared from long-chain monomers are generally more suitable than those prepared from shorter-chain monomers for use as medium viscosity synthetic lubricants.[164]

Lubricants are often formulated with additives to enhance the properties for some specific application. The synthetic lubricants may be produced by homogenous catalysts, such as BF_3 or $AlCl_3$. The homogeneous process has several significant drawbacks, the catalysts are toxic and corrosive, requiring an expensive separation procedure, formation of noticeable amounts of by-product as well as recovery and disposal of the catalyst residues, which can cause environmental problems.

In the present review attention is only paid to heterogeneous catalysts. Table 4 (page 268) constitutes a summary of heterogeneous catalyst systems in oligomerization of higher olefins.

7.1 Zeolites – The oligomerization of light olefins in the presence of zeolites has been investigated extensively. There has been little effort on studying the oligomerization of higher α-olefins using zeolite catalysts. Garwood[18] described the phenomenon of shape-selectivity in converting C_2–C_{10} olefins over HZSM-5 to higher boiling-range olefins and compared the likeness and contrast to when using amorphous silica–alumina. He concluded that the absence of the more branched oligomers and the lesser aromatics portion obtained over HZSM-5 were both indicative of shape-selective constraints.

A common intermediate step was suggested, confirmed by the similarity, C_4–C_7, of the final products from the oligomerization of ethene, propene, a mixture of internal pentene isomers, hex-1-ene, and dec-1-ene over a zeolitic catalyst.[18]

Miller[165] discovered that, under certain conditions, medium pore size molecular sieves such as ZSM-5 and crystalline metal silicates are active for the oligomerization of C_{15}–C_{20} 1-alkenes.

Minachev *et al.*[166] reported that the activity of a partially decationized zeolite for higher α-olefin oligomerization depends on the amount of catalyst, but does not depend on the strength of Brønsted and Lewis acid sites.

Synthetic lubricant with improved viscosity index was prepared by oligomerization of the C_{10}–C_{20} 1-alkenes over catalysts containing a large pore zeolite with a high silica/alumina ratio.[27] It was found that the reactivity of α-olefins followed the order of dodecene > tetradecene > hexadecene; also, the internal olefins were less reactive than the corresponding α-olefins with similar carbon numbers, as also supported by O'Connor.[9]

A catalytic distillation reactor system including three zeolite catalyst layers or their ion-exchanged varieties was used for oligomerization of all kinds of olefins, not only α-olefins.[167]

Lubricant products were produced by oligomerization of α-olefins over zeolite catalysts, and supported reduced chromium on hexagonal honeycomb microstructure, respectively.[168,169]

In general, internal alkenes are less reactive than the corresponding α-alkenes, and conversion in oligomerization decreases as the chain lengths of the feed alkenes increase. However, the zeolites reported so far are just as selective to the desired products as clay catalysts or the currently used BF_3 or $AlCl_3$ homogeneous catalysts

Table 4 *Oligomerization of higher olefins*

Year	*Catalysts*	*Reaction conditions/remarks*	*Products*	*Ref(s)*
1997	Catalyst consisting of a solid olefin polymer, BF_3, and an organomagnesium halide as reagent	20–60 °C, and reaction time 0.25–3 h	α-Olefin oligomers useful as synthetic lubricants	179
1997	Acid-treated clay catalysts	Review	C_{10-24} linear α-olefins, useful as synthetic lubricant base stocks	180
1996	A novel zeolite	NESKO process	Diesel, solvent, and lube oil	181
1996	Sulfated zirconia	Carbocation mechanism was suggested	Synthetic lubricating base oil	32
1996	HZSM-5 (SiO_2/Al_2O_3 70), and surface deactivated HZSM-5 using a sterically-hindered base (10% solution of 2,6-di-*tert*-butylpyridine in hexane) *ca.* 100 ml of solution to 100 g catalyst	Stirred autoclave (600 ml), the ratio of catalyst to olefin 1 g: 55 g, 200 and 230 °C; at 200 °C the same oligomers were obtained with or without the addition of surface deactivating solution	C_{20}+ fraction, synthetic oil with high viscosity indices (at 200 °C : VI 165 from propene, VI 154 from dec-1-ene, at 230 °C VI 134 from propene, VI 137 from isobutene). Oil obtained using surface-modified HZSM-5 consisted of linear alkanes with a low degree of methyl branching after hydrogenation	31
1995	(1) SiO_2-supported alkyl halide (*e.g.* $RAlCl_2$); (2) with other halide ligands SiO_2–$AlBr_2$	Dec-1-ene oligomerization, 0.5 eq. H_2O as promoter, olefin : catalyst 30 : 1, 24 °C, and 24 h	Catalyst (1) 70% conversion, C_{20} 22.1%, lube 77.9% (viscosity 16.2 cSt, viscosity index 144); catalyst (2) 51% conversion, C_{20} 82.4%, lube 91.3% (viscosity 18.7 cSt, viscosity index 150)	176
1994	Zeolites	Catalytic distillation reactor system, temperature range from *ca.* 150 °C to approx. 350 °C depending on catalyst type	Gasoline, diesel fuel, solvents, and lubricant stock	167
1994	Faujasite, mordenite, and ZSM-5	Zeolite catalyst samples were tested over a range of temperatures (140–300 °C) in a batch reactor for 4 h	Synthetic lubricant base stocks: faujasite > mordenite > ZSM-5 the activity of Y depends on Si/Al ratio	27

1994	Ultrastable Y zeolite with varied silica–alumina ratios, 6.5–60, faujasite, mordenite, and pentasil	Zeolite catalyst samples were tested over a range of temperature (140–300 °C) in a batch reactor for 4 h, the dealuminated Y-zeolites were evaluated in an up-flow continuous reaction system, LHSV 0.16–0.30, 1.38–2.07 MPa, 140–200 °C	For the faujasite-class, an increase in silica/alumina ratio leads to a parallel increase in both olefin conversion and in total oligomer formation, *e.g.* $SiO_2/Al_2O_3 = 60$ gives C_{14} conversion 79.8%, dimer 67.4% and trimer 12.3%. Mordenite sample exhibits low conversion, *e.g.* $SiO_2/Al_2O_3 = 10$, C_{14} conversion 18.7%, dimer 15% and trimer 3.7 %. Pentasil and ZSM-5 were nearly inactive for C_{14} oligomerization at 180 °C for 4 h, faujasite > mordenite > pentasil. Olefin oligomerization activity: $C_{12} > C_{14} > C_{16}$ and alpha > internal.	168
1993	Sulfate-activated TiO_2 or ZrO_2 or molecular sieve	Atmospheric pressure–6.99 MPa, 50–300 °C	Dimer, trimer, and higher oligomers	175
1993	Cr oxide/silica gel, silica support with pore size at least 40 Å, and commercial chrome/silica which contains 1% Cr	0.1 MPa, 90–250 °C	Liquid hydrocarbons suitable as lubricant base-stocks having a branch ratio of less than 0.19, pour point below −15 °C, and viscosity index >130. Reduced CrO_3 on silica gives a liquid lubricant composition which has a C_{30} fraction with a branch ratio below 0.19 and pour point −45 °C	177
1993	Reduced Cr oxide supported on mesoporous materials with hexagonal honeycomb microstructure, cell diameter >13 Å, and mesoporous range 20–100 Å, *e.g.* MCM-41	90–250 °C, this support enables making products of greater viscosity compared with the products from catalyst using amorphous support	Liquid hydrocarbons suitable as lubricant base-stocks having a branch ratio of less than 0.19 and high-viscosity index lubricants	169

Continued

Table 4 *Continued*

Year	*Catalysts*	*Reaction conditions/remarks*	*Products*	*Ref(s)*
1993	Activated layered silicate containing silica and Cr oxide between the layers	Oligomerization conditions	Gives mainly trimers and higher oligomers which were hydrogenated to give a liquid lubricant with pour point −36 °C	178
1993	Ti or Zr salts deposited on silicon dioxide; the titanium or zirconium oxide supported on SiO_2 is sulfate-activated and physically blended with an acidic montmorillonite clay	0.1–7 MPa, 120–250 °C	Oligomers suitable for use as synthetic lubricant base stocks	174
1993	Sulfate-activated zirconium dioxide	C_{14} as feed, olefin 100 g, catalyst 10 g, temperature 160 °C, 5 h, conversion 86.8% and the ratio of dimer/trimer was 1.68	Synthetic lubricant base-stocks	182
1992	Halogen-free Ti, Zr salts on clay	Oligomerization conditions	Synthetic lubricant base-stocks	183
1992	Supported Al_2O_3, TiO_2, SiO_2 or ZrO_2 with a heteropolyacid, *e.g.* tungstosilicic acid, molybdosilicic acid, or molybdophosphoric acid	Oligomerization conditions	Synthetic lubricant base-stocks	184
1992	Halogen-stabilized silicalite, or a zeolite	Oligomerization conditions	Gasoline with high octane no.	185
1992	Difluorophosphoric acid on montmorillonite clay	To obtain a lower dimer/trimer ratio oligomerization performed at 135–145 °C, 100 g olefins, 10 g catalyst 160 °C, 5 h conversion 82.2% and dimer/trimer = 0.80	Very high percentage of trimer and higher oligomer, very low dimer/trimer ratio	173
1992	Hexagonal honeycomb microstructure, *e.g.* MCM-41, impregnated with a group VI(B) metal, preferably Cr	90–250 °C	Liquid oligomers with a branch ratio of <0.19	186

1992	Large pore size metallosilicate structure of MCM-41with uniformly distributed honeycomb pores in the 100–200 Å range	WHSV 1, 3.45 MPa, 96 °C and conversion 84 wt% propylene, the same conditions except WHSV 0.25 conversion >98 wt%. Increasing temperature results in high MW product, increasing pressure causes increase in activity and a shift in selectivity to lower MW products, improved activity compared with ZSM-5 and ZSM-23 also provide higher C_9 and C_{12} for propylene oligomerization	For propylene WHSV 0.25, 3.45 MPa, 96 °C, $C_6 = 2$, $C_9 = 43$, $C_{12} = 34$, $C_{15} = 12$, $C_{18} = 5$, and $C_{21} = 3$ wt%, C_9 and C_{12} hydrogenated to provide clean gasoline and diesel fuel. For dec-1-ene (5 parts dec-1-ene per part catalyst and 1 h) 0.101 MPa, 180 °C, C_{10} 22.8, C_{20} 58.2, C_{30} 15.3 and C_{40} 3.7 wt%. C_{30}–C_{40} hydrogenated to provide synthetic lubricant.	81
1992	Montmorillonite clay deposited with orthophosphoric acid	Oligomerization conditions	Dimer/trimer ratio 1 : 18, after hydrogenation the product has viscosity 5.70 cSt at 38 °C, VI 134, and pour point −34 °C	172
1991	Al nitrate deposited on calcium montmorillonite clay	Oligomerization conditions	Lubricant base obtained after hydrogenation of oligomers	187
1991	Calcium montmorillonite clay (Filtrol clay-124)	Oligomerization conditions	After hydrogenation of oligomers lubricant base obtained with 100 °C viscosity 8–62 cSt, viscosity index 112, and pour point −2 °C	171
1991	Reduced chromium on porous supports (silica gel)	Tubular reactor, oligomerization conditions	C_{2-6} α-olefin oligomers	188
1990	Extruded reduced chromium oxide on a porous silica support	0.69 MPa, 123, 150, and 182 °C. As reactor temperature is raised a corresponding decrease in product viscosity was noted	High-quality lubes with high VI [according to formula $VI = 156.8 + 4.94 \times (V_{100}\,°C)^{0.5}$, where V_{100} °C is kinetic viscosity in cSt at 100 °C]	189
1989	Commercial Cr/ SiO_2 containing 1% Cr	Space velocity 1.2 h^{-1}, 1.72–2.07 MPa, 150 °C	High quality lubricants with a viscosity index of 185, and a viscosity of 266.8 cSt at 40 °C, and 36.2 cSt at 100 °C	190

Continued

Table 4 *Continued*

Year	*Catalysts*	*Reaction conditions/remarks*	*Products*	*Ref(s)*
1989	Activated Cr/SiO_2	90–250 °C	Synthetic lubricant with high viscosity index and low pour point	191
1989	HZSM-23 containing active internal Brønsted acid sites and inactivated surface acid sites neutralized with 2,4,6-collidine	175 °C	C>12 selectivity 22.0% and branching index of 30.5, *vs.* 73.7 and 51.4, respectively, for a catalyst not treated with collidine	73
1988	Siliceous zeolite catalyst	A primary fluidized bed under high temperature dehydrogenation conditions to obtain an oligomerizable product; the product then contacted with a siliceous zeolite catalyst in a secondary fluidized bed to obtain oligomers	Gasoline boiling range aliphatic and aromatic hydrocarbons	192
1988	Silicate (SiO_2/Al_2O_3 2200), ZSM-5 (SiO_2/Al_2O_3 78), amorphous SiO_2/Al_2O_3, mordenite (SiO_2/Al_2O_3 63), ultrastable Y (SiO_2/Al_2O_3 6)	(1) For 3,3-dimethylbut-1-ene, 0.33 LHSV, 0.1 and 5.59 MPa, 204 °C; (2) for 1-C_{10}=, 1-C_{14}=, 2.86 MPa, 204 °C; (3) for propylene, 0.5 LHSV, 11.1 MPa, 54 °C	(1) Dimerization of 3,3-dimethylbut-1-ene over ZSM-5 at 0.1 MPa was 1%, and at 5.59 MPa over ZSM-5 and silicate, selectivity was 95% and 80%, respectively; (2) oligomerization with 1-C_{10}=, and 1-C_{14}=, over ZSM-5 could be fit to second-order kinetics with an E_{app} of about 42 kJ mol^{-1};(3) dimerization of propylene over ZSM-5 can be improved by a factor 4 by Ni addition and operating at relatively low temperature	21
1988	ZSM-5	3–10 MPa, 200–300 °C	Distillate range olefins	22

1986	$TaCl_5$–SiO_2	LHSV 10 for 24 h, 0.345 MPa, 100–150 °C	After hydrogenation a synthetic lubricant obtained, with viscosity index of 104, a kinematic viscosity (at 100 °C) of 5 cSt, and a pour point of −3 °C	193
1985	HZSM-5	4.14–6.89 MPa, 230–235 °C	After further treatment, gasoline fraction (octane no. 79 and API 72) and a distillate (bp > 154 °C and API 8)	194
1983	ZSM-5 and crystalline metal silicate	Oligomerization conditions	Lube oil	165
1983	Zeolite ZSM-5	0.5–1 WHSV, 10.3 MPa, 200–260 °C	$\geq C_{30}$ olefins	18
1983	High SiO_2 zeolites impregnated with 0.8 wt% Zn	Fluidized catalyst bed	C_5+ liquids	195
1983	H_3PO_4 applied to activated carbon (1400 $m^2\,g^{-1}$ and 0.8 $cm^3\,g^{-1}$)	Reactant heated under N_2 and gaseous BF_3 until final pressure was 0.3 MPa, reaction temp 100 °C for 5 h; conversion was 80%	The product after purification and hydrogenation was lubricating oil, viscosity 4.24 and 20.30 cSt at 100 and 40 °C, respectively; selectivity was 84%	196
1981	BF_3–H_2O–SiO_2	7 h at 52 °C, 87.8% conversion	Selectivity for dimer, trimer, and tetramer of 19.7, 53.4, 18.9%, respectively	197
1980	BF_3–silica		C_{20-50} fractions	198
1978	Fluorinated graphite (Lewis acids are SbF_5, TaF_5, NbF_5 and Brønsted acids are HF, HSO_3F)	Used in oligomerization reactions	Isomerized *n*-heptane	199
1978	Ga_2O_3 supported by SiO_2 or ε-Al_2O_3	Residence time 5.5 s at 550 °C	Catalyst not selective for olefins and mostly aromatics formed	200
1977	Zeolite	288–482 °C	Gasoline with a good octane no.	201
1977	K-ZeoliteA, Ca- ZeoliteA, Co, K-ZeoliteA, Co, Ca- ZeoliteA		2,4,4-Trimethylpent-1-ene, 2,4,4-trimethylpent-2-ene, 2,2,4,6,6-pentamethylheptene and 1,3,3,5,5-pentamethylheptene	202

7.2 Other Solid-acid Catalysts – The literature available on oligomerization of C_{10-24} linear α-olefins contains patented information, where the catalysts used are either acid-treated calcium montmorillonites or montmorillonite clays modified by phosphorous-containing acids. C_{14} α-olefins (preferably in the presence of a tertiary hydrocarbon) treated with an acidic calcium montmorillonite clay as catalyst at >200 °C, gave oligomers of C_{10-24} linear α-olefins.[170] Lubricant base-stocks with low pour point and high viscosities were claimed to be prepared by oligomerization of 1,3-di-isopropenylbenzene and C_{10-24} α-olefins using acidic calcium montmorillonite clay catalyst, followed by hydrogenation of the oligomers.[171] A process was claimed for the preparation of oligomers by treatment of linear C_{10-24} olefins with a catalyst, obtained by treatment of phosphorous-containing acid on montmorillonite clay.[172] The same investigators also examined the oligomerization of α-tetradecene using montmorillonite clay catalyst treated with fluorophosphoric acid.[173]

ZrO_2/SiO_2 and TiO_2/SiO_2 catalysts and sulfate-activated ZrO_2/SiO_2 and TiO_2/SiO_2 were physically blended with an acidic clay and were then used in the oligomerization of higher olefins; excellent yields of suitable oligomers were reported.[174] In the case of the ZrO_2/SiO_2 and TiO_2/SiO_2 catalysts a high percentage of dimers to trimers, tetramers, *etc.* was obtained; a high portion of dimers is often desirable when preparing a synthetic lubricant from olefins having 14 or more carbon atoms.

Sulfate ion was incorporated on alumina and/or silica–alumina or transition metal oxide (iron oxide or zinc oxide); the catalysts were used in the oligomerization of oct-1-ene.[133] The results showed that these sulfated catalysts had a higher activity and selectivity in the olefin oligomerization process, while the conventional catalysts such as cobalt sulfate or magnesium sulfate supported on alumina or silica–alumina exhibited lower activity in the same process. The oligomers had a sufficiently low degree of branching and were useful as a material for, *e.g.* phthalic acid plasticizer.

Sulfate-activated TiO_2 and ZrO_2 or a molecular sieve (LZ-Y52, 4A, AW-500, and AW-300) were used in the oligomerization of C_{10-20} alkenes.[175] Olefin conversion was 64% and dimer/trimer ratio 5.57, and 86% and 1.80 for sulfated TiO_2 and ZrO_2, respectively. The AW-sulfate activated molecular sieve showed higher activity in comparison with the other molecular sieves.

Conventional Friedel–Crafts catalysts, *e.g.* $AlCl_3$ and BF_3, were replaced by those based on heterogeneous catalysis. For example, $AlCl_3$ or BF_3 was supported on amorphous silica. Presumably, these methods suffer from incomplete reaction between $AlCl_3$, HCl or Cl_2 and the support. In another invention, $AlCl_2$ was heterogenized by incorporating into a solid support. The prepared catalyst was used in the oligomerization of *n*-decene for the production of hydrocarbon lubricant base-stocks; water or methanol was added to the process as a promoter.[176]

Oligomers suitable for lubricant applications with a controlled molecular weight and low branch ratio were produced over supported chromium oxide catalysts.[177,178]

8 Dimerization of Alkenes

It is well known that olefins dimerize on contact with a nickel oxide catalyst at elevated temperature. Since the discovery of their specific activity in catalysing dimerization reactions, nickel-based catalysts have been the subject of intensive mechanistic and kinetic studies directed toward identifying and understanding the active sites responsible for this property. Earlier nickel catalysts were homogeneous; however, nickel deposited on different supports has shown high selectivity towards alkene dimerization (see Table 5, page 276).

Ethylene dimerization has been extensively reviewed by Al-Jarallah *et al.*[13] The main product from ethylene dimerization is but-1-ene, and it is useful as a co-monomer in polyethylene production. Usually, the homogeneous titanium catalytic system is established in the commercial production routes. But still, much work has been made concerning ethylene dimerization using heterogeneous catalytic systems with high activity.[203,204]

There is a close relationship between the homogeneous and heterogeneous catalyst systems used in ethylene dimerization, the support of the active sites in a heterogeneous system plays the role of ligand on a coordinated metal atom in a homogeneous system.[9,205] The change of the nature of the support alters the behaviour of an active site in the same way that the ligands in the homogeneous catalysts modify the charge and polarity of a metal atom, and its reactivity.[203]

Nickel oxide and nickel complexes supported on silica, silica–alumina, different zeolites, and polymeric materials have been reported to be active for ethylene dimerization.[2–5, 206–210] Yashima *et al.*[211] reported that ethylene dimerization can be catalysed selectively using NiY–RhY and RuY zeolites. The experimental study was conducted at less than 100 °C and 0.026 MPa; the active nickel sites had valence zero. The deposition of a nickel complex on alumina, and alumina–silica systems showed improved selectivity for ethylene oligomerization to but-1-ene. The observed kinetic parameters were comparable with those of the homogeneous system.[212]

The conventional processes used in the dimerization of isobutene are generally based on the use of an acidic catalyst, such as sulfuric acid, polyphosporic acid (in liquid or supported by an inorganic solid), heteropolyacids, *e.g.* phosphomolybdic and silicotungstic acids, ion-exchange resins of the polystyrenesulfonic or fluorosulfonic type, especially nickel salts activated with an aluminium alkyl, or an aluminium alkyl halide, and metal oxides, such as bismuth oxide optionally associated with a phosphor oxide. The conversion of C_3 or C_4 olefins into dimers using catalysts consisting of nickel deposited on silica or silica–alumina and zeolite supports is also recognized.[213,214] These nickel-supported catalysts are usually prepared by impregnation or ion-exchanged with the support. The catalysts are particularly useful for alkene dimerization. It is believed that the nature of the active sites, *i.e.* nickel, is either a coordinatively unsaturated nickel (Ni^+) ion or an acidic site which is formed during the reaction and has a higher nickel valence.[9] The dimerizing sites in the case of ethylene as feed were supposed to be a combination of nickel sites of low and high valence.[215] In another study, the $NiSO_4/\gamma\text{-}Al_2O_3$ catalyst was identified as retaining both coordination and

Table 5 *Dimerization of olefins*

Year	Catalysts	Reaction conditions/remarks	Products	Ref(s)
1997	Mo/HZSM-5, Mo–W/HZSM-5, and Mo–Zr/HZSM-5	GHSV methane 1440 ml/g h, 0.2 MPa, 600–700 °C; at 700 °C with Mo/HZSM-5, the methane conversion is 10.8%, and product distribution is 6.6% dimers and 93.4% aromatics, the reaction can be intiated at lower temperature by presence of a C_{2-3} mixture	Dimers and aromatics	220
1995	(1) NiO 50, TiO_2 13, SiO_2 27, and Al_2O_3 10 wt%; (2) NiO 50, TiO_2 13, and SiO_2 37 wt%	At 7 MPa, 190 °C, with catalyst (1) 78% butene conversion, and with catalyst (2) 53% butene conversion	Catalyst (1): 62 wt% C8 oligomers of which n-octane 21 wt%; catalyst (2): 73 wt% C8 oligomers of which *n*-octane 24 wt%	135
1992	Super-dealuminated Y-zeolite SiO_2/Al_2O_3 ratio of 40 : 1	160 °C, 5 h, yield 68.8% and conversion 76.5%	Dimers	221
1992	Ni on Si support containing Ca, Si/Ca weight ratio $\geq$ 10 : 1	Space velocity about 1.79 h^{-1}, 185 °C, 473 h	14.3% C_{12} fraction	222
1988	Impregnation of various inorganic oxides (silica, alumina, and silica–alumina) with solutions of $Ni(NO_3)_2$, $NiSO_4$, and $NiCl_2$	0.4 MPa and 40 °C: $Ni(NO_3)_2$ deposited on silica and alumina had no activity, whereas all the others were active, $NiCl_2/Al_2O_3$ was most active and selective for oligomerization	For $NiCl_2/Al_2O_3$ with reaction time 2 h the dimer product contained 6.3% 2,3-DMB, 59.0% 2-MP, 34.1% *n*-hexene and 0.6 wt% others	213
1986	Bis(1,5-cyclooctadiene) Ni(0) deposited on an omega zeolite or Y-type zeolite support	0.69–6.89 MPa, 20–70 °C	but-1-ene	223
1986	Nickelocene hydrogenated onto an omega zeolite support with promoter comprising an oxide of Ni, Fe, and/or Cu		but-1-ene	224
1985	Alumina impregnated with $Ni(NO_3)_2{\cdot}6H_2O$	LHSV of propylene 2–3 for 1385 h, 4.83 MPa, 35–50 °C, propylene conversion was 88.7–99.8%, C_6 olefin selectivity 70.7–81.5%	Dimers, with motor octane no. 80–81	216

1985	SiO_2/Al_2O_3 (mol ratio 240) was mixed with peptized and neutralized Catapol Al_2O_3 at 2 : 1 sieve/Al_2O_3 wt. ratio, impregnated with solution of Zn, and Ni	LHSV 0.5, 0.69 MPa, 65.6 °C gave C_5+ olefins with 72.7% conversion and 77% selectivity to dimers after 40 h on stream	Dimers	225
1984	Fluorided aluminum phosphate	6.89 MPa, 300–350 °C	Dimers	226
1984	$NiSO_4/Al_2O_3$ (containing 4% Ni)	WHSV 1.7, 4.14 MPa, 27 °C	>95% dimers after 47 h on stream	227
1983	Ferric sulfate or cobalt sulfate supported on Al_2O_3 or SiO_2–Al_2O_3	21 h at 65 °C	93.5 wt% dimer, composed of 90.2% EtCHMeCHMeEt	228
1977	LaX, CeX, MgX, NiY, CoY, AlY, MgX, MnY, NiX, CoX, and CaX	Fixed bed, 0.1 MPa, at 190 °C, with the exception of NiX, all the zeolites tested showed 95.5% selectivity to dimer formation	Dimers	69
1975	NiX/Li_2O	The selectivity to codimerization (heptenes) is not high due to self-dimerization of reactants	Dimers of propylene and butene	229
1975	NiX precursor catalyst impregnated with molten lithium acetate	Trickle-bed reactor, reactant in gas phase and the products in the liquid phase	Octenes	230
1975	NiY (70% exchanged, equivalent to 9 wt% Ni), RhY	0.027 MPa, 20 °C	Dimers	211
1960	Impregnation of Ni^{2+} on amorphous aluminosilicate, Ni^{2+}- exchanged zeolites	0.1 MPa, 275–300 °C; selectivities of about 80%	*n*-Butenes	231–234

acidic sites, whereas dimerization of ethylene was only attributed to the coordination sites, *i.e.* those of low valence.[83]

The obtainable dimers from oligomerization of propene and butene, such as hexenes and octenes are particularly useful for conversion by the well-known oxo alcohol process into the corresponding heptyl and nonyl alcohols which may be used in the production of plasticizers, lubricating oil additives, detergents, defoamers, and similar products. Dimers of propene are even used to improve octane rating in gasoline.

Different nickel salts impregnated on silica, alumina, and silica–alumina were used in dimerizing propene.[213] The calcination following impregnation in air at 500 °C decomposed the nitrate ions and partially removed the chloride ions, whereas the sulfate ions remained entirely in the catalyst. The activities of the Ni–alumina catalysts containing sulfate and chloride increased with increasing anion content. With nickel nitrate on alumina, calcination led to a mixture of nickel oxide and aluminate and the appearance of weak Lewis acidity. This catalyst was inactive. The addition of sulfate or chloride ions induced the appearance of strong Lewis and Brønsted acidity, the strength and population of sites increasing with increasing anion content. γ- or η-Alumina deposited with nickel sulfate dimerizes propene and butene at 60 °C, 4 MPa. Alumina was impregnated with a solution of nickel nitrate hexahydrate and then activated with an anhydrous aluminium chloride and diethyl aluminium chloride.[149,216] The alumina–nickel activated system resulted in 98% conversion of propene to dimers at 35–50 °C and 4 MPa. Replacing activation with aluminium halide by activation with aluminium alkoxide resulted in decreasing activity in propene oligomerization.

Silica–alumina–nickel oxide, amorphous to X-rays, with a high ratio of SiO_2/Al_2O_3 was found to be a suitable catalyst for dimerizing isobutene into α- and β-di-isobutene with high selectivity and was also useful for oligomerizing propene to dimers and trimers.[93]

Based on the fact that the heterogeneous nickel-containing catalysts for oligomerization of lower olefins suffer several drawbacks such as poisoning by components present in hydrocarbon feeds, *e.g.* sulfur compounds, production of a high level of saturate compounds, low conversion and relatively expensive catalyst, the non-nickel-containing catalyst was developed for dimerization.[139] The non-nickel-containing catalyst consisted of an amorphous trivalent metal oxide support which was impregnated with a monolayer of SiO_2, and a dispersed layer of TiO_2 on top of the SiO_2-monolayer. This catalyst produced the desired octene mixture and minor amounts of C_{5-7} and C_{9-16} olefins in the oligomerization of butene.

The dimers of isopentene (isoamylene) have a wide range of applications in the perfume and flavour industries. Isopentene was dimerized in the presence of acid-treated clay or an ion-exchanged resin catalyst. The reaction of isopentene dimerization was found to be highly selective towards the dimers, *e.g.*, even at a conversion level of 90% only 2–3% yields of higher oligomers were obtained.[180]

The catalytic properties of SAPO-34 in isobutene dimerization have been studied by Minachev *et al.* The properties of SAPO-34 resemble those of HZSM-5 rather than HNa-erionite.[217] Catalysts for cyclodimerization of norbornadiene have been studied.[218] Rh/Na-ZSM-5 possesses high activity in norbornadiene cyclodimerization, due to Rh interacting with the zeolite framework as an additional stabilizing ligand.[219]

In general, however, the solid catalysts reported so far are not as good as the currently used commercial systems, especially the homogeneous nickel-based catalytic systems.

The present review is more focused on oligomerization than on dimerization. However, in order to obtain a clear picture of the comparative performance of the various catalyst systems for dimerization of alkenes, a detailed and systematic study needs to be made.

9 Conclusions

The focus of the present review has been on oligomerization of alkenes using solid catalysts. Cyclic compounds such as cycloalkanes and cycloalkenes have not been considered. For a variety of olefin molecules and a variety of solid catalytic systems, the reaction selectivity to dimers, trimers, *etc.* was found to be determined by at least three factors, namely:

(1) type of active site;
(2) size of reactant molecules;
(3) reaction conditions, *e.g.* temperature, pressure, and space velocity.

For the zeolite catalysts, another factor dominates, namely shape and size of the zeolite pores. The shape of the product molecules is governed primarily by the pore structure of the zeolite catalyst.

The proposals published to date for olefin oligomerization over solid catalysts have advantages and drawbacks. Many advantages can be gained if a homogeneous catalyst can be supported on a solid phase without any significant loss in activity. But usually there is insufficient ability in these catalytic systems to control the extent of oligomerization, and there is also a lack of selectivity to the formation of desired products. A problem which is common to many conventional oligomerization catalysts lies in the corrosive action of the acids used.

In practice, the feedstock for oligomerization of low and/or high molecular weight olefins is usually some refinery cut, constituting a multicomponent system. We believe more work in this area, multicomponent feedstocks and solid catalysts, is needed.

Finally, there are some works about progress in the development of adequate equipment design for solid catalytic oligomerization, *e.g.* reactive distillation. Since a solid catalytic system often implies frequent catalyst deactivation, it seems reasonable that the development of the catalytic system should be accompanied by a more convenient process and improved upgrading of oligomerization products.

References

1 H. Pines, *The Chemistry of Catalytic Hydrocarbon Conversion*, Academic Press, New York, 1981.
2 V.N. Ipatieff and G. Egloff, *Oil and Gas J.*, 1934, **33**, 31.
3 C. Whitmore and P.L. Meunier, *J. Am. Chem. Soc.*, 1941, **63**, 2197.
4 F.C. Whitmore, W.S. Roop and N.C. Cook, *J. Am. Chem. Soc.*, 1950, **72**, 1507.
5 *Advances in Catalysis*, eds. D.D. Eley, H. Pines and P.B. Weisz, Academic Press, New York, 1982, vol. 31, p. 24.
6 S.A. Tabak, *US Patent* 4 254 295, 1981, *Chem. Abstr.*, 95, 24233, 1981.
7 A. Corma, *Chem. Rev.*, 1995, **95**, 559.
8 Kh.M. Minachev and A.A. Dergachev, *Itogi Nauki, Tekch. Ser. Kinet. Katal.*, 1990, **23**, 3.
9 C.T. O'Connor, in *Handbook of Heterogeneous Catalysis*, eds. G. Ertl, H. Knözinger and J. Weitkamp, VCH Verlagsgesellschaft mbH, Weinheim, 1997, vol. 5, p. 2380.
10 C.T. O'Connor, E. Van Steen and M.E. Dry, *Stud. Surf. Sci. Catal.*, 1996, **102**, 323.
11 C.T. O'Connor and M. Kojima, *Catal. Today*, 1990, **6**, 329.
12 J. Skupinska, *Chem. Rev.*, 1991, **91**, 613.
13 A.M. Al-Jarallah, J.A. Anabtawi, M.A.B. Siddiqui, A.M. Aitani and A.W. Al-Sa'doun, *Catal. Today*, 1992, **14**, 1.
14 Kh.M. Minachev and A.A. Dergachev, *Russ. Chem. Bull.*, 1998, **47**, 1037.
15 M.L. Occelli, J.T. Hsu and L.G. Galya, *J. Mol. Catal.*, 1985, **32**, 377.
16 S.J. Miller, *Catalysis*, 1987, **38**, 187.
17 D.B. Lukyanov, N.S. Gnep and M.R. Guisnet, *Ind. Eng. Chem. Res.*, 1994, **33**, 223.
18 W.E. Garwood, *ACS Symp. Ser.*, 1983, **218**, 383.
19 J. Datka, *J. Chem. Soc., Faraday Trans.* 1, 1981, **77**, 2633.
20 L.M. Kustov, V. Yu. Borovkov and V.B. Kazansky, *Stud. Surf. Sci. Catal.*, 1984, **18**, 241.
21 S.J. Miller, *Stud. Surf. Sci. Catal.*, 1988, **38**, 187.
22 R.J. Quann, L.A. Green, S.A. Tabak and F.J. Krambeck, *Ind. Eng. Chem. Res.*, 1988, **27**, 565.
23 A.W. Chester, F.G. Dwyer, W.E. Garwood and J.C. Vartuli, *US Patent* 5 243 112, 1993, *Chem. Abstr.*, 120, 81301, 1993.
24 P.A. Jacobs, N.I. Jaeger, P. Jiru, V.B. Kazansky and G. Schulz-Ekloff, *Stud. Surf. Sci. Catal.*, 1984, **18**, 241.
26 T.J.G. Kofke and R.J. Gorte, *J. Catal.*, 1989, **115**, 233.
27 P.E. Dai, J.R. Sanderson and J.F. Knifton, *Stud. Surf. Sci. Catal.*, 1994, **84**, 1701.
28 M.L. Occelli, J.T. Hsu and L.G. Galya, *J. Mol. Catal.*, 1985, **32**, 377.
29 L.G. Galya, M.L. Occelli and J.T. Hsu, *J. Mol. Catal.*, 1985, **32**, 391.
30 W.E. Garwood, *Prepr. ACS., Div. Pet. Chem.*, 1982, **27**, 563.
31 C.S.H. Chen and R.F. Bridger, *J. Catal.*, 1996, **161**, 687.
32 R. Sarin, D.K. Tuli, S. Sinharay, M.M. Rai, S. Ghosh and A.K. Bhatnagar, *Prepr. ACS., Div. Pet. Chem.*, 1996, **41**, 625.
33 H. Imai, T. Hasegawa and H. Uchida, *Bull. Chem. Soc. Jpn.*, 1968, **41**, 45.
34 J.P. Van den Berg, J.P. Wolthuizen and J.H.C. Van Hoof, *J. Catal.*, 1983, **80**, 139.
35 Q. Zhang, M. Kantcheva and I.G. Dalla Lana, *Ind. Eng. Chem. Res.*, 1997, **36**, 3433.
36 M.P. McDaniel, M.B. Welch and M.M. Johnson, *US Patent* 5 087 789, 1992, *Chem. Abstr.*, 116, 173546, 1992.
37 J. Heveling, *Ger. Offen. Patent* 3 741 302 A1, 1988, *Chem. Abstr.*, 109, 76461, 1988.

38 M. Yamamura, K. Chaki, T. Kazutoshi, H. Okado and K. Fujimoto, *Zeolites*, 1994, **14**, 643.
39 J. Heveling, A. van Der Beek and M. De Pender, *Appl. Catal.*, 1988, **42**, 325.
40 E.G. Derouane, C. Lefebvre and J.B. Nagy, *J. Mol. Catal.*, 1986, **38**, 387.
41 W.E. Garwood, F.J. Krambeck, J.D. Kushnerick and S.A. Tabak, *US Patent* 4 942 021, 1990, *Chem. Abstr.*, 114, 27052, 1990.
42 D. Fraenkel, *Ind. Eng. Chem. Res.* 1990, **29**, 1814.
43 D. Farcasiu, J. Hutchison and L. Luowei, *J. Catal.*, 1990, **122**, 34.
44 S. Schwartz, M. Kojima and C.T. O'Connor, *Appl. Catal.*, 1989, **56**, 263.
45 W.K. Reagan and M.P. McDaniel, *US Patent* 5 438 027, 1995, *Chem. Abstr.*, 123, 290379, 1995.
46 M. Goledzinowski and V.I. Briss, *Ind. Eng. Chem. Res.*, 1993, **32**, 1795.
47 F.T.T. Ng and D.C. Creaser, in 'Studies of Surface Science and Catalysis', eds. K.J. Smith and E.C. Sanford, *Prog. in Catal.*, 1992, **73**, 123.
48 A.M. Raspolli Galletti, G. Geri, G. Sbrana, M. Marchionna and P. Ferrarini, *J. Mol. Catal.*, 1996, **111**, 273.
49 S.-H. Chien, Y.-W. Wei and M.-C. Lin, *J. Chin. Chem. Soc. (Taipei)*, 1997, **44**, 195.
50 C.S.H. Chen, D.H. Hoskin and S.E. Schramm, *WIPO Patent* 9 306 069, 1993, *Chem. Abstr.*, 119, 142801, 1993.
51 M.N. Harandi, *US Patent* 5 202 015 A, 1993, *Chem. Abstr.*, 119, 76124, 1993.
52 R. He, in *Proc. Int. Conf. Pet. Refin. Petrochem. Process*, ed. Hou Xianglin, Int. Acad. Publ., Beijing, Peop. Rep. China, 1991, vol. 3, p. 1412.
53 R.L. Burwell, Jr., *Chemtracts: Inorg. Chem.*, 1991, **3**, 242.
54 K.P. Datema, A.K. Nowak, J. Van Bram Houckgeest and A.F.H. Wielers, *Catal. Lett.*, 1991, **11**, 267.
55 M.N. Harandi, H. Owen and S.A. Tabak, *US Patent* 5 034 565, 1991, *Chem. Abstr.*, 115, 186583, 1991.
56 P. Chu and A.B. Schwartz, *US Patent* 4 868 146, 1989, *Chem. Abstr.*, 112, 101966, 1989.
57 A.M. Gaffney, *US Patent* 4 849 571, 1989, *Chem. Abstr.*, 111, 177739, 1989.
58 H. Mazurek and J.A. Sofranko, *US Patent* 4 788 376, 1988, *Chem. Abstr.*, 110, 98578, 1988.
59 W.E. Garwood, F.J. Krambeck, J.D. Kushnerick and S.A. Tabak, *US Patent* 4 740 645, 1988, *Chem. Abstr.*, 109, 173340, 1988.
60 W.E. Garwood, J.D. Kushnerick and S.A. Tabak, *US Patent* 4 717 782, 1988, *Chem. Abstr.*, 108, 207521, 1988.
61 R.L. Espinoza, R. Snel, C.J. Korf and C.P. Nicolaides, *Appl. Catal.*, 1987, **29**, 295.
62 H. Mazurek, *US Patent* 4 605 807, 1986, *Chem. Abstr.*, 105, 229638, 1985.
63 S.A. Tabak and F.J. Krambeck, *Hydrocarbon Process.*, Int. Ed., 1985, **64**, 72.
64 S.A. Tabak, *US Patent* 4 487 985, 1984, *Chem. Abstr.*, 102, 81488, 1984.
65 S.A. Tabak, 'Mobil olefins to gasoline and distillate process', *US Natural Gas Utilization Symposium*, 7–11 February 1984, Bangkok, Thailand.
66 H. Mitchell and R.H. Waghorne, *US Patent* 4 172 810, 1979, *Chem. Abstr.*, 92, 146398, 1979.
67 S. Munemiya, *Ger. Offen. Patent* 2 228 262, 1972, *Chem. Abstr.*, 78, 72857, 1972.
68 C.T. O'Connor and M. Kojima, *S. A. J. Chem. Eng.*, 1989, **1**, 23.
69 S.M. Hassan, G.M. Panchenkov and O.I.Kuznetsov, *Bull. Chem. Soc. Jpn.*, 1977, **50**, 2597.
70 C.T. O'Connor and M. Kojima, *Catal. Today*, 1989, **6**, 329.
71 S.A. Tabak, F.J. Krambeck and W.E. Garwood, *AIChE J.*, 1986, **32**, 1526.

72 K.G. Wilshier, P. Smart, R. Western, T. Mole and T. Behrsing, *Appl. Catal.*, 1987, **31**, 339.
73 N.M. Page, L.B. Young and D.A. Blain, *US Patent* 4 870 038, 1989, *Chem. Abstr.*, 112, 99492, 1989.
74 A.V. Kucherov, A.A. Slin'kin, M.S. Kharson, T.N. Bondarenko and Kh.M. Minachev, *J. Mol. Catal.*, 1989, **53**, 293.
75 M.R. Apelian, J.R. Boulton and A.S. Fung, *US Patent* 5 284 989, 1994, *Chem. Abstr.*, 121,10213, 1994.
76 S. Han, R.H. Heck and F. Diguiseppi, *US Patent* 5 234 875, 1993, *Chem. Abstr.*, 119, 163879, 1993.
77 L.R. Martens, J.P. Verduijn and G. Mathys, *Catal. Today*, 1997, **36**, 451.
78 G.M.K. Mathys, L.M. Martens, M.A. Baes, J.P. Verduijn and D.R.C. Huybrechts, *WIPO Patent* 9 316 020 A2, 1993, *Chem. Abstr.*, 119, 202975, 1993.
79 W.H. Verrelst and L.R.M. Martens, *WIPO Patent* 9 522 516 A1, 1995, *Chem. Abstr.*, 124, 8225, 1995.
80 F.T. DiGuiseppi and S.H. Han, *US Patent* 5 264 643, 1993, *Chem. Abstr.*, 120, 58345, 1993.
81 N.A. Bhore, Q.N. Le and G. Yokomizo, *US Patent* 5 134 243, 1992, *Chem. Abstr.*, 118, 106144, 1992.
82 W.E. Garwood, G.G. Kuelh and R.B. Lapierre, *US Patent* 4 788 375, 1988, *Chem. Abstr.*, 110, 98621, 1988.
83 T. Cai, D. Cao and L. Li, *Stud. Surf. Sci. Catal.*, 1993, **75**, 2753.
84 C.D. Chang, T.J. Huang, J.G. Santiesteban and J.C. Vartuli, *US Patent* 5 608 133, 1997, *Chem. Abstr.*, 126, 214283, 1997.
85 M. Repas, *Chem. Tech.*, 1965, **17**, 222.
86 M.M. Wu and S.C. Ho, *US Patent* 5 068 476, 1991, *Chem. Abstr.*, 116, 155269, 1991.
87 S.M. Harms, M. Kojima and C.T. O'Connor, *Fuel Process. Technol.*, 1989, **21**, 231.
88 P.G. Bercik, K.J. Metzger and H.E. Swift, *I&EC Prod. Res. Dev.*, 1978, **17**, 214.
89 C.T. O'Connor, L.L. Jacobs and M. Kojima, *Appl. Catal.*, 1988, **40**, 277.
90 W.H. Verrelst, L.R.M. Martens and J.P. Verduijn, *WIPO Patent* 9 624 567 A1, 1996, *Chem. Abstr.*, 125, 226392, 1996.
91 T. Inui and J.-B. Kim, *Stud. Surf. Sci. Catal.*, 1996, **100**, 489.
92 C. Perego, S. Peratello and R. Millini, *Eur. Patent* 659 478 B1, 1995, *Chem. Abstr.*, 123, 118144, 1995.
93 M. Brunelli, B.G. Castelvetro and S. Peratello, *WIPO Patent* 9 306 926 A1, 1993, *Chem. Abstr.*, 119, 31339, 1993.
94 A.R. Bryson, R.B. Andrew and N.M. Van Niekerk, *S. African Patent* 9 101 970 A, 1992, *Chem. Abstr.*, 119, 184519, 1993.
95 J.W. Ward and D.D. Delaney, *US Patent* 5 051 386, 1991, *Chem. Abstr.*, 115, 24505, 1991.
96 F. Nierlich, J. Neumeister, T. Wildt, W. Droste and F. Obenaus, *Eur. Patent* 395 857 A1, 1990, *Chem. Abstr.*, 114, 62990, 1990.
97 S.G. Kukes, R.L. Banks and J.R. Harris, *US Patent* 4 879 425, 1989, *Chem. Abstr.*, 112, 122016, 1989.
98 A. Giusti, S. Gusi, G. Bellussi and V. Fattore, *Eur. Patent* 293 950 A1, 1988, *Chem. Abstr.*, 110, 155048, 1988.
99 A. Giusti, S. Gusi, G. Bellussi and V. Fattore, *Eur. Patent* 290 068 A1, 1988, *Chem. Abstr.*, 110, 115529, 1988.
100 P. Chu, W.E. Garwood and A.B. Schwartz, *US Patent* 4 788 374, 1988, *Chem. Abstr.*, 110, 98622, 1988.

101 A. Giusti, S. Gusi, G. Bellussi and V. Fattore, *Eur. Patent* 272 496 A1, 1988, *Chem. Abstr.*, 109, 150287, 1988.

102 C.D. Chang, J.D. Dixon, A.B. Schwartz and D.S. Shihabi, *US Patent* 4 754 096, 1988, *Chem. Abstr.*, 109, 76450, 1988.

103 G.W. Kirker, M.E. Landis and N.M. Page, *WIPO Patent* 8 800 176 A1, 1988, *Chem. Abstr.*, 108, 224135, 1988.

104 C.S.H. Chen, *US Patent* 4 658 079, 1987, *Chem. Abstr.*, 107, 42930, 1987.

105 A.K. Ghosh and R.A. Kydd, *J. Catal.*, 1986, **100**, 185.

106 J.H. Beech and H. Owen, *Brit. Patent* 2 171 717 A1, 1986, *Chem. Abstr.*, 105, 229681, 1986.

107 S.J. Miller, *US Patent* 4 608 450, 1986, *Chem. Abstr.*, 105, 156044, 1986.

108 C.D. Chang, J.N. Miale, *US Patent* 4 607 130, 1986, *Chem. Abstr.*, 106, 6716, 1986.

109 C.D. Chang and C.T.W. Chu, *US Patent* 4 554 396, 1985, *Chem. Abstr.*, 104, 53398, 1985.

110 S.A. Tabak, *US Patent* 4 547 612, 1985, *Chem. Abstr.*, 104, 36785, 1985.

111 C.D. Chang, J.E. Penick and R.F. Socha, *US Patent* 4 542 247, 1985, *Chem. Abstr.*, 103, 218152, 1985.

112 S.A. Tabak, B.S. Wright and H. Owen, *US Patent* 4 504 693, 1985, *Chem. Abstr.*, 103, 8780, 1985.

113 R.M. Dessau, *Eur. Patent* 135 385 A2, 1985, *Chem. Abstr.*, 102, 223227, 1985.

114 S.J. Miller, *US Patent* 4 423 268, 1983, *Chem. Abstr.*, 100, 102729, 1983.

115 S.J. Miller, *US Patent* 4 417 086, 1983, *Chem. Abstr.*, 100, 36879, 1983.

116 D.J. Ward, V.B. Vora, *US Patent* 4 393 259, 1983, *Chem. Abstr.*, 99, 125395, 1983.

117 S.J. Miller, *Ger. Offen. Patent* 3 229 895 A1, 1983, *Chem. Abstr.*, 99, 5183, 1983.

118 S.J. Miller, *Ger. Offen. Patent* 3 235 507 A1, 1983, *Chem. Abstr.*, 98, 215166, 1983.

119 R.M. Dessau, *Eur. Patent* 37 671 A1, 1981, *Chem. Abstr.*, 96, 85033, 1981.

120 B. Juguin, J. Cosyns, P. Le, F. Jean and J. Miguel, *Brit. Patent* 2 006 263, 1979, *Chem. Abstr.*, 92, 21696, 1979.

121 J.-Y. Ryu, *US Patent* 4 048 109, 1977, *Chem. Abstr.*, 87, 204128, 1977.

122 J.-Y. Ryu, *US Patent* 4 048 108, 1977, *Chem. Abstr.*, 87, 204127, 1977.

123 W. Schulz, H. Mix, W.F.W. Kurras, J. Reihsig, H. Fuhrmann and I. Grassert, *Ger. (East) Patent* 100478, 1973, *Chem. Abstr.*, 81, 50325, 1973.

124 R. Jacques and B. Juguin, *Ger. Offen. Patent* 2 116 922, 1971, *Chem. Abstr.*, 76, 86422, 1971.

125 J.S. Yoo, *Ger. Offen. Patent* 2 131 814, 1971, *Chem. Abstr.*, 76, 85362, 1971.

126 B. Juguin, P. Le and F. Jean, *Ger. Offen. Patent* 1 954 326, 1970, *Chem. Abstr.*, 73, 59618, 1970.

127 T. Hill, *Brit. Patent* 1 183 201, 1970, *Chem. Abstr.*, 72, 132007, 1970.

128 L.E. Addy and T. Hill, *Brit. Patent* 1124766, 1968.

129 T. Hill, *Brit. Patent* GB 1095982, 1967.

130 C.T. O'Connor, R.E. Fasol and G.A. Foulds, *Fuel Proc. Tech.*, 1986, **13**, 41.

131 M. Kojima, M.W. Rautenbach and C.T. O'Connor, *Ind. Eng. Chem. Res.*, 1988, **27**, 248.

132 Kh.M. Minachev, O.N. Usacheva, T.N. Bondarenko and N.Ya. Usachev, *Neftekhimiya*, 1994, **34**, 17 (in Russian).

133 H. Fujie, K. Imura, H. Matsumoto, T. Noh and K. Nakanishi, *Eur. Patent* 734 766 A1, 1996, *Chem. Abstr.*, 125, 333831, 1996.

134 J.P. Van den Berg and K.H.W. Roebschlaeger, *Eur. Patent* 439 865 B1, 1991, *Chem. Abstr.*, 115, 236125, 1991.

135 W.H. Verrelst, L.R.M. Martens and J.P. Verduijn., *WIPO Patent* 9 624 567 A1, 1996, *Chem. Abstr.*, 125, 226392, 1996.
136 G.G. Podrebarac, F.T.T. Ng and G.L. Rempel, *Chemtech*, 1997, **27** 37.
137 M. Tiitta, K.M. Keskinen and P. Raulo, *Finnish Patent* 963 20 B, 1996, *Chem. Abstr.*, 124, 347922, 1996.
138 M. Marchionna, M. Di Gerolamo and F. Ancillotti, *Eur. Patent* 745 576 A1, 1996, *Chem. Abstr.*, 126, 46878, 1996.
139 R.Y. Saleh and S.L. Soled, *US Patent* 5 073 658, 1991, *Chem. Abstr.*, 116, 196592, 1991.
140 R.R. Frame, *US Patent* 4 935 575, 1990, *Chem. Abstr.*, 113, 154767, 1990.
141 S.B. Gevert, P. Abrahamsson and S. Järås, *ACS Symp. Ser.* 1990, **437**, 272.
142 D.A. Young, D.E.W. Vaughan, R.L. Pruett and M.E. Tunison, *Eur. Patent* 329 305 A1, 1989, *Chem. Abstr.*, 112, 56970, 1989.
143 H.U. Hammershaimb and R.R. Frame, *US Patent* 4 835 331, 1989, *Chem. Abstr.*, 111, 154570, 1989.
144 G.M.K. Mathys, *Brit. Patent* 2 207 613 A1, 1989, *Chem. Abstr.*, 110, 213642, 1989.
145 G. Busca and G. Ramis, *J. Chem. Soc., Faraday Trans. 1*, 1989, **85**, 137.
146 A.W. Kucherov, A.A. Slinkin, M.S. Kharkov, T.N. Bondarenko and K.M. Minachev, *Kinet. Kataliz*, 1989, **30**, 193.
147 G. Ramis, G. Busca, V. Lorenzelli, P.F. Rossi, M. Bensitel, O. Saur and J.C. Lavalley, in *Proc. 9th Int. Congr. Catal.*, eds. M.J. Phillips and M. Ternan, 1988, **4**, 1874.
148 R.R. Frame and T. Imai, *US Patent* 4 737 480, 1988, *Chem. Abstr.*, 109, 55475, 1988.
149 R.R. Frame and P.T. Barger, *US Patent* 4 737 479, 1988, *Chem. Abstr.*, 109, 55474, 1988.
150 J.P. Van den Berg, G.T. Kortbeek and P. Grandvallet, *Eur. Patent* 281 208 A1, 1988, *Chem. Abstr.*, 109, 231744, 1988.
151 K.M. Minachev, T.N. Bondarenko and D.A. Kondratev, *Izv. Akad. Nauk SSSR, Ser. Khim.*, 1987, **6**, 1225.
152 S.J. Miller, *US Patent* 4 608 459, 1986, *Chem. Abstr.*, 105, 156044n, 1986.
153 H. Zimmerman and H.J. Wernicke, *Ger. Offen. Patent* 3 417 548 A1, 1985, *Chem. Abstr.*, 104, 91881, 1985.
154 T.H. Johnson, *US Patent* 4 476 343, 1984, *Chem. Abstr.*, 102, 28122, 1984.
155 M.J. O'Hara and T. Imai, *US Patent* 4 490 571, 1984, *Chem. Abstr.*, 102, 134801, 1984.
156 J.R. Sanderson and J.F. Knifton, *Canadian Patent* 2 046 374 AA, 1992, *Chem. Abstr.*, 118, 106174, 1992.
157 W.R. Cares, *US Patent* 4 065 512, 1977, *Chem. Abstr.*, 89, 6805a, 1978.
158 J. Welker, W. Hegenbarth, R. Schulz, A. Lettau, R. Voigt and R. Nauendorf, *Ger. (East) Patent* 121 630, 1976, *Chem. Abstr.*, 87, 104325, 1976.
159 J.F. Brennan and G.R. Lester, *US Patent* 3 773 853, 1973, *Chem. Abstr.*, 80, 36686, 1973.
160 O.I. Kuzenetsov, G.M. Panchenkov, A.D. Guseinov and S.G. Kiryushkin, *Neftekhimiya*, 1972, **12**, 176.
161 A.L. Lapidus, L.N. Rudakova, Y.I. Isakov, K.M. Minachev and Y.T. Eidus, *Izv. Akad. Nauk SSSR, Ser. Khim.*, 1972, 1637.
162 A.L. Lapidus, L.N. Rudakova, V.P. Chemagina, Y.I. Isakov, K.M. Minachev and Y.T. Eidus, *Izv. Akad. Nauk SSSR, Ser. Khim.*, 1973, 1261.
163 A.L. Lapidus, Y.I. Isakov, L.N. Rudakova, K.M. Minachev and Y.T. Eidus, *Proc. Symp. Mech. Hydrocarbon React.*, Hungary, June 1973, p. 60.

164 J.R. Sanderson and J.F. Knifton, *US Patent* 5 191 130, 1993, *Chem. Abstr.*, 118, 234689, 1993.

165 S.J. Miller, *US Patent* 4 417 088, 1983, *Chem. Abstr.*, 100, 36881, 1983.

166 K.M. Minachev, A.A. Dergachev and I.V. Mishin, *Neftekhimiya*, 1988, **28**, 164.

167 K.I. Keskinen, *WIPO Patent* 9 413 599 A1, 1994, *Chem. Abstr.*, 121, 304452, 1994.

168 J.F. Knifton, J.R. Sanderson and P.E. Dai, *Catal. Lett.*, 1994, **28**, 223.

169 B.P. Pelrine, K.D. Schmitt and J.C. Vartuli, *US Patent* 5 270 273, 1993, *Chem. Abstr.*, 120, 303101, 1993.

170 J.R. Sanderson, J.F. Knifton and E.T. Marquis, *Eur. Patent* 515 022, 1992, *Chem. Abstr.*, 118, 194935, 1993.

171 J.R. Sanderson and E.T. Marquis, *US Patent* 5 030 791, 1991, *Chem. Abstr.*, 115, 256934, 1991.

172 J.R. Sanderson and J. Knifton, *Eur. Patent* 466 305 A2, 1992, *Chem. Abstr.*, 116, 105607, 1992.

173 J.R. Sanderson and J.F. Knifton *US Patent* 5 191 130, 1992, *Chem. Abstr.*, 118, 234689, 1992.

174 J.R. Sanderson, J.F. Knifton and E.T. Marquis, *Eur. Patent* 549 156 A, 1993, *Chem. Abstr.*, 119, 142793, 1993.

175 J.R. Sanderson, J.F. Knifton, J.M. Larkin and E.T. Marquis, *Eur. Patent* 541 286 A, 1993, *Chem. Abstr.*, 119, 27681, 1993.

176 S.C. Ho and M.M. Wu, *US Patent* 5 451 704, 1995, *Chem. Abstr.*, 123, 318567, 1995.

177 M.M. Wu, *US Patent* 5 264 642, 1993, *Chem. Abstr.*, 120, 81287, 1993.

178 I.D. Johnson, G.W. Kirker, M.E. Landis and M.M. Wu, *US Patent* 5 243 114, 1993, *Chem. Abstr.*, 120, 34328, 1993.

179 T.-C. Chung, R. Ding and R.L. Shubkin, *WIPO Patent* 9 721 651 A1, 1997, *Chem. Abstr.*, 127, 123278, 1997.

180 S.R. Chitnis and M.M. Sharma, *React. Funct. Polym.*, 1997, **32**, 93.

181 K.M. Keskinen, *VTT Symp.*, Espoo, Finland, 1996, **163**, 111.

182 J.R. Sanderson, J.F. Knifton and J.M. Larkin, *US Patent* 5 191 139, 1993, *Chem. Abstr.*, 119, 31363, 1993.

183 J.R. Sanderson and J.F. Knifton, *US Patent* 5 146 030, 1992, *Chem. Abstr.*, 118, 106157, 1992.

184 J.R. Sanderson and J.F. Knifton, *Canadian Patent* 2 046 374 AA, 1992, *Chem. Abstr.*, 118, 106174, 1992.

185 G.L.G. Debras, R.M. Cahen and G.E.M.J. De Clippeleir, *US Patent* 5 171 331, 1992, *Chem. Abstr.*, 118, 257880, 1992.

186 B.P. Pelrine, K.D. Schmitt and J.C. Vartuli, *US Patent* 5 105 051, 1992, *Chem. Abstr.*, 117, 154327, 1992.

187 E.T. Marquis and J.R. Sanderson, *Eur. Patent* 449 453 A2, 1991, *Chem. Abstr.*, 116, 24536, 1991.

188 B.P. Pelrine and M.M. Wu, *US Patent* 4 996 384, 1991, *Chem. Abstr.*, 115, 50509, 1991.

189 B.P. Pelrine, *US Patent* 4 914 254, 1990, *Chem. Abstr.*, 113, 26733, 1990.

190 M.M. Wu, *US Patent* 4 827 073, 1989, *Chem. Abstr.*, 112, 80802, 1989.

191 M.M. Wu, *US Patent* 4 827 064, 1989, *Chem. Abstr.*, 112, 80801, 1989.

192 M. Harandi, *US Patent* 4 788 364, 1988, *Chem. Abstr.*, 110, 138489, 1988.

193 T.H. Johnson, *US Patent* 4 615 790, 1986, *Chem. Abstr.*, 105, 229666, 1986.

194 H. Owen, S.A. Tabak and B.S. Wright, *US Patent* 4 520 215, 1985, *Chem. Abstr.*, 103, 24731, 1985.

195 S.J. Miller, *US Patent* 4 417 087, 1983, *Chem. Abstr.*, 100, 36880, 1983.

196 A. Priola, V. Fattore, S. Arrighetti and G. Mancini, *Eur. Patent* 68 554 A1, 1983, *Chem. Abstr.*, 98, 163757, 1983.
197 A.M. Madgavkar and H.E. Swift, *US Patent* 4 308 414, 1981, *Chem. Abstr.*, 96, 68302, 1981.
198 E. Harold, *US Patent* 4 213 001, 1980, *Chem. Abstr.*, 93, 220368, 1980.
199 G.A. Olah, *US Patent* 4 116 880, 1978, *Chem. Abstr.*, 89, 221652, 1978.
200 A.J. Kolombos, *Brit. Patent* 1 533 169, 1978, *Chem. Abstr.*, 91, 57833, 1978.
201 C.J. Plank and E.J. Rosinski, *US Patent* 4 021 502, 1977, *Chem. Abstr.*, 87, 25804, 1977.
202 E. Detrekoy and D. Kallo, *ACS Symposium Ser.* 1977, **40**, 549.
203 D. Ballard 'Catalysis – Homogeneous and Heterogeneous', *Proc. Int. Symp. on the Relations between Heterogeneous and Homogeneous Catalytic Phenomena, Bruxelles 1974*, eds. B. Delmon and G. James, Elsevier, Amsterdam, 1975, p. 253.
204 P.D. Smith, D.D. Klendworth and M.P. McDaniel, *J. Catal.*, 1987, **105**, 187.
205 G. Lefebvre and Y. Chauvin, in *Aspects of Homogeneous Catalysis,* eds. R. Ugo, C. Manfredi, Milan, 1979, vol. 1, p.107.
206 T. Matsuda, H. Miura and K. Sugiyama, *J. Chem. Soc., Faraday Trans. 1*, 1979, **75**, 1513.
207 J.R. Shon, H.W. Kim and J.T. Kim, *J. Mol. Catal.*, 1987, **41**, 375.
208 J.R. Shon and A. Ozaki, *J. Catal.*, 1980, **61**, 29.
209 R. Ceder, G. Muller, J. Saleh and J. Vidal, *J. Mol. Catal.*, 1991, **68**, 23.
210 J.C. Carlu and C. Caze, *Reactive Polymers*, 1990, **13**, 153.
211 T. Yashima, Y. Ushida, M. Ebisawa and N. Hara, *J. Catal.*, 1975, **36**, 320.
212 E. Angelescu and M. Udrea, *Proc. VIth Int. Symp.*, Heterogeneous Catalysis, Sofia, 1987, p. 55.
213 Y. Chauvin, D. Commereuc, F. Hugues and J. Thivolle-Cazat, *Appl. Catal.*, 1988, **42**, 205.
214 H. Knut, D. Kiesling, G. Wendt and A. Schollner, *Chem. Tech.*, 1998, **8**, 40.
215 K. Kimura, A.-I. Hideo and A. Ozaki, *J. Catal.*, 1970, **18**, 271.
216 R.R. Frame, *US Patent* 4 613 580, 1986, *Chem. Abstr.*, 105, 211537, 1985.
217 M.S. Dadashov, A.V. Kazakov, G.I. Kapustin, A.A. Dergachev and E.S. Shpiro, in *Proc. Int. Meet. Zeolite Catal. Solution Environ. Probl.*, ed. K.G. Ione, Nova Science Publishers, Commack, NY, 1995.
218 B. Azbel, N.F. Goldshleger, A.A Grigorev, Ya.I. Isakov, Kh.M. Minachev, I.G. Sirotina and M.L. Khidekel, *SU Patent* 1 015 525 A1, 1994, *Chem. Abstr.*, 123, 285323, 1994.
219 N.F. Goldshleger, B.I. Azbel, Y.I. Isakov, E.S. Shpiro and Kh.M. Minachev, *J. Mol. Cat. A.*, 1996, **106**, 159.
220 L. Wang, Y. Xu and L. Tao, *Sci. China, Ser. B: Chem.*, 1997, **40**, 161.
221 J.R. Sanderson, P.S.E. Dai and J.F. Knifton *US Patent* 5 120 891 A, 1992, *Chem. Abstr.*, 117, 194997, 1992.
222 N.A. De Munck and R.Garton, *Brit. Patent* 2 246 365 A1, 1992, *Chem. Abstr.*, 117, 91007, 1992.
223 K.W. Barnett, *US Patent* 4 628 139, 1986, *Chem. Abstr.*, 106, 120389, 1986.
224 K.W. Barnett and M.J. Brown, *US Patent* 4 628 138, 1986, *Chem. Abstr.*, 106, 85236, 1986.
225 S.J. Miller, *US Patent* 4 538 012, 1985, *Chem. Abstr.*, 104, 52440, 1985.
226 M.J. O'Hara and T. Imai, *US Patent* 4 476 342, 1984, *Chem. Abstr.*, 102, 28123, 1984.
227 S.J. Miller, *US Patent* 4 465 788, 1984, *Chem. Abstr.*, 101, 174405, 1984.

228 R.M. Dowling and D.P. Higley, *US Patent* 4 423 267, 1983, *Chem. Abstr.*, 100, 102731, 1983.
229 L. Forni, L. Van Mao and R. Invernizzi, *Chim. Ind.*, 1975, **57**, 669.
230 L. Forni, L.R. Invernizzi and L. Van Mao, *Chim. Ind.*, 1975, **57**, 577.
231 Y.T. Eidus, N.I. Ershov, K.V. Puzitskii and B.A. Kazanskii, *Izv. Akad. Nauk SSSR, Ser. Khim.*, 1960, 114.
232 Y.T. Eidus, N.I. Ershov, K.V. Puzitskii and B.A. Kazanskii, *Izv. Akad. Nauk SSSR, Ser. Khim.*, 1960, 703.
233 Y.T. Eidus, N.I. Ershov, K.V. Puzitskii and B.A. Kazanskii, *Izv. Akad. Nauk SSSR, Ser. Khim.*, 1960, 920.
234 Y.T. Eidus, N.I. Ershov, K.V. Puzitskii and B.A. Kazanskii, *Izv. Akad. Nauk SSSR, Ser. Khim.*, 1960, 1291.